A HANDBOOK OF NUCLEAR MAGNETIC RESONANCE

A Handbook of

NUCLEAR MAGNETIC RESONANCE

Ray Freeman
Oxford University

Copublished in the United States with
John Wiley & Sons, Inc., New York

Longman Scientific & Technical,
Longman Group UK Limited,
Longman House, Burnt Mill, Harlow,
Essex CM20 2JE, England
and Associated Companies throughout the world.

Copublished in the United States with
John Wiley & Sons, Inc., 605 Third Avenue, New York, NY 10158

First Published 1987

British Library Cataloguing in Publication Data

Freeman, Ray, *1932–*
A handbook of nuclear magnetic resonance
1. Nuclear magnetic resonance spectroscopy
I. Title
538′.362 QC762

ISBN 0-582-00390-3 CSD
ISBN 0-582-00574-4 PPR

Library of Congress Cataloging-in-Publication Data
Freeman, Ray, 1932–
A handbook of nuclear magnetic resonance.
Bibliography: p.
Includes index.
1. Nuclear magnetic resonance spectroscopy.
I. Title.
QD96.N8F74 1988 543′.0877 86-34419
ISBN 0-470-20812-0 (USA only)

Set in 10 point Times Roman
Printed and bound in Great Britain at the Bath Press, Avon

Dedicated to my wife Anne-Marie

Contents

Preface

Those of us who practise the art of high resolution NMR spectroscopy usually have a background in chemistry or biochemistry, yet the technique itself has a strong physical bias and its description is steeped in the vocabulary of quantum mechanics. Characteristically, the chemist's reaction has been to use NMR in a largely empirical fashion, interpreting the spectra according to familiar chemical principles, and avoiding too close a brush with the heavy physics. This approach has been remarkably successful and nowadays no self-respecting chemistry department can operate without an NMR spectrometer. However, in recent years a bewildering array of specialized NMR techniques has been introduced, each presenting an interesting problem in spin physics, and each requiring its own theoretical treatment. This leaves the hard-working spectroscopist struggling a little to keep pace with new ideas like multiple-quantum coherence and two-dimensional spectroscopy. Worse still, the new experiments come with their own protective armour of specialist terminology, discouraging to the beginner.

This book starts from the premise that it is possible to explain some of these new concepts (and some older ones) in a pictorial and largely non-mathematical manner, in the belief that this is the approach favoured by most practitioners of NMR. (After all, they have chemistry, biochemistry or even medicine as their prime concern, not the spin Hamiltonian.) It is not a treatise for the complete novice in the field – there are no entries under chemical shift or spin–spin coupling – but it is written for those who already have a grounding in practical NMR spectroscopy and who would like to understand it better. In order to focus on the difficult bits instead of embedding them in a bland matrix like a plum pudding, I have adopted an unusual format. There are about sixty separate entries arranged alphabetically, rather like a very specialized encyclopaedia. These sections are self-contained, but they are cross-referenced to related sections (indicated by an asterisk (*)) and there is also an index to help the reader to find topics that are nested within one of the main entries. This form owes a considerable debt to Peter Atkins' book *Quanta* (1) which I have always admired. It has the advantage that the busy spectroscopist can go directly to the subject in question without having to search through a lot of extraneous material. Indeed, the handbook will have achieved its aim if it sits on the spectrometer console next to the coffee, rather than on the library shelf. There are even a few cartoons to help while away the time during long signal accumulations.

The late Andy Warhol has taught us that in the future everyone will be famous for 15 minutes. Failing that, everyone can aspire to inventing a pulse sequence and bestowing on it a suitably outrageous acronym. My attempts to compile a catalogue of pulse sequences and their mnemonics proved to be a disaster and had to be abandoned. The entries were too numerous and perhaps too ephemeral. In the final version of the book, the use of acronyms has been kept to an absolute minimum; they are employed where a full description of the experiment would be cumbersome and repetitive.

In a certain sense, this is a specialist glossary or lexicon, one definition of which is a 'vocabulary characteristic of a particular group of people'. NMR spectroscopists certainly constitute such a group, constantly coining new terminology (not to say jargon) and protecting the exclusiveness of their club. Without wishing to perpetuate any of this, I have nevertheless tried to retain the most commonly used form of words, for example *solvent suppression* and *saturation transfer*, although these are really just codes for more complex ideas. The real danger with jargon lies in its very *familiarity*; this can lull us into a false sense of security about our understanding of the terms.

The scope of this book had to be restricted, and I have excluded solid-state NMR, nuclear quadrupole resonance, magnetic resonance imaging and most of the experiments performed by physicists – they have their own club anyway. If this leaves an incongruous collection of topics, this may well be unavoidable in anything remotely resembling a dictionary. This is a book for dipping into when there is nothing more exciting to do on the spectrometer. The desire to keep each entry reasonably self-contained has led to repetition of certain ideas; the alternative was a rather tedious cross-referencing scheme that made the sections hard to read. Controversy has been positively encouraged, hence the space devoted to subjects like *maximum entropy* and *zero filling* which are guaranteed to arouse the passions of most NMR spectroscopists with their hint of witchcraft and something-for-nothing.

For a work largely concerned with *explanations* of NMR phenomena, it seemed reasonable to adopt the viewpoint taken in Professor Abragam's excellent book (2) and refrain from quoting the literature simply to give credit for a particular piece of work. The references are intended more as a guide to further reading; many are to general articles and reviews. An exception had to be made when it came to acknowledging the source of spectra used as examples. Unfortunately, since I took the path of least resistance and used the spectra closest to hand, there is a discernible narcissistic tendency in these references. For both these reasons it seemed wiser not to compile a name index.

Where does the reader start in a book of this kind? Probably not with the index, nor with the list of contents, but more likely with one of the 'cartoons'. Actually, these are not mere cartoons; they are intended to highlight some of the anecdotes used in the text to illustrate NMR concepts. The drawings themselves are the work of a young Italian art student, Valeria Petrone, who is also a close family friend. They have given me a lot of pleasure, and I am most grateful to Valeria for her excellent work. Some connoisseurs may well want the book for the illustrations rather than the undoubtedly limited appeal of the magnetic resonance ideas.

In the early years of this battle with words and word processors I was greatly helped by the drive and enthusiasm of my co-author, Gareth Morris, but he was unfortunately forced to withdraw through ill-health. Several colleagues were kind enough to read the final manuscript and made many valuable comments – Geoffrey Bodenhausen, James Keeler, Peter Bloch, Simon Davies, Jan Friedrich, Sally Davenport and Patrick Cook – they have saved me much embarrassment and I am greatly in their debt. Many of the spectra have been taken from the work of these and other Oxford students and I am grateful for their permission to use this material. My colleague Peter Hore was kind enough to perform several computer simulations specifically for illustrations in this book. Finally, I would also like to thank Bernhard Blümich, Andrew Derome and Tom Frenkiel for providing original spectra.

Magdalen College, Oxford. September 1986

REFERENCES

1. P. W. Atkins, *Quanta: A Handbook of Concepts.* Oxford University Press, 1974.
2. A. Abragam, *The Principles of Nuclear Magnetism.* Oxford University Press, 1961.

Acknowledgements

I am grateful to the following for permission to reproduce diagrams:

Academic Press for Fig. 1 p. 22, Fig. 2 p. 24, Fig. 3 p. 25 (from Shaka, Keeler & Freeman 1983); Fig. 2 p. 68 (from Freeman, Mareci & Morris 1981); Fig. 2 p. 94 (from Freeman, Kempsell & Levitt 1979); Fig. 1 p. 103 (from Levitt, Freeman & Frenkiel 1983); Fig. 1 p. 106 (from Freeman & Hill 1975); Fig. 2 p. 107 (from Bodenhausen *et al.* 1977); Fig. 3 p. 108 (from Levitt & Freeman 1979); Fig. 3 p. 115 (from Keeler & Neuhaus 1985); Fig. 4 p. 116 (from Bodenhausen *et al.* 1977); Fig. 2 p. 130 (from Freeman, Frenkiel & Levitt 1981); Fig. 3 p. 131 (from Shaka & Freeman 1982); Fig. 1 p. 135 (from Freeman & Morris 1978); Fig. 3 p. 138 (from Pei & Freeman 1982); Fig. 1 p. 149 (from Freeman & Morris 1978); Fig. 1 p. 152 (from Bodenhausen, Freeman & Turner 1977); Fig. 1 p. 188 (from Bax & Freeman 1980); Fig. 2 p. 201 (from Morris & Freeman 1978); Fig. 1 p. 208 and Fig. 3 p. 212 (From Bauer *et al.* 1984); Fig. 2 p. 209 (from Morris & Freeman 1978); Fig. 4 p. 234 (from Bax & Freeman 1981); Fig. 1 p. 240 (from Hore 1983); Fig. 4 p. 247 (adapted from Pei & Freeman 1982) and Fig. 1 p. 277 (adapted from Blümlich & Kaiser 1984); the American Chemical Society for Fig. 3 p. 41 (from Freeman *et al.* 1978), © 1978 American Chemical Society and Fig. 2 p. 159 and Fig. 4 p. 161 (from Morris & Freeman 1979), © 1979 American Chemical Society; the American Institute of Physics for Fig. 1 p. 260 (adapted from Freeman & Hill 1971) and Fig. 1 p. 290 (adapted from Ferretti & Freeman 1966); Franklin Institute Press for Fig. 2 p. 136 (adapted from Freeman & Morris 1979); the Royal Society of London for Fig. 1 p. 112 and Fig. 3 p. 246 (from Freeman 1980).

My thanks also to Carl Hanser Verlag, München, for the cartoon on p. 293 which is an adaptation of an illustration from 'Halbritters Waffenarsenal' copyright 1977 Carl Hanser Verlag, München, Wien.

Adiabatic Fast Passage

In most cases of interest to the NMR spectroscopist the interaction of nuclear magnetization with a radiofrequency field can be described in terms of the steady-state solutions of the Bloch equations for continuous-wave spectrometers or of the transient impulse response for Fourier transform machines. There is, however, another mode of excitation known as *adiabatic fast passage*. In this mode, the radiofrequency is applied continuously and its intensity is increased well above the normal level for continuous-wave NMR and it is swept through resonance rapidly compared with the slow-passage condition. Two inequalities must be satisfied.

$$\gamma B_1 \gg \frac{1}{B_1}\frac{dB_0}{dt} \gg T_1^{-1}, T_2^{-1} \qquad [1]$$

We shall see below that the linewidth of the NMR response is of the order of B_1, so the expression in the centre is the inverse of the time taken to sweep through the NMR response. Consequently the left-hand inequality requires that the sweep rate be slow with respect to γB_1; this is called the *adiabatic condition*. The right-hand inequality requires the sweep rate to be fast in comparison with the rates of relaxation T_1^{-1} and T_2^{-1}, so that the nuclear magnetization does not decay during passage through resonance.

Consider the motion of the nuclear magnetization vector M in a reference frame rotating about the Z axis in synchronism with the radiofrequency field, such that B_1 is a static field along the X axis, and the spins experience an effective field B_{eff} given by

$$\begin{aligned} B_{eff}^2 &= B_1^2 + \Delta B^2 \\ \tan\theta &= \Delta B/B_1, \end{aligned} \qquad [2]$$

where θ is the inclination of B_{eff} with respect to the X axis in the X–Z plane. During an adiabatic fast passage, the magnetization vector M follows the effective field B_{eff} exactly. It starts aligned along the +Z axis (Fig. 1) and follows a semicircular path in the X–Z plane, ending up aligned along the −Z axis. The maximum induced signal is proportional to M_0 and occurs at the exact resonance condition, but the vector is aligned along the X axis rather than along the Y axis, so the receiver must be adjusted as if it were to detect the dispersion-mode component. Nevertheless, the general form of the signal resembles the well-known absorption-mode lineshape, passing through a maximum at resonance and falling asymptotically to zero far from resonance. The linewidth is, however, very much greater than that in normal high-resolution work; the signal reaches half its maximum height at $\Delta B = \pm 3^{1/2} B_1$, where $\theta = \pm 60°$. Since B_1 must be quite strong in order to satisfy the adiabatic condition, the technique is not suitable for high-resolution work.

There is a second unusual property of the adiabatic fast passage signal. Provided that the inequalities [1] are satisfied, the signal does not exhibit saturation effects as B_1 is increased, it remains always proportional to M_0. For the early field-sweep spectrometers this was a very important asset, making it possible to search for an unknown resonance that might have a very long spin–lattice relaxation* time, for example silicon-29.

Failure to sweep sufficiently fast in comparison with the relaxation rates causes some loss of magnetization due to spin–spin and spin–lattice relaxation. Infringement of the adiabatic condition allows the vector M to lag behind the effective field B_{eff} and to begin a forced precession around B_{eff}, which leads to

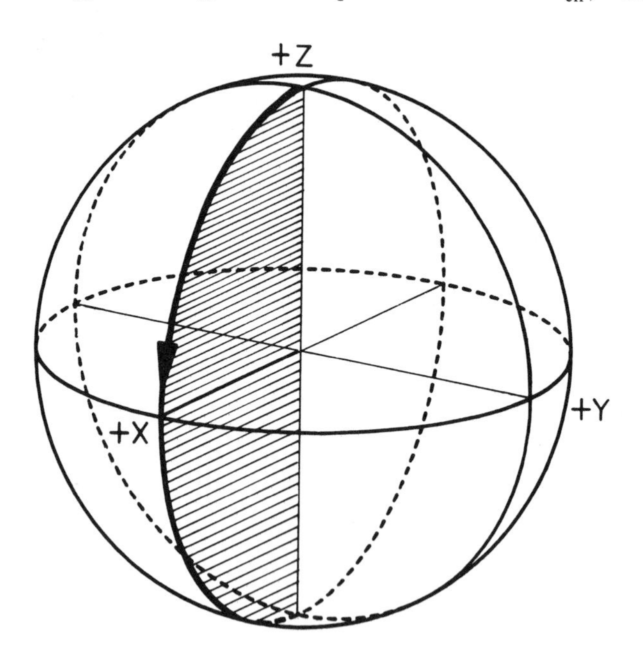

Fig. 1. Trajectory followed by the effective field B_{eff} during an adiabatic fast passage through resonance. Far below resonance B_{eff} is aligned along the +Z axis and it moves in the (shaded) Z–X plane, passing through the +X axis at exact resonance, reaching the −Z axis far above resonance. The magnetization vector remains always parallel to B_{eff} if the adiabatic condition is satisfied.

'wiggles' on the trailing edge of the observed response. This approaches the regime used in a technique (1) known as *rapid-scan correlation spectroscopy* (not to be confused with chemical shift correlation*). This is an alternative to pulse-excited Fourier transform spectroscopy, where the high-resolution spectrum is scanned rapidly with a sawtooth sweep. The transient 'wiggles' on the trailing edge of the resonances interfere with each other and produce complicated beat frequencies, and the linewidths are much broader than in slow-passage spectra. However, the broadening and beat patterns can be removed by cross-correlation with the lineshape and wiggle pattern from an isolated single line. After this data processing stage, the original high-resolution spectrum can be recovered. Two practical advantages are claimed for rapid-scan correlation spectroscopy. First, it achieves a uniform excitation of the nuclear spins over a wide spectral width with a relatively low-intensity B_1 field. (This is a more efficient use of radiofrequency power than in a Fourier transform spectrometer.) Second, it circumvents the problem of very intense solvent peaks, since it can be set up so that the sweep ramp does not pass through the solvent resonance. (*See* Solvent suppression*.) The sensitivity of correlation spectroscopy is considerably higher than that of continuous-wave slow passage methods and approaches that of Fourier transform spectroscopy. However, time must be allowed between consecutive sweeps for transverse magnetization to decay essentially completely, otherwise complicated steady-state effects* would ensue.

Adiabatic passage through resonance produces a population inversion (M aligned along the $-Z$ axis). If a return sweep is undertaken before any appreciable relaxation has occurred, the process is reversed and a normal Boltzmann population is restored. Alternatively, some delay may be introduced to allow spin–lattice relaxation and this can be made the basis for a method for measuring the relaxation time (2) by varying the time for which the spin populations are inverted between two consecutive adiabatic fast passages.

An adiabatic fast passage which is interrupted at exact resonance provides a simple method for aligning M along the radiofrequency field direction (X axis); this is the first step in a spin-locking* experiment, thus avoiding the necessity for a 90° radiofrequency phase shift. Recently, there have been some new techniques for wideband spin inversion related to adiabatic fast passage (3). Another innovation has been to perform adiabatic fast passage with a nonlinear frequency sweep (4).

REFERENCES

1. J. Dadok and R. F. Sprecher, *J. Magn. Reson.* **13**, 243 (1974).
2. A. Abragam, *The Principles of Nuclear Magnetism.* Oxford University Press, 1961.
3. J. Baum, R. Tycko and A. Pines, *J. Chem. Phys.* **79**, 4643 (1983).
4. C. J. Hardy, W. A. Edelstein and D. Vatis, *J. Magn. Reson.* **66**, 470 (1986).

Cross-references

Continuous-wave spectroscopy
Fourier transformation
Rotating frame
Shift correlation
Solvent suppression
Spin–lattice relaxation
Spin-locking
Steady-state effects
Vector model

Alignment of Molecules

High resolution NMR spectra are normally obtained from mobile liquids in which molecules tumble rapidly and isotropically: molecular motion is sufficiently fast and random that the observed resonant frequencies of individual nuclei represent an average over all possible orientations. For example, the chemical shift is known to be a function of the orientation of the molecule with respect to the applied magnetic field direction, but in a liquid we observe an average value: the 'isotropic' chemical shift. The field B_{DD} due to a dipole of strength μ at a distance r, subtending an angle θ with respect to the applied magnetic field direction, is given by the formula

$$B_{DD} = \pm(\mu_0/4\pi)\mu(3\cos^2\theta - 1)/r^3 \quad [1]$$

Here, the term $(\mu_0/4\pi)$ is the conversion factor for SI units. In an isotropic liquid the mean value of $(3\cos^2\theta - 1)$ vanishes and no dipole–dipole splittings are observed. Similarly, nuclear quadrupole splittings do not appear in conventional high-resolution spectra of liquids. In complete contrast, solid-state NMR spectra exhibit very broad lines since dipolar couplings, chemical shift anisotropy and quadrupolar coupling exert their full effect.

It can, however, be useful to restrict molecular motion so that the dipolar coupling is only partly averaged. Although this can be achieved for polar molecules by applying a very large electric field gradient, the principal method used is to dissolve small molecules in a nematic liquid crystal. This is an intermediate state of matter which behaves like a liquid except that the rod-shaped molecules exert anisotropic forces on one another, inducing a degree of alignment of their long axes. This is shown schematically in Fig. 1. These local domains of one-dimensional order would normally be oriented at random, but in the intense magnetic field of a high-resolution spectrometer they all acquire the same preferred direction. Usually, the liquid crystal molecules are not themselves observed, but a small solute molecule is added in low concentration. Although the small solute molecule still tumbles rapidly, it acquires a slight net orientation through its interactions with the neighbouring liquid crystal molecules. The degree of ordering is measured by an *order parameter* S, which is normally very much less than unity. As a result, dipole–dipole splittings appear in the high-resolution spectrum, but they are scaled down considerably by the factor S.

The great advantage of this controlled reintroduction of dipolar couplings is that, unlike the solid case, only intramolecular couplings are seen. Thus, for solute molecules with comparatively few magnetic nuclei, it is usually possible to resolve all the solute lines just as in an isotropic spectrum, although the very large number of lines that can result often makes analysis difficult. (The signals from the liquid crystal itself are usually broad and unresolved, since such systems tend to be tumbling rather slowly and they contain large numbers of protons.) Iterative computer analysis may be employed to extract accurate dipolar coupling constants from experimental spectra. From the known orientational dependence of the dipolar coupling it is then possible to derive very accurate information on molecular geometry. In general, the order parameter S is not known; however it is possible to take one internuclear distance as the standard for measuring all the others, thus bypassing the need to know S. In this manner the *relative* internuclear distances can be determined with high accuracy provided that the molecule is rigid.

The alignment methods described above rely on a cooperative effect between molecules with large anisotropies in their bulk magnetic susceptibilities. In the very intense magnetic fields now being introduced for high-resolution spectrometers (protons at 600 MHz) some *simple* molecules begin to show a small alignment, sometimes aided by dimerization. It is reported, for example, that many deuterated aromatic compounds (1) and monodeuterobenzene (2) at high field (14.1 tesla) show splittings attributable to the interaction of the deuteron nuclear quadrupole with the electric field gradient of the molecule. This interaction would average to zero in a strictly isotropic liquid. This raises important questions about the suitability of some deuterated reference materials (e.g. chloroform-d) for field/frequency regulation*.

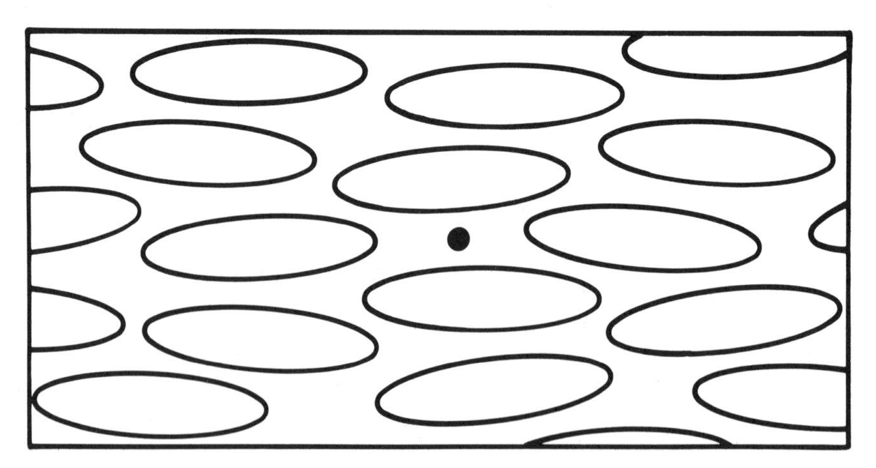

Fig. 1. Schematic representation of a solute molecule (black circle) dissolved in a nematic liquid crystal matrix where the long axes of the molecules are aligned.

REFERENCES

1. J. A. B. Lohman and C. MacLean, *Chem. Phys. Lett.* **58**, 483 (1978).
2. A. A. Bothner-By, C. Gayathri, P. C. M. van Zijl and C. MacLean, *J. Magn. Reson.* **56**, 456 (1984).

Cross-reference

Field/frequency regulation

Apodization

Apodization is a term borrowed from infra-red Fourier transform spectroscopy, with the literal meaning 'removing the feet'. The feet in question are the set of small sidelobes which appear on each side of the resonances in the frequency domain; they are derived from the sinc function character imposed on the lineshape by truncation of the time-domain signal before it has completely decayed.

Although an ideal free induction signal* in Fourier transform NMR would decay smoothly to zero, in practice the process of acquisition is usually terminated before the decay is complete, introducing a step function at the end. Often the data table is then completed by zero-filling*. The signal fed into the Fourier transformation* program may thus be regarded as the 'ideal' free induction decay multiplied by a step function. The convolution* theorem tells us that this corresponds to the convolution of the ideal frequency-domain lineshape with a sinc function; the more severe the discontinuity in the time-domain signal, the more marked are the sinc function 'wiggles'.

The effect may be removed by the application of a suitably chosen sensitivity enhancement* weighting function to the free induction signal, but this sacrifices resolution. Apodization is used to describe attempts to remove the sinc function 'wiggles' without significantly altering the resolution or signal-to-noise ratio. The idea is to apply a particular type of weighting function to the last 20–30% of the time-domain signal in order to bring it smoothly to zero at the end of the acquisition period. A widely used apodization function is made up from a half cycle of a cosine wave, $\frac{1}{2}(1 + \cos x)$. In fact, this only alleviates the problem and does not eliminate the wiggles (Fig. 1). Only a weighting function which operates on the entire time-domain signal can eradicate the sinc function artifacts entirely.

The frequency-domain spectrum obtained by digital Fourier transformation is of course discrete. It turns out that a step function at the very end of the free induction decay (with no zero filling) leads to a spectrum where the data points fall at the crossing points of the sinc function, disguising the effect. The transform of the corresponding zero-filled signal interpolates data points and reveals the underlying oscillatory structure.

Apodization is particularly important when a strong resolution enhancement* function has been used to emphasize the later parts of the free induction signal. Although many computer programs incorporate a mild apodization routine, it is preferable in these cases to include a component in the time-domain weighting function which brings the signal smoothly to zero at the end of acquisition. The Lorentzian-to-Gaussian function accomplishes this quite effectively. In two-dimensional spectroscopy*, apodization functions may be required in both the t_1 and t_2 dimensions.

In many types of two-dimensional Fourier transform experiment, early truncation of the signal in the t_1 time domain is unavoidable if the total experimental time is to be kept within reasonable limits. There are also some related 'interferogram' experiments where echo modulation patterns are used to distinguish CH_3, CH_2, CH and quaternary carbon sites, and where the total experiment time must be limited. In such cases, some rather more sophisticated methods may be used to eliminate the sinc function artifacts (1). Parametric fitting of the time-domain signal is one approach, allowing for the fact that the signal is truncated. Alternatively, the maximum entropy method* (2) may be applied; it is particularly effective for removing known artifacts of this kind (3).

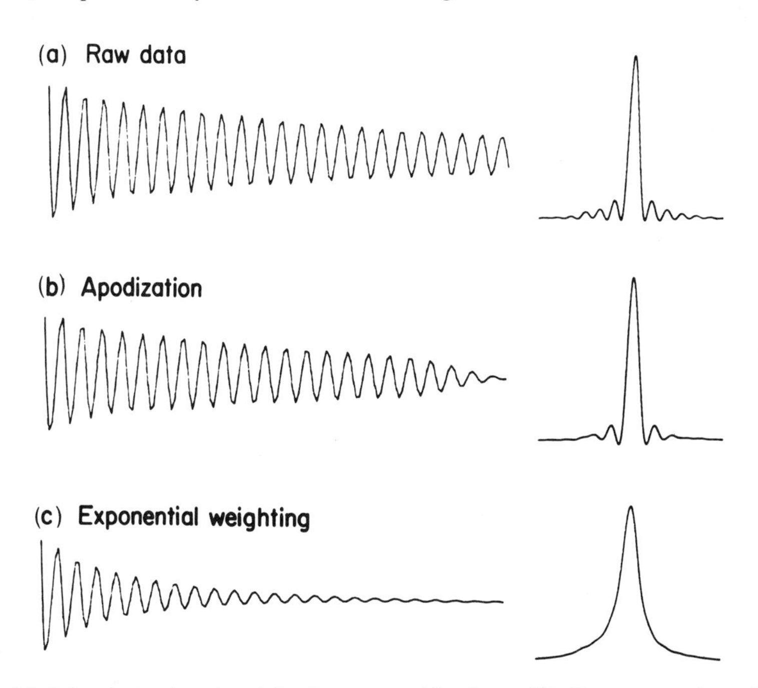

Fig. 1. (a) A free induction signal that is truncated (and zero filled) corresponds to a line in the frequency domain that has sinc function wiggles. (b) Apodization with a $0.5\,(1 + \cos\theta)$ function on the last 25% of the data attenuates but does not eliminate these wiggles. (c) An exponentially decaying weighting function removes the wiggles but broadens the line significantly.

REFERENCES

1. J. Keeler, *J. Magn. Reson.* **56**, 463 (1984).
2. S. F. Gull and G. J. Daniell, *Nature (Lond.)* **272**, 686 (1978).
3. P. J. Hore, *J. Magn. Reson.* **62**, 561 (1985).

Cross-references

Free induction decay
Fourier transformation
Convolution
Maximum entropy method
Resolution enhancement
Sensitivity enhancement
Two-dimensional spectroscopy
Zero-filling

Baseline Correction

The quantitative use of high resolution NMR spectroscopy requires accurate integration of the area under the resonance line since this is directly proportional to the number of nuclei giving rise to that resonance. The integral is very sensitive to small changes in the baseline of the spectrum, d.c. offset, tilt, hump or 'rolling baseline'. Gross disturbances can even give rise to perceptible errors in determining resonance frequencies.

These problems can arise from several sources, some purely instrumental, others with an NMR origin. Perhaps the most common of all is the effect of transmitter breakthrough – the fact that the receiver takes a short time to recover from the effect of the intense transmitter pulse, despite being gated off while the pulse is applied. This can falsify the first few data points of the free precession signal, particularly the first point. The incorrect first point transforms into a d.c. offset of the baseline of the spectrum; if the first two points are falsified there is also a bowing of the baseline (Fig. 1). If several signal ordinates are distorted at the beginning of the free induction decay, then higher frequency components are introduced – the phenomenon of the 'rolling baseline'.

The Fourier transform program treats the incoming free induction decay* as a periodically repeated function (whether or not it is actually repeated). There is thus a sudden jump in the signal voltage between the last point of one free induction decay (normally zero volts) and the first point of the next. The transformation program only works properly if this discontinuity is smoothed out, either by an analogue filter in the receiver or by halving the first ordinate of the free induction decay. Some spectrometer systems do not do this and the transformation program then interprets the first data point as being too large, resulting in an offset of the baseline in the frequency domain (1). This effect is usually made less severe because the audio bandpass filter tends to smooth out the effect of the discontinuity before the Fourier transformation* stage. In two-dimensional spectroscopy*, however, there is no equivalent filter for the t_1 dimension, and any discontinuity at $t_1 = 0$ results in prominent ridges parallel to the F_1 axis whenever there is a strong NMR response. (See t_1 Noise*.)

Alternatively, the baseline may be distorted by a genuine NMR response, one which decays very rapidly (short T_2) because the spins lack mobility (in solids, glasses, liquid crystals, viscous liquids). Probe construction techniques usually require a small amount of adhesive and, although this is kept to an absolute minimum in the vicinity of the receiver coil, it eventually contributes a broad background signal when working at high sensitivity. Other undesirable background signals can originate in the sample itself, for example with liquid crystals or with very large biological macromolecules. Spurious humps may even arise from the protons in water picked up by the leads to the receiver coil, where the local B_0 field is non-uniform and differs in intensity from the B_0 field at the receiver coil. High-resolution phosphorus-31 *in vivo* spectroscopy of animals commonly displays a broad background hump underneath the sharp lines of interest. This is partly due to the technique for spatial localization, which often employs B_0 field gradients specifically designed to produce a flat region of homogeneous field surrounded by regions of very steep gradients which give rise to a very broad signal. Immobile phosphorus atoms in bone also contribute a broad background signal. Many systems of biological interest are grossly heterogeneous and give rise to both narrow and broad resonances in the proton NMR spectra.

Genuine as these effects may be, it is often necessary to suppress the broad background in order to obtain quantitative intensities for the well-resolved resonances. A simple remedy is to delay acquisition of the free induction signal until after the first few data points which carry the broad signal component. The main consequence is to introduce a (calculable) linear frequency-dependent phase shift into the spectrum, which is easily corrected automatically or manually during presentation of the spectrum. Note, however, that if any broad lines remain in the spectrum, a frequency-dependent phase shift will distort them; each individual resonance should really have a *single* phase angle associated with it – not one which varies across the width of the line. A rather more elegant solution is to employ a suitable time-domain weighting function which de-emphasizes the first few sampling points of the free precession signal.

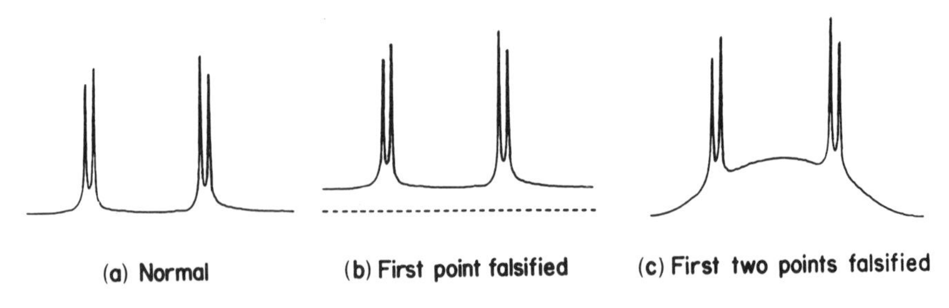

Fig. 1. The effect on the spectrum of falsifying the first data point or the first two data points of the free induction decay. The first gives rise to a baseline offset which may not be noticed until an integral is computed. The second gives rise to the effect known as 'rolling baseline'. The magnitude of the perturbation has been exaggerated for the purpose of illustration.

One that seems particularly appropriate in this context is the 'convolution difference' method (2). This applies a strong exponential sensitivity enhancement* function which broadens all the lines; this is then scaled down by a suitable factor and is then subtracted from the unprocessed free induction decay. The procedure is most simply visualized in the frequency domain. The spectrum contains some narrow lines and some very broad lines. The former are appreciably broadened by the Lorentzian convolution*, but the latter are scarcely affected. Subtraction cancels the broad resonances and leaves the sharp lines superimposed on a weak broad inverted pedestal, hardly apparent in the final spectrum. Several other possible weighting functions can be used; they should be designed so as to de-emphasize the early section of the free induction decay.

Baseline correction can also be carried out in the frequency domain. Baseline offset and tilt are easily corrected by the computer, the only problem is to indicate suitable regions of the spectrum (usually the high- and low-field edges) which are known to contain no genuine signals. More complicated baseline distortions can be handled by fitting the baseline to a polynomial which may then be subtracted out. This requires that the operator indicate several points throughout the spectral width corresponding to true baseline with no signal components.

REFERENCES

1. G. Otting, H. Widmer, G. Wagner and K. Wüthrich, *J. Magn. Reson.* **66**, 187 (1986).
2. I. D. Campbell, C. M. Dobson, R. J. P. Williams and A. V. Xavier, *J. Magn. Reson.* **11**, 172 (1973).

Cross-references

Convolution
Fourier transformation
Free induction decay
Sensitivity enhancement
t_1 Noise
Two-dimensional spectroscopy

Bloch–Siegert Effect

Almost all NMR experiments use an excitation which is a linearly oscillating radiofrequency field. This can be thought of as the superposition of two counter-rotating fields of the same frequency (Fig. 1), and the nuclear spins are principally affected by the component which rotates in the same sense as their precession. Thus protons would be sensitive to (say) the clockwise component, whereas nitrogen-15, which has a negative magnetogyric ratio, would react to the counter-clockwise component of radiation at the appropriate frequency. Bloch and Siegert (1) calculated the effect of the other 'unused' rotating component and found that its main effect was a very slight shift of the observed resonance line. To a good approximation, this *Bloch–Siegert shift* is given by

$$\Delta f = \tfrac{1}{2}(\gamma B_1/2\pi)^2/2F, \qquad [1]$$

where B_1 is the intensity of one rotating component of the transmitter field and

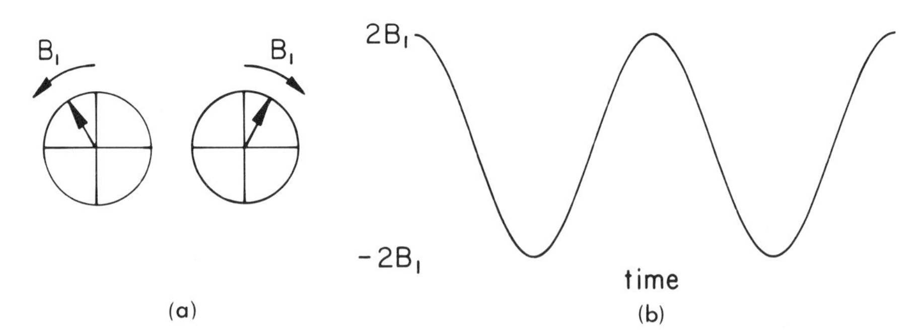

Fig. 1. NMR experiments are normally performed with a linearly oscillating field (b) which can be decomposed into two equal counter-rotating fields (a). Only the component that rotates in the same sense as the nuclear precession gives rise to resonance, but the other component causes the Bloch–Siegert effect.

F is the Larmor precession frequency, making a resonant offset of 2F Hz for the unused rotating component of the field. The shift is in a direction *away* from the offending non-resonant rotating component. In most practical high resolution cases, where $\gamma B_1/2\pi$ would seldom exceed 10 kHz, and 4F might be as high as 1000 MHz, the Bloch–Siegert shift is very small (in this example only 0.1 Hz). The shift disappears as soon as B_1 is extinguished, so that in pulsed NMR we are seldom concerned with this type of Bloch–Siegert effect.

An equivalent effect occurs whenever there is another radiofrequency field B_2 applied away from resonance for the nuclear spins. It was first described by Ramsey (2) but is still universally known as the Bloch–Siegert effect. It can be visualized using the vector model for nuclear magnetism in the rotating reference frame*. In the coordinate frame rotating in synchronism with the appropriate rotating component of this *second* radiofrequency field B_2, the spins 'see' an effective field which is the resultant of B_2 and the resonant offset ΔB. Their precession frequency is thus increased slightly by an amount $\gamma(B_{eff} - \Delta B)/2\pi$. For the usual cases where $\Delta B \gg B_2$ this gives a Bloch–Siegert shift of approximately

$$\Delta f = \tfrac{1}{2}(\gamma B_2/2\pi)^2/(\gamma \Delta B/2\pi) \qquad [2]$$

in analogy with eqn [1]. Since B_2 fields employed for decoupling are usually quite intense, and since ΔB can now be very much smaller than the Larmor frequency, this type of Bloch–Siegert shift can be appreciable. For example, in the determination of nuclear Overhauser effects*, where B_2 is applied during the observation of the free induction decay, these shifts can be an undesirable complication if difference spectroscopy* is employed. Much better difference spectra are obtained if B_2 is switched off during the acquisition of the free induction decay.

The ΔB term in the denominator of eqn [2] suggests that significant Bloch–Siegert shifts will normally only appear in *homonuclear* double resonance experiments. However, the effect cannot be entirely ruled out in certain heteronuclear experiments where the B_2 field is applied to another nuclear species. One example is the relatively unusual case of broadband decoupling of carbon-13 (proton observation), where a strong irradiation field B_2 can give rise to an appreciable Bloch–Siegert shift of the deuterium resonance at a distance of several tens of MHz. The process of switching the decoupler on or off may then seriously interfere with the operation of the field-frequency regulation*.

The Bloch–Siegert effect is quite general. The expression for the shift, eqn [2], can be used to predict coherent decoupling* and spin tickling* effects in double resonance by transforming the energy-level diagram into the rotating frame of reference and then calculating the Bloch–Siegert shifts of the energy levels due to the presence of the effective field B_{eff}, the resultant of B_2 and the resonance offset of the X spins. This accounts for the coalescence of spin multiplet structure on the A spin resonance, the appearance of weak satellite lines or (under the appropriate conditions) the spin tickling effect.

In situations where there is an appreciable spatial inhomogeneity in the radiofrequency field B_2, the Bloch–Siegert effect can give rise to line broadening as well as a shift. Different volume elements of the sample experience different Bloch–Siegert shifts; the consequent broadening is usually accompanied by a marked distortion of the lineshape.

When homonuclear decoupling experiments are used for determining chemical shifts, accurate measurements require correction for the Bloch–Siegert shifts. There is also a second correction to be applied which is twice as large and in the opposite sense. It arises because the optimum decoupling condition requires the quantization axes of the observed and irradiated spins to be orthogonal (3). Optimum decoupling occurs when B_2 is slightly displaced from exact resonance in a direction towards the observed resonance by an amount

$$\Delta f \approx (\gamma B_2/2\pi)^2/(\delta_A - \delta_X), \quad [3]$$

where $\delta_A - \delta_X$ is the chemical shift difference.

REFERENCES

1. F. Bloch and A. Siegert, *Phys. Rev.* **57**, 522 (1940).
2. N. F. Ramsey, *Phys. Rev.* **100**, 1191 (1955).
3. W. A. Anderson and R. Freeman, *J. Chem. Phys.* **37**, 85 (1962).

Cross-references

Coherent decoupling
Difference spectroscopy
Field/frequency regulation
Nuclear Overhauser effect
Rotating frame
Spin tickling

Broadband Decoupling

Carbon-13 spectra which include all the proton–carbon splittings are normally too complicated for easy interpretation, and are of low sensitivity because each carbon resonance is split into many multiplet components. Spectral simplicity thus requires a technique for strong irradiation over the entire range of proton resonance frequencies in order to decouple all CH spin–spin interactions. This is called broadband decoupling. The technique also provides the broadband saturation of the proton spins necessary for the generation of a nuclear Overhauser effect*.

Continuous monochromatic irradiation of protons is effective only over a very restricted frequency range. Consider a typical experimental situation where $\gamma_H B_2/2\pi$ might be of the order 8 kHz, J_{CH} of the order 200 Hz, and the maximum acceptable splitting of the carbon-13 resonances of the order 1 Hz. Application of the coherent decoupling* theory derived by Anderson (1) indicates that this limit is reached for a proton offset of only 40 Hz.

One of the earliest attempts to broaden the effective range of the B_2 field involved coherent phase modulation of B_2 at an audiofrequency related to the frequency $\gamma B_2/2\pi$ (2). More recently, some success has been achieved with square-wave phase modulation (inversion of the radiofrequency phase at a fixed rate) employing a modulation frequency of a few hundred hertz (3). It is tempting to visualize this experiment in terms of spreading out the B_2 field into an extended array of modulation sidebands. Unfortunately, such a picture does not lend itself to a detailed analysis and can give quite misleading conclusions.

A widely used method of broadband decoupling employs noise modulation (4). If two nuclei A and X are coupled and if the atom carrying X exchanges chemically at a rate fast compared with J_{AX}, then the multiplet on the A resonance coalesces to a single line. The exchange is a random process and thus differs in this respect from the coherent process of spin decoupling. Noise-modulated decoupling may be regarded as an attempt to introduce a similar incoherence into the irradiating field B_2. In practice, noise modulation is usually achieved by inverting the phase of a monochromatic radiofrequency source according to a pseudo-random sequence which is generated digitally. The mean rate of phase inversion acts as a bandwidth control, fast inversion corresponding to an irradiation over a broad band. The sequence repeats itself periodically, but it is a simple matter to make the number of steps in the period very large, so that the behaviour is almost indistinguishable from purely random modulation.

Chemical exchange at an intermediate rate gives a broad line for the observed A resonance. In much the same way, noise decoupling at a reduced level of B_2 can broaden the observed lines and introduce spurious noise components. Even with high-level noise decoupling it is sometimes possible to detect additional noise in the steep flanks of the decoupled carbon-13 resonance. Worse still, there may be a perceptible broadening of the decoupled carbon-13 resonances, attributable to unresolved residual splittings, and this translates into a sensitivity loss which can be ill afforded. Experiments which involve the generation of carbon-13 spin echoes* seem particularly sensitive to phase incoherence transmitted from the noise-irradiated proton spins (5).

This broadening at reduced levels of irradiation has been turned to good advantage for the identification of quaternary carbon sites, exploiting the fact that these resonances require only weak decoupling fields (small J_{CH}), while all resonances with one-bond couplings are broadened and usually disappear from the spectrum. Note, however, that in this experiment, part of the central line of a CH_2 group with equivalent protons also remains sharp, since the antisymmetric state is unaffected by CH decoupling (6).

Another effective way of spreading the influence of B_2 over a wide range of proton shifts is by 'chirp modulation', a term borrowed from radar technology. This involves a rapid repetitive sawtooth sweep of the irradiation frequency through all the proton resonance frequencies (7). It is tempting to speculate that this method works by inverting the proton spins by a form of adiabatic fast passage*, but the values of the experimental parameters used do not appear to support such a hypothesis.

All these techniques for broadband decoupling work better the higher the intensity of the B_2 field, but it is not usually possible to increase this much beyond the levels now used (typically $\gamma_H B_2/2\pi \approx 4$ kHz) because of sample heating by the radiofrequency field. Indeed, for aqueous samples containing ions, this limit has already been greatly exceeded. For any sample that might be altered or damaged by heating, the operator should always check that radiofrequency heating is at a tolerable level before embarking on a lengthy decoupling experiment. These problems can be somewhat alleviated by computer control of the decoupler level, so that a relatively low-level irradiation is used in the preparation period just before carbon-13 excitation, and a high level is used during acquisition of the free induction decay. This is because the requirements for the establishment of the nuclear Overhauser enhancement are rather less demanding than those for efficient broadband decoupling. In critical cases no preirradiation is used at all (sacrificing the nuclear Overhauser enhancement), high flow rates of cooling gas are used, and the preparation period is made long compared with the acquisition time (low duty cycle).

When it is necessary to decouple from a nuclear species with a very wide range of chemical shifts, for example fluorine, or to remove dipolar interactions from spectra of partially oriented samples, the present technology breaks down. Similar difficulties can be foreseen for proton decoupling in any future very high field spectrometers. There has therefore been a serious search for alternative schemes for broadband decoupling. This culminated in the discovery of a whole family of *deterministic* pulse sequences (8–10) which replace the *pseudo-random* sequences that held sway for so many years. These are repeated sequences of radiofrequency pulses. Each pulse is actually a composite pulse*, designed to have a built-in compensation for resonance offset effects or other imperfections. Reviews of the new broadband decoupling sequences have been written by Levitt *et al.* (11) and by Shaka and Keeler (12).

THEORIES OF BROADBAND DECOUPLING

The first step is to abandon the frequency-domain picture entirely and to examine the phenomena in the time domain as the carbon-13 free induction decay is being acquired. There are two principal theoretical treatments – coherent averaging theory (average Hamiltonian theory) and the exact theory due to Waugh (13). Average Hamiltonian theory is an approximate treatment which has proved most successful in solid-state multiple-pulse NMR. It assumes that the perturbation is *periodic* and that the sampling operation is synchronized with this period. It further assumes that the sequence is *cyclic* in that it returns the system to the same state once each period (relaxation is neglected). Just as a disco dancer viewed in a suitable strobe light might appear motionless although actually performing quite complex gyrations, so the observed carbon-13 signal appears to execute a deceptively simple time evolution.

The actual Hamiltonian describing the complex periodic motion is replaced by a time-independent *average* Hamiltonian $\overline{\mathcal{H}}$, expressed as an infinite series (the Magnus expansion). Provided that the cycle time is short enough, this power series converges rapidly and the higher-order terms can be neglected. In fact, for the carbon-13 case all odd-order terms vanish anyway; it is only the zero-order term $\overline{\mathcal{H}}^{(0)}$ which is of practical importance, the higher-order terms giving rise only to weak 'cycling sidebands' which can usually be disregarded. The decoupling problem is thus reduced to the consideration of the motion of the proton spins over one period, a considerable simplification. The drawback is that the treatment is only an approximation; nevertheless much of the initial progress in the development of the new broadband decoupling sequences was achieved by average Hamiltonian theory (11).

The Waugh theory (13) is simpler and more direct. It starts with the assumption that the sequence is periodic but does not necessarily require that it be cyclic. Initially, for the sake of simplicity, it assumes that the carbon-13 free induction signal is sampled in synchronism with the end of the decoupler period, although later it proves possible to relax this condition. Attention is focused on the two satellite resonances H(+) and H(−) flanking the main proton resonance, and corresponding to the two possible spin states of carbon-13. At any particular stage of the decoupling sequence H(+) and H(−) experience different effective fields B_{eff} owing to their different resonance offsets. Consider the motion during the first whole period t_s of the decoupling sequence. Each satellite performs a series of pure rotations, one for each radiofrequency pulse. The resultant of a succession of rotations is also a pure rotation since the trajectory is a series of linked arcs on the surface of a unit sphere, and the resultant is merely another such arc. We may then define an *overall* rotation

$$R(+) = \exp(-i\beta(+)\mathbf{n}(+)\cdot\mathbf{I}), \quad [1]$$

where $\mathbf{n}(+)$ is a unit vector defining the rotation axis and $\beta(+)$ is the angle of rotation for the H(+) satellite. A similar expression can be written for the other satellite in terms of $\mathbf{n}(-)$ and $\beta(-)$. Viewed in this manner, the problem takes on a remarkable similarity to continuous-wave coherent decoupling, the two vectors representing H(+) and H(−) appearing to execute a uniform rotational motion (at least under this 'stroboscopic' observation). Indeed, when the evolution of the carbon-13 free induction signal is calculated, the similarity is further emphasized because identical equations emerge

$$S(t) = 0.5[1 + \mathbf{n}(+)\cdot\mathbf{n}(-)]\cos(\Omega(-)t) + 0.5[1 - \mathbf{n}(+)\cdot\mathbf{n}(-)]\cos(\Omega(+)t). \quad [2]$$

Here $\Omega(+)$ is a fictitious frequency representing the sum of the precession rates of H(+) and H(−) under the influence of the decoupling sequence, while $\Omega(-)$ represents the difference.

$$\Omega(+) = [\beta(+) + \beta(-)]/2t_s$$
$$\Omega(-) = [\beta(+) - \beta(-)]/2t_s \quad [3]$$

This means that the carbon-13 free induction signal S(t) is amplitude modulated at $\Omega(+)$ and at $\Omega(-)$. Fourier transformation of S(t) gives a spectrum of four lines symmetrically disposed about the carbon-13 chemical shift frequency δ_c, just as in coherent decoupling. We focus attention on the two lines which have frequencies $\delta_c \pm \Omega(-)$ and relative intensities

$$I(-) = 0.5[1 + \mathbf{n}(+)\cdot\mathbf{n}(-)]. \quad [4]$$

There are two other lines at frequencies $\delta_c \pm \Omega(+)$ with relative intensities

$$I(+) = 0.5[1 - \mathbf{n}(+)\cdot\mathbf{n}(-)]. \quad [5]$$

Now, if we assume that $\gamma B_2 \gg 2\pi|J_{CH}|$ then both H(+) and H(−) experience very similar overall rotations (the condition for good decoupling) and $\mathbf{n}(+)$ is nearly parallel to $\mathbf{n}(-)$ so the first set of lines carry nearly all the intensity; the other two lines may be disregarded. The carbon-13 resonance is said to have a *residual splitting* $2\Omega(-)$ which is very much smaller than J_{CH}. Usually this is presented in terms of a *scaling factor*

$$\lambda = 2\Omega(-)/J_{CH}. \quad [6]$$

The aim is to make λ as small as possible over a wide range of proton resonance offsets so that the residual splittings are lost within the carbon-13 linewidth. One way to achieve this is to force the proton satellite vectors to follow closely similar trajectories so that $\beta(+) - \beta(-)$ is always small and $\mathbf{n}(+)$ and $\mathbf{n}(-)$

remain closely collinear. We might say that in this regime the two proton satellites behave as if there were no coupling to carbon-13, and conversely the carbon-13 resonance behaves as if it were not coupled to the protons. A more certain method of achieving the same goal is to make the decoupling sequence as nearly as possible an exact cycle; the residual overall rotations $\beta(+)$ and $\beta(-)$ are then both very small and so are their difference and their sum. It does not then matter what the rotation axes $\mathbf{n}(+)$ and $\mathbf{n}(-)$ are, and for exact cyclicity they become undefined. The new decoupling sequences seek to establish good cyclicity and to improve it in a systematic way by concatenating decoupler cycles into *supercycles*.

Evolution of the carbon-13 signal over the entire free induction decay need not be calculated explicitly, the results follow from the behaviour over the first period t_s. It is not even necessary to restrict sampling to the end of the period. More rapid sampling merely introduces very weak additional modulation components which appear in the spectrum as 'cycling sidebands'. If the carbon-13 free induction decay is deliberately sampled out of synchronism with the proton decoupling sequence, some of these cycling sidebands cancel because they have random phases. Some other sidebands have constant phase and persist, but they can be dispersed into a large number of much weaker components by stringing together different permutations of a given supercycle.

THE NEW DECOUPLING SEQUENCES

We start with a very simple-minded experiment. Although this is an impractical method in itself, it helps us to understand the way the ideas evolved that eventually led to the new decoupling sequences. The proton spins are periodically inverted by radiofrequency pulses while the carbon-13 free induction signal is sampled at the midpoint of the 'window' between adjacent pulses. Because of the scalar coupling, the individual carbon-13 spin multiplet vectors diverge in the X–Y plane of the rotating frame at $2\pi m_Z J_{CH}$, where m_Z is the proton spin quantum number (that is $\pm 1/2$ for a CH spin system). The proton pulses reverse the sign of m_Z, causing the spin multiplet vectors to coalesce at the time of signal acquisition. Sampled in this manner, the carbon-13 free induction signal appears to evolve only under the influence of its chemical shift with no effect of scalar coupling. (Note how this picture immediately suggests the use of average Hamiltonian theory.)

The catch is that the proton pulse repetition rate determines the carbon-13 sampling rate, and if the method is to be of practical utility these would have to be high, of the order of 20 kHz. The pulse widths are in this case no longer negligible in comparison with the windows between them. The simple picture breaks down because the proton motion during the pulse assumes increasing importance, and this can hardly represent an ideal 180° rotation for off-resonance spins. It is not feasible to increase the intensity of the pulses (and to decrease their widths) because this would soon exceed the high-voltage limits of the probe circuitry.

The breakthrough came with the discovery (8) of composite 180° pulses which provide accurate spin inversion over an appreciable range of resonance offsets. For example, the three-pulse sandwich

$$R = 90°(+X)\ 180°(+Y)\ 90°(+X) \qquad [7]$$

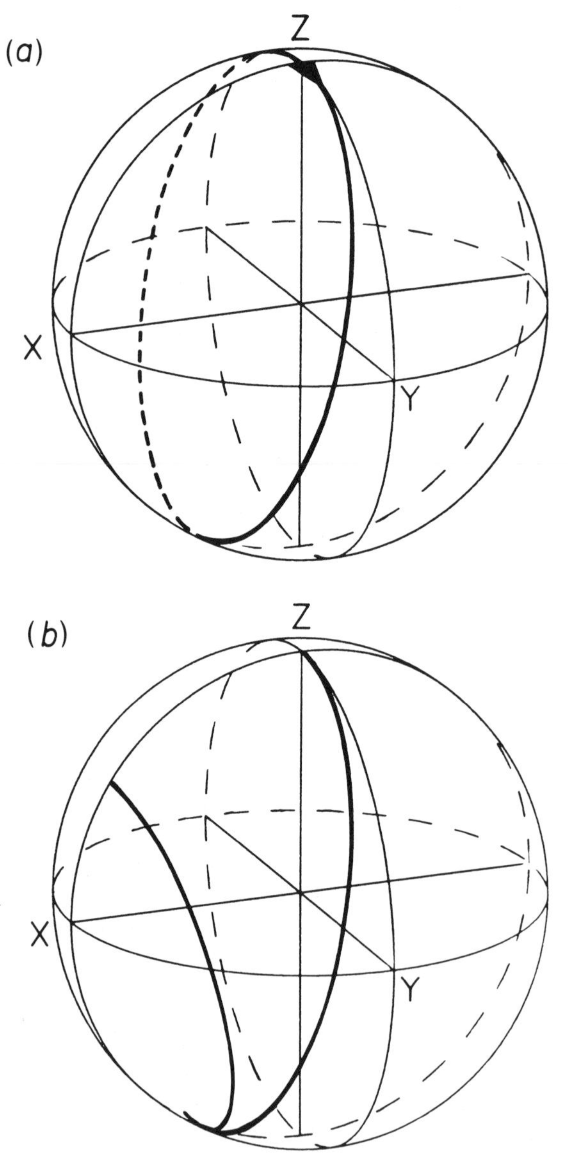

Fig. 1. (a) A repeated off-resonance inversion pulse RR has only small cumulative errors (the length of the arrowhead). (b) When R is combined with its phase-inverted counterpart $\bar{R}$ the cumulative errors are large. The combination R R $\bar{R}$ $\bar{R}$ is largely self-compensating.

provides reasonably good spin inversion over a range of $\Delta B = \pm B_1$. The idea is to assemble these composite pulses into a sequence with no windows between pulses. However, at finite resonance offsets, the repeated sequence $[R]_n$ would have cumulative errors (and so too would $[R\bar{R}]_n$, the sequence in which every other composite pulse has the phases of all its constituent pulses inverted (Fig. 1)). The first viable *magic cycle* is MLEV-4

$$[RR\bar{R}\bar{R}]_n, \tag{8}$$

where the subscript indicates that the sequence is repeated indefinitely. Clearly, if R is close to a 180° pulse, RR is nearly cyclic, the residual error being largely cancelled by the rest of the cycle $\bar{R}\bar{R}$. MLEV-4 already provides better decoupling than pseudo-random noise modulation, and this is only the first step.

Now, there are two important properties of such periodic decoupling sequences. First, their performance is unaffected by phase inversion. $RR\bar{R}\bar{R}$ has the same efficiency as $\bar{R}\bar{R}RR$. Second, they are unaffected by cyclic permutation of some part of the sequence, for example $RR\bar{R}\bar{R} \rightarrow \bar{R}RR\bar{R}$. However, if we examine in detail the nature of the residual imperfections (the deviations from exact cyclicity), these do change with phase inversion and cyclic permutation. For example, the residual overall rotation of $RR\bar{R}\bar{R}$ is a small rotation about an axis near +Z. It can be shown that the permuted sequence $\bar{R}RR\bar{R}$ has a residual rotation of similar amplitude about an axis near −Z. Consequently, when the two are combined into the MLEV-8 *supercycle*

$$RR\bar{R}\bar{R}\ \bar{R}RR\bar{R} \tag{9}$$

the two constituent imperfections cancel to a large degree.

Now, it turns out that the overall rotation which measures the residual imperfection of MLEV-8 is a small rotation about an axis near the X–Y plane. This can be largely cancelled by combining [9] with its phase-inverted counterpart to give MLEV-16

$$RR\bar{R}\bar{R}\ \bar{R}RR\bar{R}\ \bar{R}\bar{R}RR\ R\bar{R}\bar{R}R. \tag{10}$$

The compensation can be visualized by drawing the four trajectories representing the overall rotations for each of the four MLEV-4 subcycles which make up MLEV-16. They are four small arcs near the Z axis linked together and forming an almost closed loop; the small gap represents the residual imperfection of the MLEV-16 supercycle (Fig. 2). There is one such pattern for each proton satellite H(+) and H(−) and the residual splitting of the carbon-13 line depends on the difference between the small overall residual rotations.

These principles of cyclic permutation and phase inversion form the basis of all the later *expansion procedures* used for broadband decoupling. The cyclic permutation rule can be stated rather more generally: if a pulse (or combination of pulses) representing a rotation $\theta(X)$ is cyclically permuted, then the axis of the residual rotation of the sequence is itself rotated about the X axis through an angle θ. Thus, for the formation of the MLEV-8 sequence above, an approximately 180° pulse was permuted in MLEV-4, taking the axis of the residual rotation of $RR\bar{R}\bar{R}$ from near +Z to near −Z for $\bar{R}RR\bar{R}$. Now, of course, this means that the permuted segment should itself be well-behaved as a function of proton resonance offset; if not, the compensation will deteriorate with large offsets. It turns out that one of the best single pulses in this regard is a 90° pulse, for this is largely self-compensating with respect to resonance offset. This has been made the basis of a more efficient expansion procedure (9). Suppose we discover a cycle whose residual rotation is about an axis near +Z. Permutation of a 90° pulse rotates this axis so that it lies near the X–Y plane; we may call this permuted cycle P. Its residual imperfections are largely compensated by combining it with $\overline{P}$, whose residual rotation is also about an axis near the X–Y plane, but with the X and Y coordinates changed in sign.

This is the strategy adopted in the WALTZ family of decoupling sequences (10). These were devised when it was discovered that the MLEV sequences were extremely sensitive to small errors in the radiofrequency phase shifts. Now, it can be shown that sequences which only use radiofrequency phases of 0° and 180° are remarkably insensitive to errors in phase shift, an error as large as 10° having very little effect. The WALTZ sequences start with the composite inversion pulse

$$R = 90°(+X)\ 180°(-X)\ 270°(+X). \qquad [11]$$

In a convenient shorthand notation introduced by Waugh, where all pulses are expressed as multiples of 90°, this becomes $1\bar{2}3$ (hence the name WALTZ). The primitive sequence is WALTZ-4

$$RR\bar{R}\bar{R} = 1\bar{2}3\ 1\bar{2}3\ \bar{1}2\bar{3}\ \bar{1}2\bar{3}. \qquad [12]$$

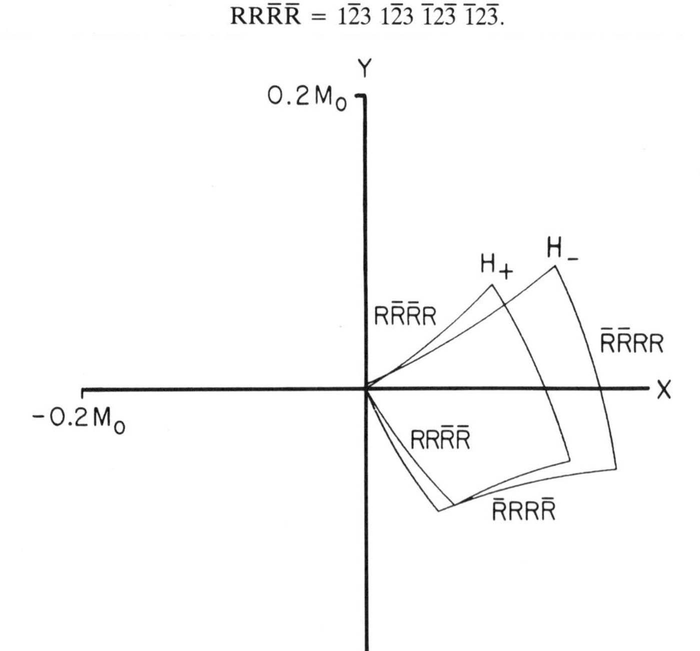

Fig. 2. The net overall rotations of a series of four MLEV-4 sequences shown in projection in the X–Y plane. Note that the excursions are very small compared with the radius M_0. Although the trajectories for the two proton satellites H_+ and H_- are different, they both form almost closed loops, indicating that the net rotations for the MLEV-16 sequence are both extremely small.

The residual overall rotation axis is near +Z. Permutation of a 90° pulse from left to right and combination with the phase inverted counterpart gives WALTZ-8

$$K\bar{K}\bar{K}K = \bar{2}4\bar{2}3\bar{1}\ 2\bar{4}2\bar{3}1\ 2\bar{4}2\bar{3}1\ \bar{2}4\bar{2}3\bar{1}. \qquad [13]$$

Note that if we allow for the trivial cyclic permutation $K\bar{K}\bar{K}K \rightarrow KK\bar{K}\bar{K}$, WALTZ-4 and WALTZ-8 have the same underlying structure.

The next stage of expansion permutes a 90° pulse cyclically from right to left and combines the result with the phase-inverted sequence. This gives the widely-used WALTZ-16 supercycle

$$Q\bar{Q}\bar{Q}Q = \bar{3}4\bar{2}3\bar{1}2\bar{4}2\bar{3}\ 3\bar{4}2\bar{3}1\bar{2}4\bar{2}3\ 3\bar{4}2\bar{3}1\bar{2}4\bar{2}3\ \bar{3}4\bar{2}3\bar{1}2\bar{4}2\bar{3} \qquad [14]$$

The scaling factor for WALTZ-16 is so small over the operating bandwidth that there is little point in continuing the expansion procedure. In fact, in practice, other sources of imperfection make their appearance and they may not be compensated for by further expansion. For example, because the sample is spinning, the B_2 intensity is modulated owing to the spatial inhomogeneity of the decoupler field, and this may be a problem for extended supercycles. The performance can also be degraded by pulse-width miscalibration or amplitude imbalance between the 0° and 180° channels.

WALTZ-16 appears adequate for most present-day applications of carbon-13 spectroscopy. Figure 3 illustrates the decoupling performance, as a function

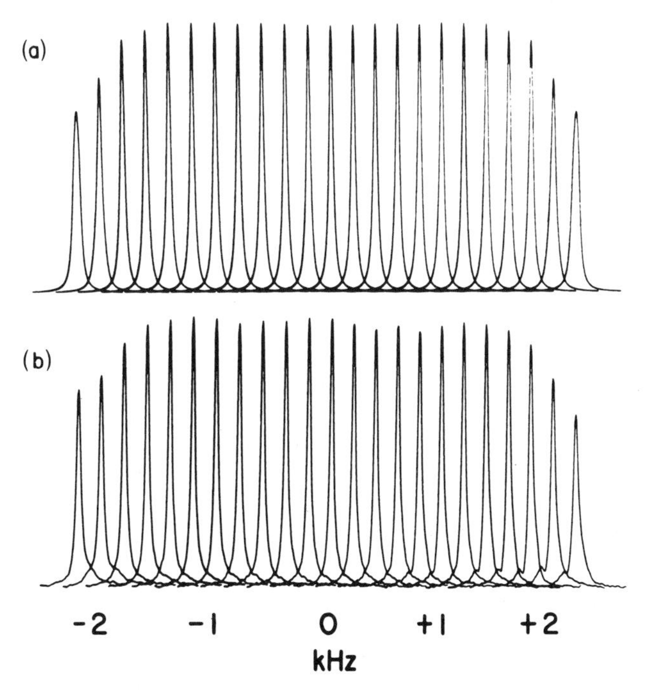

Fig. 3. (a) Computer simulated carbon-13 resonances for WALTZ-16 decoupling calculated at proton offsets which increase in 200 Hz steps, and taking into account the measured spatial inhomogeneity of the B_2 field. (b) The corresponding experimental carbon-13 resonances observed in a sample of formic acid. $\gamma B_2/2\pi$ = 2k Hz.

of proton resonance offset, for the carbon-13 resonance of formic acid (J_{CH} = 221 Hz). It has been demonstrated that, with suitable precautions, carbon-13 linewidths of the order of 0.02 Hz can be obtained (14). However, when there are two or more coupled protons with different chemical shifts, there can be residual line broadenings of the order of a few tenths of a hertz (15). The operating proton bandwidth is roughly $\Delta B = \pm B_2$. Wider bandwidths can be achieved by using pulses that are not restricted to multiples of 90° and by accepting somewhat larger residual splittings in the observed spectrum (usually masked by line broadening due to sensitivity enhancement). One such sequence is GARP-1, which achieves a bandwidth $\Delta B = \pm 2.4\ B_2$ (16).

REFERENCES

1. W. A. Anderson and R. Freeman, *J. Chem. Phys.* **37**, 85 (1962).
2. W. A. Anderson and F. A. Nelson, *J. Chem. Phys.* **39**, 183 (1963).
3. J. B. Grutzner and R. E. Santini, *J. Magn. Reson.* **19**, 173 (1975).
4. R. R. Ernst, *J. Chem. Phys.* **45**, 3845 (1966).
5. R. Freeman and H. D. W. Hill, *Dynamic Nuclear Magnetic Resonance Spectroscopy*, Eds.: L. M. Jackman and F. A. Cotton . Academic Press: New York, 1975, Ch. 5.
6. W. A. Anderson, *Phys. Rev.* **102**, 151 (1956).
7. V. J. Basus, P. D. Ellis, H. D. W. Hill and J. S. Waugh, *J. Chem. Phys.* **35**, 19 (1979).
8. M. H. Levitt and R. Freeman, *J. Magn. Reson.* **43**, 502 (1981).
9. J. S. Waugh, *J. Magn. Reson.* **49**, 517 (1982).
10. A. J. Shaka, J. Keeler and R. Freeman, *J. Magn. Reson.* **53**, 313 (1983).
11. M. H. Levitt, R. Freeman and T. Frenkiel, *Advances in Magnetic Resonance*, Ed.: J. S. Waugh. Academic Press: New York, 1983, Vol. 11.
12. A. J. Shaka and J. Keeler, *Progress in NMR Spectroscopy*, Eds.: J. W. Emsley, J. Feeney and J. Sutcliffe. Pergamon Press: Oxford, 1987, Vol. 19, p 47.
13. J. S. Waugh, *J. Magn. Reson.* **50**, 30 (1982).
14. A. Allerhand, R. E. Addleman, D. Osman and M. Dohrenwend, *J. Magn. Reson.* **65**, 361 (1985).
15. P. B. Barker, A. J. Shaka and R. Freeman, *J. Magn. Reson.* **65**, 535 (1985).
16. A. J. Shaka, P. B. Barker and R. Freeman, *J. Magn. Reson.* **64**, 547 (1985).

Cross-references

Adiabatic fast passage
Coherent decoupling
Composite pulses
Nuclear Overhauser effect
Spin echoes

Chemical Exchange

Historically, the first substance to have its high resolution spectrum investigated, ethanol, showed clear evidence of chemical exchange of the hydroxyl proton, and the physics of this process was analysed in a classic paper by Arnold (1). In effect, NMR provides a completely innocuous label for monitoring the exchange of an atom between two chemically distinguishable sites. They may be in different molecules or in different conformers of the same molecule. Whereas almost all other methods of following rate processes involve displacing the system from equilibrium, the NMR method applies to systems actually in an equilibrium state.

The way chemical exchange affects the NMR spectrum depends very much on the rate of exchange, and there are two regimes that are important – the slow exchange limit and the fast exchange limit.

SLOW EXCHANGE

When the rate of exchange is slow in comparison with the chemical shift difference between the two sites, the only observable effect in the NMR spectrum is a line broadening of the resonances involved in the exchange. Figure 1 illustrates how an exchange event interrupts the regular precession at each site, so that the frequencies are less well defined – a line broadening effect. We may think of T_2 the spin–spin relaxation time* as a *phase memory time*. If two spins are precessing exactly in phase at time zero, local magnetic fields from surrounding spins slowly upset their timekeeping and they lose phase coherence with a time constant T_2. If one of the spins is removed physically and replaced by another spin of random phase, the phase memory time of the ensemble is thereby shortened. Exchange at a rate $1/\tau$ thus makes an additional contribution to the natural line width

$$(T_2')^{-1} = (T_2)^{-1} + (\tau)^{-1}, \qquad [1]$$

where T_2 is the spin–spin relaxation time in the absence of exchange. In practical situations we seldom observe the natural linewidth $(T_2)^{-1}$ rad sec^{-1}, but rather a broader instrumental linewidth $(T_2^*)^{-1}$ rad sec^{-1} so that in the slow exchange limit the effect becomes observable only when the broadening exceeds the instrumental linewidth. Slower rates of exchange can be detected by a spin echo* experiment, by the saturation transfer* technique (2) or by two-dimensional spectroscopy*.

FAST EXCHANGE

When the rate of exchange is fast in comparison with the chemical shift difference $\omega_A - \omega_X$ between the two sites, quite different effects are observed. In the general case, the populations of the two sites p_A and p_X will be different. We must now fix our attention on a given spin, which hops rapidly between sites A and X, spending such a short time at each site that the 'phase jump' $(\omega_A - \omega_X)\tau$ is always very much less than 1 radian. The precession is thus a

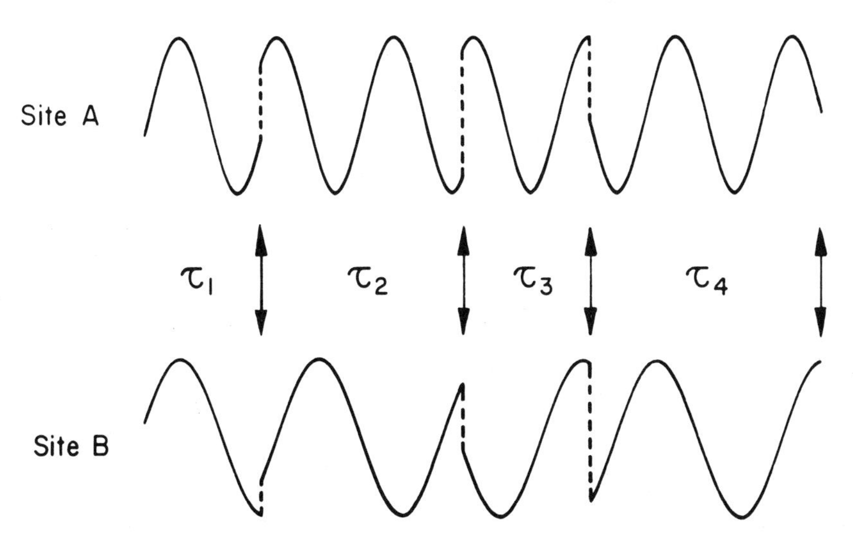

Fig. 1. Chemical exchange interrupts the free precession of a spin at site A by replacing it with another spin from site B that has a random phase. Random phase jumps lead to line broadening which increases as the reciprocal of the mean time between exchange events τ.

very jerky motion, made up of segments at frequency ω_A alternating with segments at frequency ω_X and with many such segments per revolution. The motion closely approximates precession at the weighted mean frequency

$$\Omega = \frac{\omega_A p_A + \omega_X p_X}{p_A + p_X}. \qquad [2]$$

Instead of the expected resonances at ω_A and ω_X, we observe only a single coalesced line at Ω, the weighted mean frequency or 'centre of gravity' of the spectrum. As the exchange rate increases, the phase jumps become smaller, making the precessional motion less jerky, so that the coalesced line becomes narrower. Note that this is the opposite of the behaviour observed in the slow exchange limit. The general expression for observed linewidth in the fast exchange limit may be written (3)

$$(T_2')^{-1} = (T_2)^{-1} + M_2\tau, \qquad [3]$$

where M_2 is the second moment of the lines in the absence of chemical exchange, defined by

$$M_2 = \sum_j p_j(\omega_j - \Omega)^2. \qquad [4]$$

The frequencies are all measured with respect to the centre of gravity of the spectrum Ω where

$$\sum_j p_j(\omega_j - \Omega) = 0. \qquad [5]$$

For the simplest case of only two sites, equally populated, eqn [3] reduces to

$$(T_2')^{-1} = (T_2)^{-1} + \left[\frac{\omega_A - \omega_X}{2}\right]^2 \tau. \qquad [6]$$

At very fast exchange rates the last term becomes negligible and the coalesced line eventually reaches the natural linewidth $(T_2)^{-1}$.

In many fast-exchanging systems, the relative populations are also changing, causing the coalesced line to shift dramatically within the spectrum. For example, in a two-site problem, if site A becomes more highly populated than site X, the coalesced line moves towards the chemical shift of site A, the 'centre of gravity' having been shifted according to eqn [5]. A good example is provided by the proton spectrum of ethanol, where the hydroxyl resonance is normally in fast exchange between hydrogen-bonded and 'free' sites. Inert solvents such as carbon tetrachloride discourage hydrogen bonding and the hydroxyl resonance moves to high field (between the methylene and methyl resonances). On the other hand, in neat ethanol, where hydrogen bonding is favoured, the hydroxyl resonance moves to low field. When a resonance line changes its frequency over a wide range as a result of solvent or temperature effects, this is often evidence for hydrogen bonding.

SPIN-ECHO METHODS

In order to avoid the effects of instrumental line broadening, investigations of chemical exchange often employ the spin echo method. These are usually studies of simple spin systems in the fast exchange limit, where the decay rate of the spin echoes yields the modified relaxation time T_2'. A surprising effect is observed if the repetition rate ($1/t_R$) of the spin-echo experiment is increased (4). The apparent relaxation time T_2' (normally much shorter than the 'true' spin–spin relaxation time T_2) increases and eventually approaches T_2 as the repetition rate becomes high in comparison with the exchange rate. The nuclei appear to behave as if no chemical exchange were taking place. This can be understood when it is realized that the line broadening in the fast exchange limit is proportional to the amplitude of the phase error introduced by a chemical exchange event. In the case of low pulse repetition rates the mean phase error due to an exchange event would be $M_2\tau$ radians, where M_2 is the second moment of the spectrum, defined in eqn [4]. However, we must consider the case where the pulse repetition rate becomes comparable with the exchange rate. The echo peak intensity is only affected by phase errors which build up during the time $t_R/2$ between a refocusing pulse and the echo peak. Consequently, the effective phase error is reduced, and at sufficiently high repetition rates the spin echo decay rate becomes independent of chemical exchange (4). This experimental approach allows both T_2' and T_2 to be evaluated.

INTERMEDIATE EXCHANGE RATES

When the exchange rate is comparable to the chemical shift difference between the two sites, these simple pictures break down. For two-site exchange, increasing exchange rate first broadens the individual lines, then draws them together so that they eventually coalesce into a single peak, which then gets narrower as the exchange rate increases. A proper analysis of line profiles in the case of intermediate exchange between several sites requires numerical methods. However, this is where the NMR spectrum is most strongly affected by chemical exchange, so reliable values of the exchange rates can be extracted.

Chemical exchange can also influence spin multiplet structure. If a resonance A exhibits a spin multiplet pattern due to coupling to another group of spins X, then if X undergoes chemical exchange at a rate which is fast compared with $|J_{AX}|$, the A multiplet will coalesce into a single line. This is why the hydroxyl resonance of ethanol is usually a singlet and the nearby methylene group shows no doublet structure attributable to coupling to the OH group. If, however, the sample is purified, cutting down the rate of chemical exchange of the hydroxyl proton, the OH resonance becomes a 1:2:1 triplet and the methylene group acquires an additional doublet splitting.

The possibility of chemical exchange can inhibit the use of the NMR method for certain structural determinations. Suppose that a molecule is suspected of existing in two conformers. The fact that only one spectrum is observed does not mean that only one conformer exists: it may merely indicate that the two forms are interchanging at a rate which is fast compared with the appropriate chemical shift difference.

REFERENCES

1. J. T. Arnold, *Phys. Rev.* **102**, 136 (1956).
2. S. Forsén and R. A. Hoffman, *J. Chem. Phys.* **39**, 2892 (1963).
3. L. H. Piette and W. A. Anderson, *J. Chem. Phys.* **30**, 899 (1959).
4. Z. Luz and S. Meiboom, *J. Chem. Phys.* **39**, 366 (1963).

Cross-references

Saturation transfer
Spin echoes
Spin–spin relaxation
Two-dimensional spectroscopy

Chemically Induced Nuclear Polarization

An NMR spectrum measures the absorption of radiofrequency energy, and net emission only occurs in very special circumstances, for example if some kind of pumping of spin populations occurs as in selective population transfer or some kinds of Overhauser experiments (see Nuclear Overhauser effect*). It was therefore quite a surprise when it was discovered that certain free radical reactions yielded products with strong (transient) emission and absorption lines. These 'anomalous' lines decayed to near-zero intensity by spin–lattice relaxation*, suggesting that they were the result of a non-Boltzmann population distribution. The presence of unpaired electron spins in the free radical stage led the initial investigators to the erroneous conclusion that this must be some kind of electron–nucleus Overhauser or dynamic polarization effect and hence used the name *chemically induced dynamic nuclear polarization* (CIDNP).

The actual mechanism (1–3) is quite remarkable, for it proposes that the nuclear spin can influence the course of a chemical reaction despite the fact that the energy gap between nuclear spin states in a magnetic field is tiny by comparison with bond dissociation energies. Consider the simplest possible case of a molecule ABH. In order to keep the model simple, only one nuclear spin (H) has been assumed, but more complicated extensions follow the same principles. Ultraviolet irradiation raises this to an excited singlet state ABH* which usually undergoes rapid intersystem crossing to the triplet state and then dissociates into radicals

$$ABH \rightarrow ABH^{*}\ (\text{singlet}) \rightarrow ABH^{*}\ (\text{triplet}) \rightarrow A\bullet + \bullet BH$$

Thus the two electrons in the radical pair have *parallel* spins, a triplet state. The reaction takes place in solution, usually in a deuterated solvent, so the nascent radical pair is trapped in a solvent cage for a time that is sufficiently long that many recollisions occur between the two radicals. As long as the two radicals remain within interaction range of each other they are referred to as a *geminate pair* and if they recombine to form a molecule this process is called *geminate recombination*. Although the two radicals may collide many times inside the solvent cage, they cannot recombine as long as the electron spins remain in a triplet configuration, for this represents antibonding. Bond formation can only occur if there is some triplet to singlet conversion before one of the radicals escapes from the cage.

There are three triplet states of different energy in a magnetic field (Fig. 1) called the T_0, T_+ and T_- states. All that is required for interconversion between the T_0 and S states is that one electron precess faster than the other for a sufficiently long period that a 180° phase difference builds up. This differential precession occurs because of the presence of the nuclear spin (H) operating through its hyperfine coupling to the electron on the •BH radical.

We might visualize this process by sketching the hypothetical electron spin resonance spectra of the two radicals (Fig. 2a). This is a single line at the g-value for the A radical, but is a doublet of splitting equal to the hyperfine coupling a_H for the B radical. For any given radical pair *only* the α nuclear spin state *or* the β nuclear spin state is applicable; somewhere else in the solution there will be another radical pair with the opposite label. Suppose it is in fact an α state, then for this radical pair the two electron precession frequencies are very close and there is only a low probability of triplet–singlet conversion. The two radicals may not form a chemical bond and the most likely fate is escape from the solvent cage followed by eventual reaction with the solvent (deuteron extraction). Note that escape of an A• radical and subsequent reaction with some distant •BH radical is *not* the same as geminate recombination: all phase memory is lost when one of the radicals escapes from its partner. By contrast, if we had considered a different radical pair where the proton was in a β state, then the two electrons would have had appreciably different precession frequencies, there would have been a high probability of triplet–singlet conversion, and the two radicals would be quite likely to form a bond (geminate recombination).

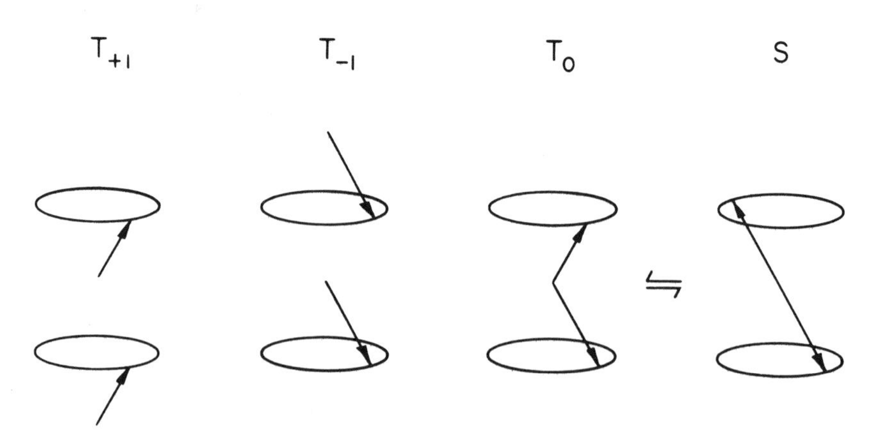

Fig. 1. Representation of the triplet and singlet states of a radical pair trapped in a solvent cage. In a magnetic field the T_0 and S states may interchange if one of the unpaired electron spins precesses faster than the other.

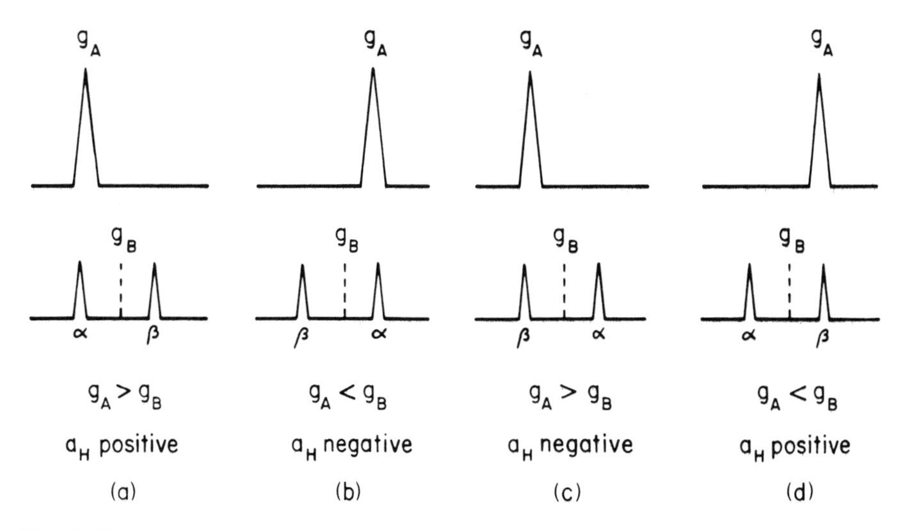

Fig. 2. Hypothetical ESR spectra of a radical A· and a radical ·BH in a magnetic field. The important parameters are the sign of the difference in g values and the sign of the hyperfine coupling a_H. Cases (a) and (b) favour triplet to singlet conversion for radical pairs with a β proton, whereas cases (c) and (d) favour conversion for radical pairs with an α proton.

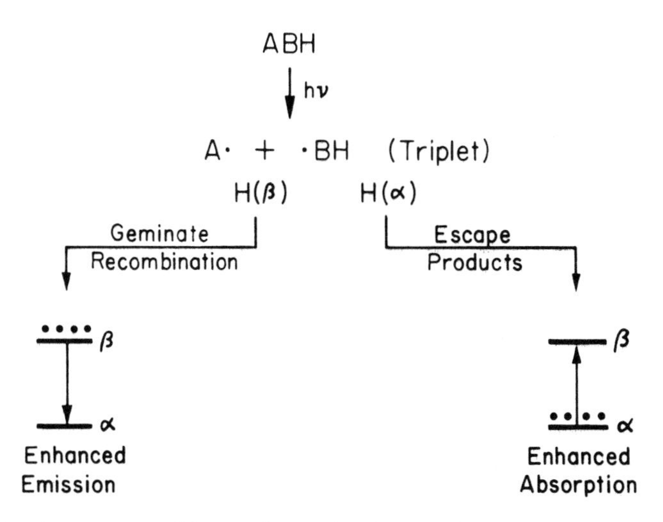

Fig. 3. Sorting into two possible reaction paths after a radical pair has been generated. If radical pairs with a β proton undergo triplet to singlet conversion, they are more likely to recombine within the solvent cage, producing ABH molecules with enhanced NMR emission. Radical pairs with an α proton are unlikely to recombine; they eventually escape from the cage, attacking the solvent and giving enhanced NMR absorption.

Geminate recombination thus leads to an ABH molecule with an enormous preponderance of protons in the β state, far in excess of the usual Boltzmann population. The corresponding proton resonance is therefore in emission and is extremely intense considering the small number of such product molecules (the free radical concentration is never very high). This is not a stable situation and at a rate determined by the spin–lattice relaxation* time almost 50% of this excess population relaxes by dropping down to the α level. The corresponding equilibrium NMR signal is very weak indeed and is often undetectable.

An escape product, perhaps a DBH molecule, has a large excess of protons in the α state, giving a very intense line in *absorption*. Since its resonance will in general be at a different frequency from the ABH molecule cited above, the two signals do not overlap and cancel. Spin–lattice relaxation causes a decay of this signal as proton spins move up to the β state (Fig. 3).

The sorting of the chemical reaction path is not of course as complete as inferred in this argument; it is possible that the radical pair makes two conversions: triplet–singlet–triplet, but usually the lifetime for trapping in the solvent cage is short compared with the inverse frequency difference for electron precession.

By now the reader will be objecting that some quite *ad hoc* assumptions were made in sketching the hypothetical electron spin resonance spectra used in this treatment. Indeed, it was assumed that $g_A > g_B$ and that the hyperfine coupling constant a_H was positive (otherwise the α and β labels would need to be interchanged). Reversal of one of these assumptions would have reversed the argument: geminate recombination generating an absorption spectrum and escape products an emission spectrum. Reversal of *both* assumptions brings us back to the result of Fig. 3. The four different cases are set out in Fig. 2. These sign reversals for the CIDNP effect are embodied in Kaptein's Rules (3).

The beauty of this experiment is that it provides good evidence for a free radical mechanism in the reaction under investigation, and the form of the enhanced NMR spectrum gives information about the structure of the radicals involved. The presence of two or more nuclei in a given radical gives rise to a new phenomenon – the *multiplet effect* – in which the relative intensities are perturbed within the spin multiplet. The mechanism for this follows along similar lines.

Historically, the discovery of chemically induced nuclear polarization provides a cautionary tale. It appears that several spectroscopists had noticed occasional inverted resonance lines long before the effect was reported and characterized. They had dismissed them as instrumental artifacts because when they tried to repeat the experiment (a second scan through the spectrum, for this was in the days of continuous-wave spectroscopy*) the anomalies disappeared. What they did not realize was that a free radical reaction was involved, initiated by ultraviolet irradiation (sunlight). To repeat the experiment properly would have required taking the sample out of the probe and re-exposing it to sunlight; then the emission lines would have been reproducible. They were not simply things that went bump in the night.

REFERENCES

1. G. L. Closs, *J. Am. Chem. Soc.* **91**, 4552 (1969).
2. G. L. Closs and A. D. Trifunac, *J. Am. Chem. Soc.* **91**, 4554 (1969).
3. R. Kaptein and L. J. Oosterhoff, *Chem. Phys. Letters* **4**, 195, 214 (1969).

Cross-references

Continuous-wave spectroscopy
Nuclear Overhauser effect
Polarization transfer
Spin–lattice relaxation

Coherent Decoupling

One of the first indications that NMR might provide scope for sophisticated new techniques was the early introduction of double resonance methods (1). Many of the new laser experiments involving double irradiation have their counterparts in some of these early NMR experiments. Included under this heading are the related fields of broadband decoupling*, off-resonance decoupling*, gated decoupling*, the nuclear Overhauser effect*, spin tickling*, polarization transfer*, the Hartmann–Hahn experiment*, saturation transfer*, and selective decoupling*. The simplest of all these techniques is coherent decoupling (2,3).

Coherent decoupling involves the irradiation of one group of nuclear spins X with a monochromatic source of intensity B_2, such that γB_2 is of the same order as $2\pi J_{AX}$, so as to modify the spin multiplet structure of any group of coupled spins A. In a Fourier transform spectrometer decoupling effects are only observed if the B_2 field is applied while the free induction signal is being acquired. If the B_2 irradiation is centred on the chemical shift frequency of X, and is sufficiently far away from the A resonance, then the A multiplet coalesces to a single line. On the other hand, if the irradiation is offset somewhat from this condition then the A spin multiplet splitting is scaled down, the apparent splitting increasing with decoupler offset and decreasing with the intensity B_2. Some new lines also make their appearance in each flank of the A spin multiplet; they are generally weak and their positions are quite sensitive to the intensity of B_2, so they are severely broadened by B_2 inhomogeneity. These *satellite lines* are accounted for in terms of multiple-quantum effects where at least one quantum comes from the irradiation field B_2.

Coherent decoupling is mainly used to establish which resonances are coupled together in spectra too complicated for this to be decided by inspection. The irradiation field B_2 can also be used to search for resonances that are hidden by overlapping lines (in homonuclear systems) or resonances that are not directly observable in the spectrometer (in heteronuclear systems). Heteronuclear coherent off-resonance decoupling is widely used to determine multiplicities of carbon-13 resonances. When it is necessary to eliminate all CH splittings from a carbon-13 spectrum, broadband decoupling methods are used.

The form of a decoupled spectrum can be calculated from a relatively simple theory (3). For convenience, we may neglect all population disturbances and concentrate our attention on the line positions and their intrinsic intensities. In fact, in a Fourier transform spectrometer this simplification is readily achieved if B_2 is only applied during acquisition of the free induction signal. Consider the simple case of a homonuclear AX spin system where the X region is irradiated with the B_2 field while the A region of the spectrum is investigated. First-order coupling is assumed. Figure 1 shows the energy level diagram appropriate to this system. In order to calculate the effect of introducing the radiofrequency field B_2, the energy level diagram is transformed into a rotating frame* synchronized with f_2, the frequency of the B_2 irradiation. This transformation allows B_2 to be represented as a *static* field in this frame. Energies and magnetic fields are both expressed in frequency units (hertz). The new energies after the transformation are easily calculated since all the allowed transitions are diminished in frequency by f_2 Hz. Some energy levels become nearly degenerate in this frame (1 and 2, for example) and are particularly sensitive to the introduction of the static field B_2 because a non-crossing rule comes into play, which holds these levels apart.

The amount of this perturbation is calculated according to the effective magnetic fields E_1 and E_2 acting on the X spins in the rotating frame, the resultants of their offset from resonance D_1 and D_2 and the perturbing field expressed in frequency units $\gamma B_2/2\pi$.

$$E_1 = [D_1^2 + (\gamma B_2/2\pi)^2]^{1/2} \quad E_2 = [D_2^2 + (\gamma B_2/2\pi)^2]^{1/2} \qquad [1]$$

$$D_1 = \delta_X + \tfrac{1}{2}J - f_2 \qquad D_2 = \delta_X - \tfrac{1}{2}J - f_2 \qquad [2]$$

Energy levels 1 and 2 are forced apart from D_1 to E_1, while levels 3 and 4 are forced apart from D_2 to E_2 (Fig. 1c). As a direct consequence, the A lines are

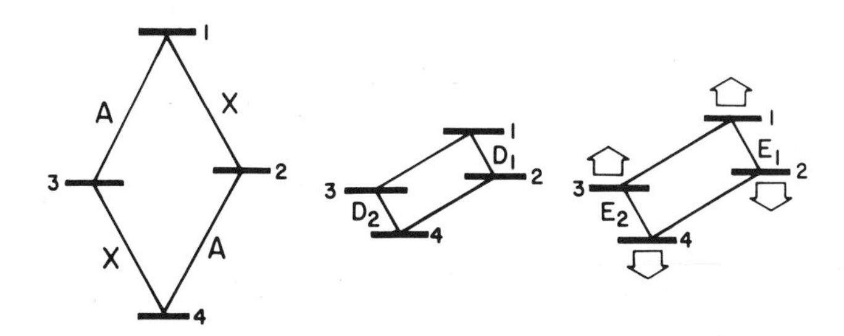

Fig. 1. Transformation from the laboratory frame (a) into the rotating frame (b) reduces all transition frequencies by the rotation frequency of the frame; consequently, it lowers energy level 1 and raises energy level 4 by the same amount (not to scale). The introduction (c) of the decoupling field B_2 near to resonance for the X spins increases the separation of levels 1 and 2 ($D_1 \rightarrow E_1$) and the separation of levels 3 and 4 ($D_2 \rightarrow E_2$). This diagram may then be used to predict the spectrum observed for the A spins.

shifted in frequency by an amount which depends on the decoupler offset. They remain symmetrically disposed about the chemical shift δ_A but move inwards to give a splitting $E_1 - E_2$ rather than J. This is illustrated for several different offsets in Fig. 2. Note that for irradiation at exact resonance, $f_2 = \delta_X$, the two lines coalesce into a single line at δ_A. The behaviour is symmetrical for positive and negative offsets.

The relative intensities of the lines are also a function of the decoupler offset. The intrinsic intensity is given by $\frac{1}{2}(1 + \cos\theta)$, where θ is the angle between the two effective field vectors E_1 and E_2. Note that if $\gamma B_2/2\pi > |J|$, then θ is never a very large angle and the intensities remain not far below unity. But where does the remaining intensity go? It reappears in weak satellite lines symmetrically placed outside the A multiplet with a separation $E_1 + E_2$. They can be considered as the two previously forbidden transitions: the zero-quantum line 2–3 and the double-quantum line 1–4, the selection rules being broken because of the perturbation by the B_2 irradiation. They are most prominent when B_2 is at exact resonance (Fig. 2), while for large decoupler offsets they move apart and lose intensity.

Analogous results apply to AX_2 and AX_3 spin systems where the X resonance is irradiated. The decoupled A spectrum remains a triplet or quartet with a reduced splitting given by $E_1 - E_2$, although a more complicated set of satellite lines can be observed. The AX_2 case has the interesting feature that one half of the intensity of the centre line of the A triplet is quite unaffected by irradiation of the X spins – it represents a transition between antisymmetric A-spin energy levels not shared by any X transitions. Thus the 1:2:1 intensity ratio is not strictly maintained. In the AX_3 case the quartet intensity ratio remains 1:3:3:1 throughout. The positions of the outer lines, however, are three times as

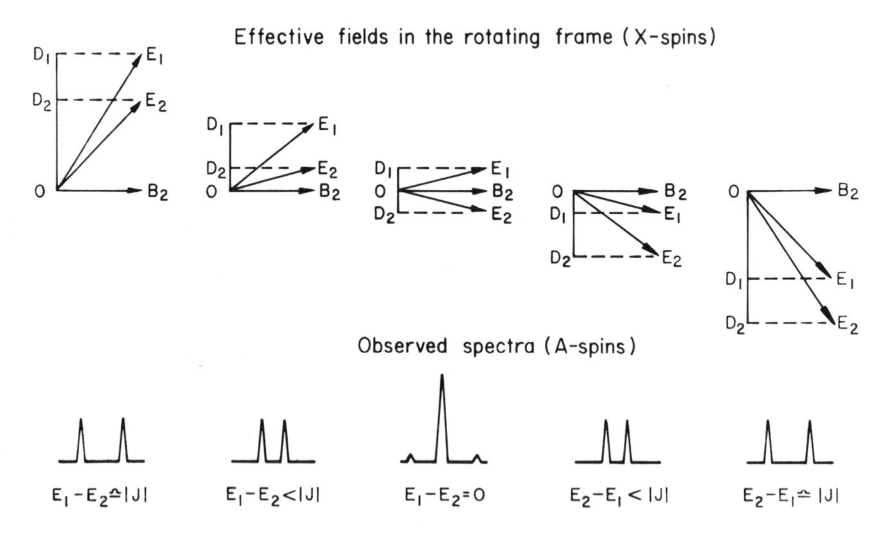

Fig. 2. Decoupling in an AX system with the B_2 field applied to the X spins. The A-spin resonance frequencies show a splitting $|E_1 - E_2|$ that is close to $|J|$ when B_2 is far from the X resonance but reduces to zero at exact X resonance. The weak satellite lines have a separation $|E_1 + E_2|$ and lose intensity as the decoupler goes off resonance.

sensitive to the strength of B_2 as the inner lines, so that in a spatially inhomogeneous irradiation field these lines are preferentially broadened. Where off-resonance proton decoupling is used to determine multiplicities of carbon-13 resonances, this has occasionally led to the misassignment of a quartet as a doublet. Figure 3 shows the dependence of a carbon-13 quartet from a methyl group as a function of proton resonance offset, taking into account the broadening due to the spatial inhomogeneity of B_2, and shows this preferential broadening of the outer lines (4).

When the sense of decoupling experiment is reversed, so that we are observing the A resonance of an A_2X or A_3X system while irradiating the X spins, the behaviour is different. It is no longer possible to find an irradiation frequency f_2 which causes complete coalescence of the A spectrum into a single line. Consider the A_3X case. It is now necessary to define four effective fields acting on the X spin, E_1, E_2, E_3 and E_4, and corresponding to the four possible combinations of the A spin states. Part of the A doublet intensity coalesces when $E_1 = E_2$, another part when $E_2 = E_3$, and yet another part when $E_3 = E_4$. There is said to be a *residual splitting* whatever the offset $f_2 - \delta_X$. However, by increasing the intensity $\gamma B_2/2\pi$ compared with J, this residual splitting can be reduced to any desired extent.

Coherent decoupling is often used to locate hidden resonance lines in homonuclear systems by searching for the frequency f_2 which gives optimum coalescence of an observed multiplet. In most cases it is perfectly permissible to take this as a measure of the chemical shift of the hidden resonance. However, the treatment given above makes the assumption that the A spins are not directly influenced by the irradiation field B_2, that is to say $|f_2 - \delta_A| \gg \gamma B_2/2\pi$. For relatively small chemical shift differences this condition may not be satisfied, and a small correction must then be applied. For the purpose of evaluating this correction we may neglect the small differences in resonance

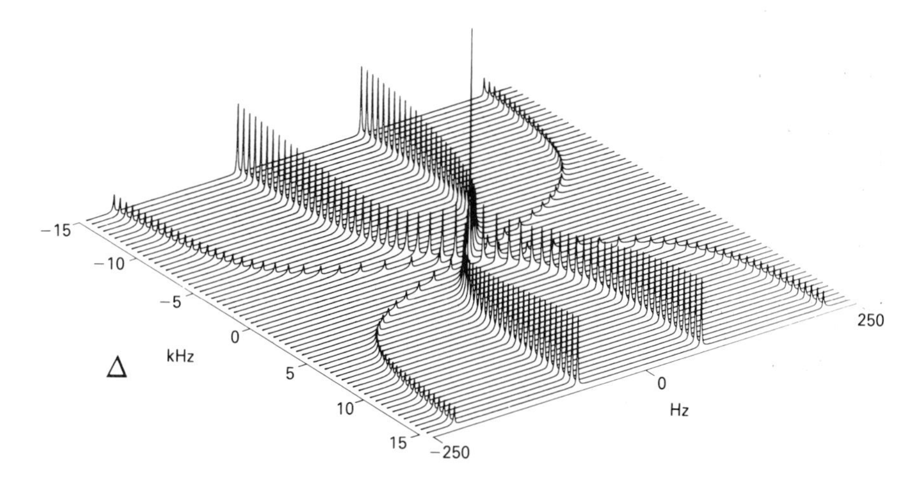

Fig. 3. Simulated A-spin spectra of an AX_3 system with the decoupler applied to the X spins, as a function of decoupler resonance offset Δ. This simulation takes into account the spatial inhomogeneity of the B_2 field which broadens the lines and distorts the lineshapes.

frequencies caused by J coupling and define an average effective field E_A acting on the A spins and an average effective field E_X acting on the X spins. The condition for exact decoupling is that E_A and E_X should be orthogonal. Since E_A is not quite aligned along the Z axis, f_2 must be moved slightly away from δ_X towards the A resonance in order to achieve this condition. The correction is therefore approximately given by

$$C = (\gamma B_2/2\pi)^2/(\delta_A - f_2) \text{ Hz}. \quad [3]$$

This effect thus slightly underestimates the chemical shift difference; on top of this, there is a Bloch–Siegert shift* of the observed A resonance due to the B_2 field; it acts in the opposite sense to the correction above and has half the magnitude.

REFERENCES

1. V. Royden, *Phys. Rev.* **96**, 543 (1954).
2. A. L. Bloom and J. N. Shoolery, *Phys. Rev.* **97**, 1261 (1955).
3. R. Freeman and W. A. Anderson, *J. Chem. Phys.* **37**, 85 (1962).
4. R. Freeman, J. B. Grutzner, G. A. Morris and D. L. Turner, *J. Am. Chem. Soc.* **100**, 5637 (1978).

Cross-references

Bloch–Siegert effect
Broadband decoupling
Gated decoupling
Hartmann–Hahn experiment
Nuclear Overhauser effect
Off-resonance decoupling
Polarization transfer
Rotating frame
Saturation transfer
Selective decoupling
Spin tickling

Composite Pulses

The pendulum is an excellent timekeeper because the period of the swing is virtually independent of its amplitude, and in any case, when it is used in a clock the amplitude of swing remains constant. The period is, however, dependent on the length, and small changes in length due to thermal expansion have a significant effect on the timekeeping. Hence the invention of the compensated pendulum in which the thermal expansion of one metal is arranged to counteract the different thermal expansion of another metal.

A similar principle has been applied to radiofrequency pulses.* When a single pulse does not achieve the desired effect, a composite 'sandwich' of pulses may often be devised to produce a better result. The possible applications of this idea cover a wide field, from solvent suppression* (1) to new methods of broadband decoupling*. Only two examples are discussed here – compensation for the effects of B_1 inhomogeneity and resonance offset errors. These are the two most serious pulse imperfections in high-resolution work with liquids.

COMPOSITE 90° PULSES

Suppose it is required to convert Z magnetization into transverse magnetization. Certain spin–lattice relaxation* methods, for example, demand an initial condition where there is no Z component from any part of the sample (not merely that residual Z components cancel). A simple 90°(X) pulse subject to spatial inhomogeneity in the B_1 field would not accomplish this, since magnetization from some volume elements of the sample would either undershoot or overshoot the equatorial plane. A composite 90°(X) 90°(Y) pulse helps considerably, by converting the spread in the Z–Y plane to a corresponding spread in the X–Y plane (Fig. 1). For certain relaxation applications this would suffice, but it does introduce phase errors, which in extreme cases could give rise to a serious loss in total transverse signal.

There is a more complicated composite sequence which takes a magnetization vector from the Z axis to the Y axis with good compensation for pulse length errors (2). This is the sequence

$$45°(-Y)\ 90°(X)\ 90°(Y)\ 45°(X) \qquad [1]$$

Figure 2 illustrates this sequence of four rotations. Note first that trajectories (b) and (c) are of equal length and are opposed tangentially, so that an increase in (b) is compensated by an equal increase in (c). More subtly, an increase in trajectory (a) is converted by the changes in (b) and (c) into a situation in which it is compensated by an equal increase in the length of trajectory (d). Such a composite pulse sequence might be used to initiate a spin–spin relaxation experiment in which it is important to have the magnetization accurately aligned along the Y axis of the rotating frame* (as in spin-locking*).

When resonance offset effects have to be taken into account, the situation becomes more complicated because rotations are now about an axis tilted in the X–Z plane through an angle θ, where

$$\tan \theta = \Delta B/B_1 \qquad [2]$$

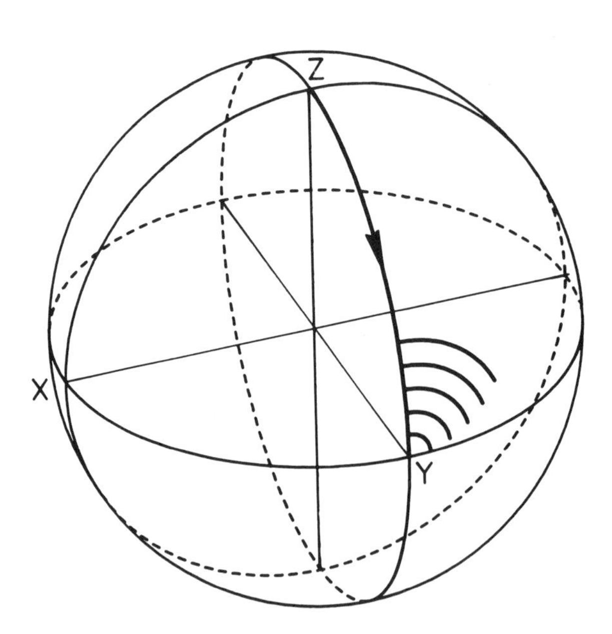

Fig. 1. A composite 90° pulse sequence 90°(X) 90°(Y) which compensates for small errors in the pulse length due to miscalibration or spatial inhomogeneity. The trajectories end close to the X–Y plane, but a significant phase error is introduced. Pulse length errors of 0°, 5°, 10°, 15°, 20° and 25° are shown.

and the flip angle is increased from $\gamma B_1 t$ to $\gamma B_{eff} t$, where

$$B_{eff} = [\Delta B^2 + B_1^2]^{1/2}. \quad [3]$$

An interesting self-compensation occurs for a 90° pulse because these two effects cancel to a large extent (3) and magnetization vectors initially along the Z axis are taken very close to the X–Y plane over a wide range of resonance offsets. Furthermore, the phase errors introduced by this single 90° pulse are essentially linearly dependent on offset and are hence easily corrected by a simple phase adjustment. Such self compensation appears to be a relatively rare event.

It is important to note that this type of compensation is not particularly demanding: there are some applications in which it would be necessary to take all Z magnetization vectors to the Y axis rather than simply disperse them in the X–Y plane. More difficult still is the general problem of constructing a compensated pulse which effects a pure 90° rotation *whatever the initial conditions* of the magnetization vectors.

COMPOSITE 180° PULSES

Population inversion pulses, which are widely used in spin–lattice relaxation measurements, may also be replaced by a composite sequence in order to

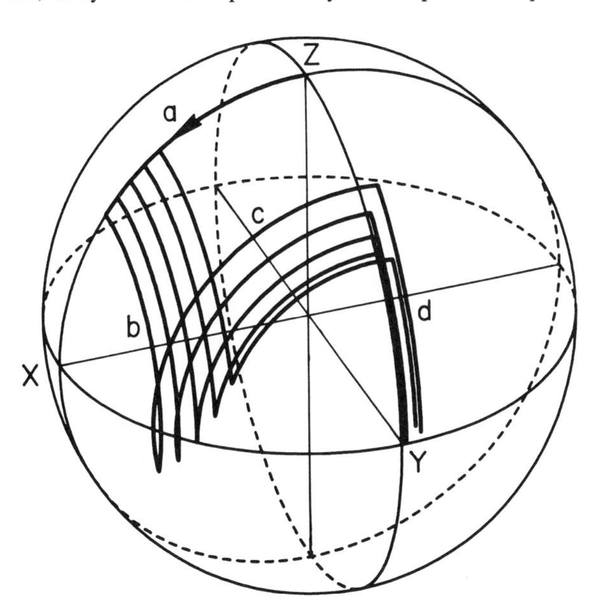

Fig. 2. A composite 90° pulse sequence 45°(−Y) 90°(+X) 90°(+Y) 45°(+X) designed to compensate pulse length errors. A family of five trajectories is shown for initial pulse lengths of 35°, 40°, 45°, 50° and 55° with the other pulses increased in proportion. The compensation relies on the approximate 'antitangential' condition of the curves b and c. All five trajectories terminate close to the +Y axis.

compensate B_1 inhomogeneity or miscalibration. A simple and effective sequence which is equivalent to 180°(Y) is

$$90°(X)\ 180°(Y)\ 90°(X). \qquad [4]$$

It is rather easy to see that if the first pulse is short, leaving the magnetization vector above the equatorial plane, the 180°(Y) pulse translates it to a point an equal distance below the equatorial plane, so that the last pulse carries the vector down to the −Z axis. Small errors in the 180°(Y) pulse have only very small second-order effects because the corresponding trajectory is so short (4).

This same sequence, or the slightly modified sequence 90°(X) 240°(Y) 90°(X), can also compensate for the effects of finite offset from resonance, although the two types of compensation do not mix – at significant offsets, pulse length errors are not corrected. Symmetry arguments are no longer useful, and the manner in which the compensation arises is best appreciated from the magnetization trajectories (Fig. 3). A family of such trajectories is shown covering the offset range where $\Delta B/B_1$ goes from 0.4 to 0.6. Figure 4 shows the corresponding trajectories for a simple 180° pulse. Spin–lattice relaxation measurements by inversion-recovery would benefit considerably from this type of offset compensation. Recently (5), sequences of this type have been used for a new method of broadband decoupling, allowing effective decoupling to be achieved at significantly lower radiofrequency powers.

A pulse which inverts longitudinal magnetization effectively must necessarily rotate X–Y magnetization back into the X–Y plane, that is to say it must behave

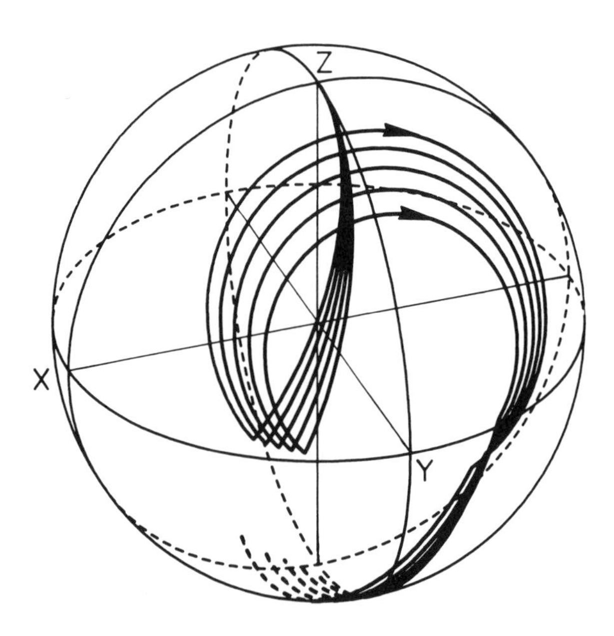

Fig. 3. A composite 180° pulse sequence 90°(X) 240°(Y) 90°(X) which compensates either for resonance offset (shown) or for pulse length error (not shown). Five trajectories are illustrated corresponding to resonance offsets between $\Delta B/B_1 = 0.4$ and 0.6. All five trajectories terminate near the −Z axis.

as a 180° rotation about some axis in the X–Y plane. Any residual errors due to resonance offset effects are translated into phase shifts of the rotation axis rather than into any tilting of this axis out of the equatorial plane. For this reason, the 90°(X) 240°(Y) 90°(X) sequence acts as an efficient refocusing pulse in spin echo* experiments. Although pulse imperfections introduce a phase error on odd-numbered echoes, these are cancelled on the even-numbered echoes, and the compensation is not affected by echo-modulation effects (6). Once again, a distinction can be drawn between different classes of compensation (7). Full compensation would take vectors through essentially the same angle as the corresponding ideal pulse, irrespective of initial conditions, provided we remain within the compensation range. Partial compensation achieves a similar result except for the introduction of a phase shift which depends on the extent of the imperfections.

EFFICIENCY OF COMPOSITE PULSES

The effectiveness of a given pulse for spin inversion may be measured in terms of the magnetization along the $-Z$ axis compared with M_0. An ideal pulse would therefore give -1. This allows the assessment of simultaneous imperfections due to pulse length error and resonance offset effects in the form of a

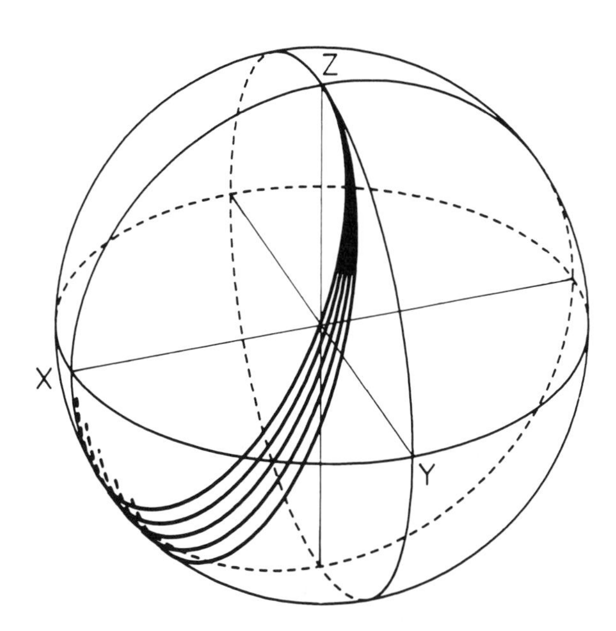

Fig. 4. The five trajectories of Fig. 3 for a simple 180°(X) pulse, indicating that resonance offset effects carry the magnetization vectors far from the $-Z$ axis.

contour diagram. Figure 5 shows such a diagram for the sequence 90°(X) 240°(Y) 90°(X), emphasizing that resonance offset effects can be compensated for provided there are not appreciable pulse length errors, or that compensation can be achieved for short pulses (not long pulses) provided there are no appreciable resonance offset effects.

Simultaneous compensation for both kinds of pulse imperfection is more difficult to achieve although there are many situations where both are likely to occur, for example with phosphorus-31 *in vivo* high resolution spectroscopy using a surface coil. Good compensation for pulse length error and resonance offset at the same time infers a contour diagram with an approximately rectangular region in which there are only contours near −1. The GROPE-16 sequence (8) makes a valuable contribution here. It was constructed by employing an important general principle known as *inverse rotation*. For a given nuclear species, the sign of the gyromagnetic ratio determines the sense of the rotation about the B_1 field. An inverse rotation is one which would turn magnetization vectors in the opposite sense. When resonance offset effects can be neglected, the inverse sequence can be constructed by reversing the chronological order of the pulses and inverting all the radiofrequency phases. In the more general case of appreciable resonance offsets, it is not possible to form the inverse sequence exactly, but a good approximation can be achieved by starting with a sequence which is a *cycle*. A cycle (9) is a set of rotations that

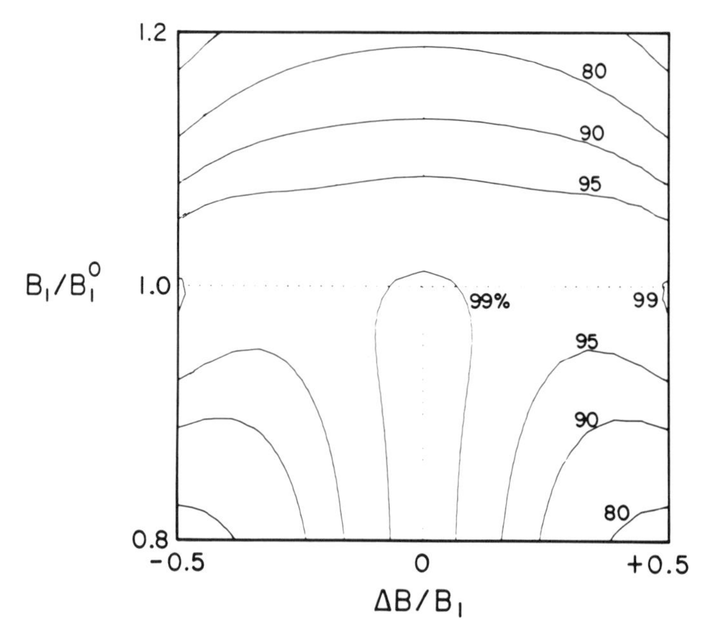

Fig. 5. Contour diagram plotting the efficiency for spin inversion for the 90°(X) 240°(Y) 90°(X) composite pulse as a function of resonance offset $\Delta B/B_1$ and misset of the radiofrequency field intensity $B_1/B_1°$. The sequence compensates for low values of B_1 near exact resonance, or for offset effects when B_1 is correctly set, but does not compensate for both types of error simultaneously.

has the overall effect of a rotation angle very close to zero. Now, if a sequence S has an inverse S^{-1}, then SS^{-1} must be a cycle. Consequently, if a broadband cycle has been found with a 90°(X) pulse at the beginning or at the end, then the remainder of the cycle (minus the 90°(X) pulse) is a good approximation to the inverse of a 90°(X) pulse. Because a 90° pulse takes magnetization efficiently from +Z to the X–Y plane, the combination $90°(X)\ 90°(-X)^{-1}$ behaves like a 180° pulse with good compensation. An inverse rotation pulse based on these principles was used to construct the GROPE-16 sequence.

PHASE SHIFTS

Suppose we take a pulse sequence S which has the overall effect of a rotation through an angle β about the −Y axis of the rotating frame, and sandwich it between two ideal 90° pulses

$$90°(X)\ \beta(-Y)\ 90°(-X). \qquad [5]$$

The result is an equal rotation β about the +Z axis of the rotating frame. That is to say the bracketing pulses 90°(X) and 90°(−X) have had the effect of rotating the rotation axis through +90° (from −Y to +Z) without altering β. This is known as a *similarity transformation*. The method can be used in order to achieve the effect of an arbitrary radiofrequency phase shift in equipment that has only hardware for 90° phase shifts (10).

PHASE DISTORTION

A given 90° pulse may provide good compensation for pulse imperfections when judged by its ability to take magnetization from the Z axis into the X–Y plane, but it still may not be an *ideal* 90° pulse. Often the imperfections appear in the form of a phase shift of the signal which varies as a function of resonance offset or pulse length error. Similar considerations apply to composite 180° pulses, limiting certain applications, for example refocusing in spin echo experiments. This is known as *phase distortion* (11). All the composite pulses mentioned so far suffer from this problem.

One important consequence of phase distortion is that it is not normally feasible to 'clean up' a complicated pulse sequence merely by replacing each pulse with its composite pulse equivalent, since the phase errors are likely to be cumulative. Levitt and Ernst (12) have demonstrated that by careful analysis of the effect of pulse imperfections in the 'INADEQUATE' experiment it is possible to incorporate composite pulses in such a manner that phase distortions are essentially self-compensating, whereas the indiscriminate substitution of composite pulses leads to disaster. Tycko *et al.* (11) have shown that a method based on the Magnus expansion can lead to the discovery of composite pulses that do not cause any significant phase distortion. Such pulse clusters can be substituted directly into a complex pulse sequence without fear of further complications due to cumulative phase errors. They are also useful where it is necessary to compensate for a *distribution* of values of some pulse parameter, for example gross spatial inhomogeneity of the radiofrequency field. If we wished to employ a surface coil to observe the *in vivo* NMR signal from an extensive sample volume, phase distortion caused by the distribution of B_1 values could lead to considerable signal cancellation.

HOW ARE COMPOSITE PULSES DEVISED?

The discovery of the first composite 180° pulse by Levitt (4) seems to have been inspired principally by intuition guided by visualization of the magnetization trajectories on the unit sphere (the vector model*). Sometimes, an analytical solution is possible, based on trigonometrical requirements or symmetry considerations. It helps enormously to have the trajectories drawn out by computer simulation as in Figs 1–4. These calculations are usually carried out using rotation operators, a powerful method for analysing the effect of a composite pulse (13). Geometrical methods remain the simplest and possibly the most satisfying approach, and have been used extensively in this section. Nevertheless, the field of composite pulses has grown enormously (even into laser spectroscopy) and many different methods of analysis have been used. Mention has already been made of the concept of *inverse rotation* where magnetization vectors rotate about an effective field in the opposite sense to that predicted by the sign of the gyromagnetic ratio.

Spin inversion can be achieved through adiabatic fast passage*, including versions where the radiofrequency amplitude or the sweep rate vary during passage through resonance. Composite pulses have been devised by approximating these frequency-sweep experiments by a limited number of radiofrequency pulses of the appropriate phases (14).

While tinkering with someone else's composite pulse sequence is frowned upon, it is often possible to take an existing composite pulse and to improve it by an iterative expansion procedure (15), in which better compensation is achieved at the expense of a more complicated sequence. Numerical optimization by computer can be a very powerful technique for this purpose (16). Finally, as mentioned above, Tycko *et al.* (11) have demonstrated the utility of a theoretical approach based on the Magnus expansion, which allows phase distortion to be compensated at the same time as the other imperfections.

We have certainly not heard the last word on composite pulses.

REFERENCES

1. A. G. Redfield, S. D. Kunz and E. K. Ralph, *J. Magn. Reson.* **19**, 114 (1975).
2. M. H. Levitt, *J. Magn. Reson.* **48**, 234 (1982).
3. R. Freeman and H. D. W. Hill, *J. Chem. Phys.* **54**, 3367 (1971).
4. M. H. Levitt and R. Freeman, *J. Magn. Reson.* **33**, 473 (1979).
5. A. J. Shaka, J. Keeler and R. Freeman, *J. Magn. Reson.* **53**, 313 (1983).
6. M. H. Levitt and R. Freeman, *J. Magn. Reson.* **43**, 65 (1981).
7. M. H. Levitt, *Progress in NMR Spectroscopy* **18**, 61 (1986).
8. A. J. Shaka and R. Freeman, *J. Magn. Reson.* **55**, 487 (1983).
9. J. S. Waugh, *J. Magn. Reson.* **50**, 30 (1982).
10. R. Freeman, T. Frenkiel and M. H. Levitt, *J. Magn. Reson.* **44**, 409 (1981).
11. R. Tycko, H. M. Cho, E. Schneider and A. Pines, *J. Magn. Reson.* **61**, 90 (1985).
12. M. H. Levitt and R. R. Ernst, *Mol. Phys.* **50**, 1109 (1983).
13. M. H. Levitt, *J. Magn. Reson.* **50**, 95 (1982).
14. J. Baum, R. Tycko and A. Pines, *J. Chem. Phys.* **79**, 4643 (1983).
15. M. H. Levitt and R. R. Ernst, *J. Magn. Reson.* **55**, 247 (1983).
16. A. J. Shaka, *Chem. Phys. Lett.* **120**, 201 (1985).

Cross-references

Adiabatic fast passage
Broadband decoupling
Radiofrequency pulses
Rotating frame
Solvent suppression
Spin echoes
Spin–lattice relaxation
Spin-locking
Vector model

Continuous-Wave Spectroscopy

Almost all high resolution NMR spectrometers before about 1970 employed continuous-wave irradiation and swept either the applied field or the radiofrequency. This mode of operation seemed natural to the early NMR workers since it represented a direct parallel with the familiar optical spectroscopy techniques. Although it had been realized for a long time that the transient response to a radiofrequency pulse gave the Fourier transform of the slow-passage spectrum, it was not until the pioneering work of Ernst and Anderson (1) that the enormous sensitivity advantage of the Fourier transform mode was properly appreciated. Implementation of this principle on commercial spectrometers was further delayed until inexpensive digital minicomputers were readily available and the Cooley–Tukey algorithm* introduced (2).

The Fourier transform revolution was soon complete. It is now difficult to find any definite advantage of the continuous-wave method for high-resolution work, except the absence of dynamic range problems and the doubtful economic argument that a digital computer is not strictly necessary (although time averaging* would require one). A Fourier transform spectrometer can perform all the same functions and at the same time delivers one or two orders of magnitude better sensitivity and provides access to time-dependent phenomena such as relaxation and chemical exchange. Continuous-wave spectroscopy has never been satisfactory for low-sensitivity nuclei such as carbon-13 or nitrogen-15. Nevertheless it is still used, particularly for simple proton spectrometers of moderate cost.

The majority of continuous-wave spectrometers sweep the irradiation frequency (or field) sufficiently slowly that the slow-passage condition is approached. This requires that the sweep rate be reduced to the point that the time taken to sweep through a resonance line be of the order of the spin–spin relaxation time T_2, if the line has its natural width $(\pi T_2)^{-1}$ Hz. In practice, linewidths are more usually determined by instrumental effects $(\pi T_2^*)^{-1}$ and the sweep rate may be increased to the point where the time taken to traverse

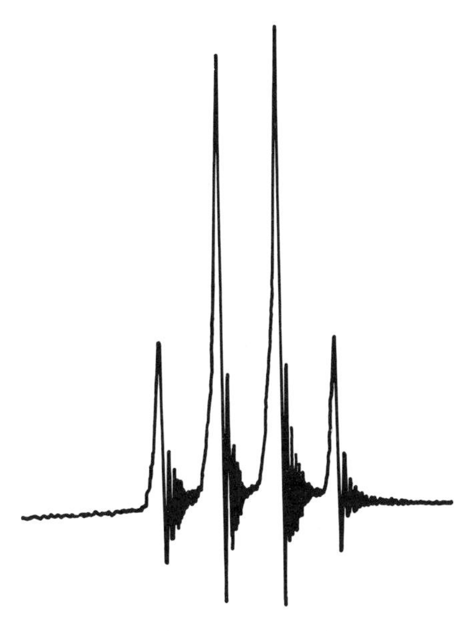

Fig. 1. Typical continuous-wave NMR spectrum (quartet of acetaldehyde) showing the wiggles on the trailing edge of the lines. This is a result of failure to meet the slow-passage condition.

resonance is of the order of T_2^*. Violation of the slow-passage criterion broadens and distorts the observed resonance, introducing the well known ringing pattern on the trailing edge of the line, the 'wiggles' (Fig. 1). Most spectra obtained by continuous-wave methods show some trace of wiggles, and this may thus be used to distinguish such spectra from those obtained by Fourier transformation* where sweep artifacts are absent. Rapid-scan correlation spectroscopy seeks to occupy the middle ground, sweeping rapidly through resonance in order to improve sensitivity* but later removing the intense wiggle–beat pattern by cross-correlation with the response from a single, isolated resonance line.

STEADY-STATE SOLUTIONS TO THE BLOCH EQUATIONS

A slow-passage continuous-wave spectrometer excites an NMR response which can be calculated from the steady-state solutions to the Bloch equations (3, 4) in the rotating frame*. It is assumed that the radiofrequency field B_1 is at a low level and has been applied for a sufficiently long time that any transient effects have all died away. The steady-state signal may then be represented by a vector M (see Vector model*) having three components, each of which is a function of Δf the offset from resonance.

$$M_X = M_0 2\pi\Delta f\gamma B_1 T_2^2/[1 + (2\pi\Delta f T_2)^2 + \gamma^2 B_1^2 T_1 T_2] \quad [1]$$

$$M_Y = M_0\gamma B_1 T_2/[1 + (2\pi\Delta f T_2)^2 + \gamma^2 B_1^2 T_1 T_2] \quad [2]$$

$$M_Z = M_0[1 + (2\pi\Delta f T_2)^2]/[1 + (2\pi\Delta f T_2)^2 + \gamma^2 B_1^2 T_1 T_2] \quad [3]$$

The signal is thus always proportional to M_0, the initial polarization at Boltzmann equilibrium. In order to gain more insight into the meaning of these equations it is convenient to make the simplification that $T_1 = T_2$, and to rewrite with the substitutions $a = 2\pi\Delta f T_2$ and $b^2 = \gamma^2 B_1^2 T_1 T_2$. Thus a is the resonance offset parameter expressed in multiples of the natural line width, and b^2 is the saturation parameter.

$$M_X = M_0 ab/(1 + a^2 + b^2) \quad \text{(dispersion-mode)} \quad [4]$$

$$M_Y = M_0 b/(1 + a^2 + b^2) \quad \text{(absorption-mode)} \quad [5]$$

$$M_Z = M_0(1 + a^2)/(1 + a^2 + b^2) \quad \text{(population difference)} \quad [6]$$

The general vector M is constrained by these equations to move so that its tip is always on the surface of a hemisphere centred at a point $+\frac{1}{2}M_0$ on the Z axis, with a radius $\frac{1}{2}M_0$ and allowing only positive values of M_Y. This is readily proved by some straightforward trigonometry.

A passage through resonance corresponds to the motion of the vector M over the surface of this hemisphere (Fig. 2) starting and finishing at the point Z where a is $\pm\infty$. In the process, an absorption mode component is induced in the receiver coil proportional to M_Y and a dispersion mode component is detected which is proportional to M_X. In the laboratory frame both components are oscillating at the frequency of the rotating frame, which is the frequency f_1 of the irradiation field B_1. In a continuous-wave spectrometer this frequency f_1 may be swept slowly through resonance at constant B_0 intensity, or the static field B_0 may be swept slowly while the frequency f_1 is held constant. Figure 2 shows a family of possible trajectories of M during slow passage through resonance for various different choices of the saturation parameter b^2. Each trajectory is a circular loop starting from Z and finishing at Z, the rate at which the vector M moves being greatest for the exact resonance condition.

Several useful deductions can be made from the pictorial description of the steady-state Bloch equations shown in Fig. 2. First, both the dispersion-mode and absorption-mode signals are very weak far from resonance when $a^2 \gg 1 + b^2$, and only reach a significant level reasonably close to resonance. It is also clear that the tails of the dispersion-mode signal M_X fall off more slowly with offset (as M_0/a) than the absorption-mode signal M_Y, which falls off as M_0/a^2. This is the reason for using the absorption mode for high-resolution work.

ABSORPTION-MODE SIGNALS

For negligible saturation, where B_1 has been reduced to a sufficiently low level that $b^2 \ll 1$, the absorption-mode response has the form of a pure Lorentzian line

$$M_Y = M_0 b/(1 + a^2) \tag{7}$$

The half width at half height occurs when $a = \pm 1$, that is to say, $2\pi\Delta f = 1/T_2$. The inverse of the linewidth is thus a measure of the spin–spin relaxation time provided there are no additional broadening influences such as B_0 inhomogeneity. A typical trajectory for M in this case of negligible saturation makes a small circular excursion near the top of the hemisphere close to point Z (Fig. 2). The induced signal is small compared with $\frac{1}{2}M_0$ but increases linearly with B_1 intensity until the onset of saturation. The optimum peak absorption-mode signal occurs for the condition $b = 1$, that is to say

$$\gamma^2 B_1^2 T_1 T_2 = 1. \tag{8}$$

Slow passage through resonance then corresponds to trajectory ii of Fig. 2 which just touches the equatorial plane of the hemisphere for $a = 0$, giving a peak signal of $M_Y = \frac{1}{2}M_0$. The linewidth is increased by a factor $2^{1/2}$, indicating that the price for optimum signal strength is a loss in resolution. This broadening occurs because saturation is most effective at exact resonance and is least effective in the tails. For strongly saturating B_1 fields, a typical trajectory iii

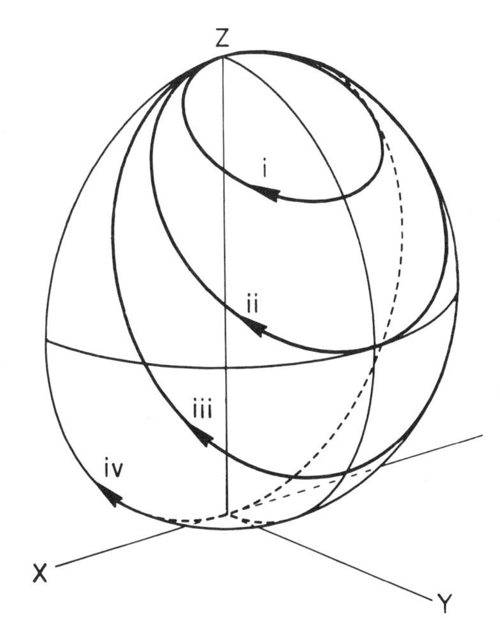

Fig. 2. The steady-state solutions of the Bloch equations (for $T_1 = T_2$) constrain the magnetization vector to follow a circular trajectory on the surface of a hemisphere centered at $Z = 0.5M_0$ on the Z axis. All trajectories start at Z far from resonance and induce a maximum Y component (absorption mode) as they pass through exact resonance, returning to Z as the sweep carries them far from resonance again. Trajectory (i) is for negligible saturation ($b^2 = 0.5$), while trajectory (ii) corresponds to $b^2 = 1$ (maximum absorption signal), trajectory (iii) is for appreciable saturation ($b^2 = 2$), and trajectory (iv) for complete saturation ($b^2 = \infty$). The dispersion-mode signal is represented by the X component of magnetization.

loops down below the bulge of the hemisphere, inducing a smaller-than-optimum peak absorption-mode signal and broadening the line by a factor $(1 + b^2)^{1/2}$. This situation is thus to be avoided on both sensitivity and resolution grounds. In the limit of extreme saturation, the vector M follows trajectory iv and the absorption-mode signal vanishes. The behaviour of the absorption-mode signals is shown in Fig. 3(a).

DISPERSION-MODE SIGNALS

The dispersion-mode signal has a different form and behaves differently under saturation. As the radiofrequency f_1 is swept through resonance, the dispersion-mode signal passes through a minimum, goes through zero at exact resonance ($a = 0$) and passes through a maximum before falling asymptotically to zero in the tail. It is thus an odd rather than even function of offset. The maximum and minimum coincide with the points of maximum slope on the absorption-mode signal, and can thus be used to measure the linewidth. For negligible saturation, these extrema occur when $a = \pm 1$. In this limit, the strength of the dispersion-mode signal increases with the strength of the radiofrequency field B_1, but

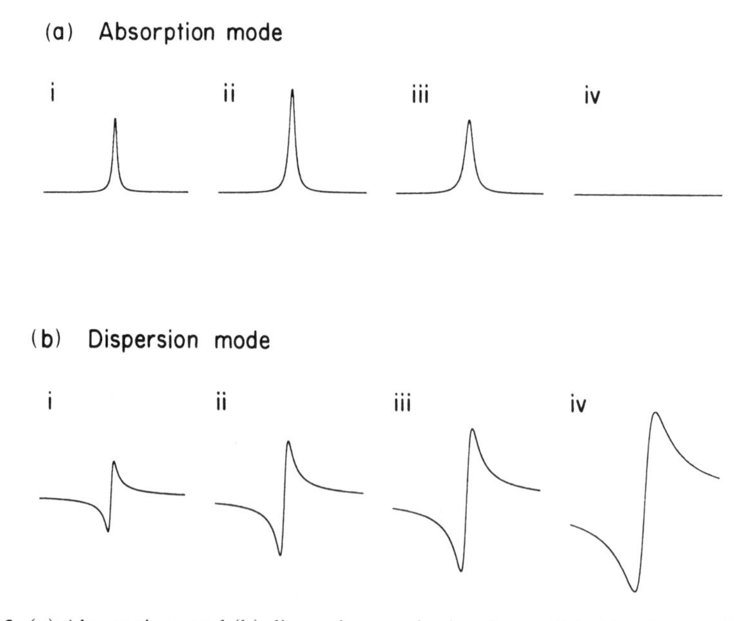

Fig. 3. (a) Absorption- and (b) dispersion-mode signals predicted by the steady-state solutions of the Bloch equations corresponding to the four trajectories of Fig. 2. The absorption-mode signal passes through a maximum for $b^2 = 1$ (trajectory ii) but the dispersion-mode signal approaches an asymptotic value when $b^2 = \infty$ (trajectory iv).

unlike the absorption-mode signal, it does not reach a maximum for $b^2 = 1$ but continues to increase with the degree of saturation until it reaches $\pm\frac{1}{2}M_0$ asymptotically for trajectory iv of Fig. 2. For this reason, a search for a completely unknown resonance is often carried out with a strong B_1 field and with the receiver adjusted to the dispersion mode, since this requires no estimate of the relaxation times T_1 and T_2. The linewidth, measured between the points of maximum and minimum signal, increases as $(1 + b^2)^{1/2}$. These signals are illustrated in Fig. 3(b).

POPULATIONS

The component M_Z is a measure of the difference in spin populations between the two energy levels involved in the NMR transition. For negligible saturation, the difference from the Boltzmann population difference depends on the saturation parameter only to second order. For the optimum absorption-mode signal ($a = 0, b = 1$) this population difference is reduced to one half as f_1 sweeps through the resonance peak. For very strong saturation ($b^2 \gg 1$) there is no population difference at exact resonance, but it recovers again in the tail of the line.

The above description assumed $T_1 = T_2$. For the general case where $T_2 < T_1$, the picture is modified, the locus for the tip of the vector M becoming an ellipsoid with major axis in the Z direction unchanged at $\frac{1}{2}M_0$ but with the two minor axes reduced by the factor $(T_2/T_1)^{1/2}$. This has the effect that the expressions for signal intensity are reduced by $(T_2/T_1)^{1/2}$, giving an optimum peak absorption-mode of $\frac{1}{2}M_0(T_2/T_1)^{1/2}$ and an asymptotic dispersion-mode of $\pm\frac{1}{2}M_0(T_2/T_1)^{1/2}$. For large molecules which are tumbling relatively slowly in solution this can mean a significant loss in sensitivity.

Many practical cases of high-resolution spectroscopy involve a significant broadening of the line by B_0 inhomogeneity, often discussed in terms of an instrumental linewidth $1/T_2^*$ corresponding to a fast decay of the free induction signal with time constant T_2^*. In the steady-state solutions of the Bloch equations discussed above it is not correct to substitute T_2^* in place of the spin–spin relaxation time T_2. In particular, the saturation behaviour of an inhomogeneously broadened line is quite different: it is possible to burn a hole in the line profile by selective saturation. The linewidth does not change with degree of saturation until the broadening of the natural linewidth by saturation becomes comparable with the inhomogeneity broadening. The peak height of the absorption-mode signal does not pass through the usual maximum at $b = 1$ but continues to increase until the saturated natural linewidth $(1 + b^2)^{1/2}/T_2$ approaches the inhomogeneous linewidth $1/T_2^*$.

In the absence of selective saturation effects, the observed resonance response should be thought of as the natural linewidth convoluted with the distribution function representing the inhomogeneity of the applied B_0 field. Since the area under the line remains constant (representing the intensity), broadening by field inhomogeneity severely reduces the peak height and hence the sensitivity. Optimum absorption-mode signals are then considerably weaker than $\frac{1}{2}M_0(T_2/T_1)^{1/2}$. This loss can be redeemed by reducing T_1 and T_2 by the addition of a relaxation reagent ('doping') until the linewidth is again determined by T_2 rather than by T_2^* and the optimum absorption-mode signal again approaches $\frac{1}{2}M_0$. Too much doping causes an unnecessary broadening of the lines, so a high-resolution spectroscopist uses this method with circumspection. Dissolved oxygen from the air is a rather mild paramagnetic relaxation reagent and a significant shortening of very long relaxation times can be achieved by bubbling pure oxygen through the sample solution and capping the sample tube.

REFERENCES

1. R. R. Ernst and W. A. Anderson, *Rev. Sci. Instr.* **37**, 93 (1966).
2. J. W. Cooley and J. W. Tukey, *Math. Comput.* **19**, 297 (1965).
3. F. Bloch, *Phys. Rev.* **102**, 104 (1956).
4. A. Abragam, *The Principles of Nuclear Magnetism.* Oxford University Press, 1961.

Cross-references

Cooley–Tukey algorithm
Fourier transformation
Rotating frame
Sensitivity
Time averaging
Vector model

Convolution

The problem of distinguishing between a true signal and random noise has proved a source of fascination and led to several misconceptions. In the stock market, the price of a given stock fluctuates with time and it would be useful and profitable to discover just what part of this fluctuation is a 'real' trend and what is just meaningless erratic behaviour. On the assumption that the longer-term fluctuations are 'real', a 30-day moving average is calculated; this is 30 times less sensitive to the day-to-day fluctuations, but shows up the longer-term trends. Mathematically, this is a convolution of the graph of daily stock price with a rectangular function spanning 30 days, a very crude convolution function indeed. One might well make an argument that the month-old reports are less significant than today's price and therefore introduce a convoluting function shaped to reflect the diminishing importance of the early data.

Almost the same problem is posed by an NMR spectrum where there are noisy fluctuations associated with the desired signal. In the stock market example, the future stock prices were of course unknown. However, in modern NMR spectrometers the entire spectrum may be stored and recalled for later processing so we can look both forward and backward in time. It is then possible to use a smooth symmetrical convolution function such as a Lorentzian or Gaussian curve in the frequency domain. Figure 1 illustrates this process for a very simple 1 : 2 : 1 convoluting function acting on a noisy input signal; the rapid fluctuations are smoothed out.

Normally the NMR spectrum is sampled at regular frequency intervals, and a digital computer is used to multiply each ordinate on the raw spectrum by the corresponding ordinate on the convoluting function. The sum of all these products is then normalized and plotted. The convolution function is then displaced one step to the right and the entire process repeated, giving another point on the smoothed spectrum. Allowance should be made for discontinuities at the ends of the unprocessed spectrum. Clearly, the degree of smoothing is a function of the width of the convoluting function. How far should this process go? The useful limit occurs when the linewidth of the convoluting function matches the NMR linewidth; beyond this the signal peak heights would be reduced faster than the noise. This is equivalent to the *matched filter* condition (1). In the frequency domain, convolution of a spectrum S(f) with a smoothing function B(f) is equivalent (in the time domain) to multiplication of the respective Fourier transforms S(t) and B(t). Fourier transform spectrometers achieve the effect of convolution by multiplying the free induction signal by a sensitivity enhancement* function, often a decaying exponential, corresponding to a Lorentzian convolution function (see Fourier transformation*).

It is quite feasible to reverse the process of sensitivity enhancement by employing a convolution function which has negative lobes on each side of the central maximum. This is sometimes called *deconvolution* since it can in principle be used to remove the effects of instrumental broadening when the form of this distribution is known. Deconvolution enhances resolution at the expense of signal-to-noise ratio. Again, the convolution theorem shows that this is equivalent to the well-known resolution enhancement* procedures used in the Fourier transform mode.

Biochemists concerned with the problem of emphasizing narrow features of the NMR spectrum at the expense of broad underlying resonances, introduced the process of *convolution difference* (2). A relatively severe broadening function is applied, calculated to affect the narrow lines much more than the broad lines; this is then subtracted from the unperturbed spectrum, largely cancelling the broad components and leaving each narrow line sitting in a shallow depression. (Note that the overall integral of such a convolution difference spectrum must be zero.) In practice, this procedure is also used as a rather mild resolution enhancement technique, and is usually implemented in the time domain. Sometimes the broadened spectrum is scaled down with respect to the unprocessed spectrum, before subtraction.

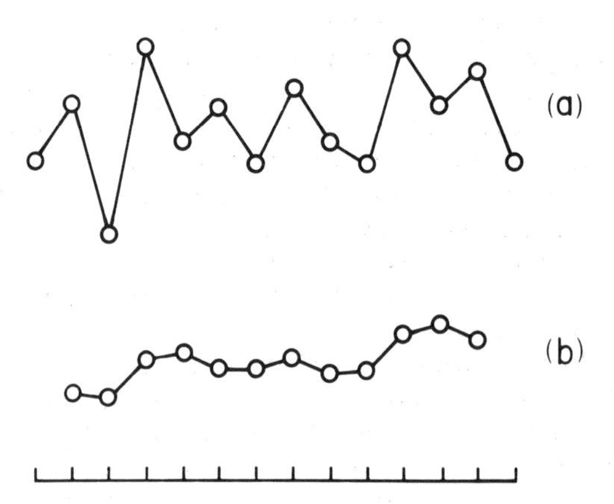

Fig. 1. Smoothing (b) of random noise by a 1:2:1 convolution function. At each point of the raw data (a), 50% of that ordinate is added to 25% of the preceding ordinate and 25% of the following ordinate. As with all convolutions, there are end-effects which must be dealt with; here the first and last points are omitted.

REFERENCES

1. R. R. Ernst, Sensitivity Enhancement in Magnetic Resonance, *Advances in Magnetic Resonance*, Ed.: J. S. Waugh. Academic Press: New York, 1966, Vol. 2.
2. I. D. Campbell, C. M. Dobson, R. J. P. Williams and A. V. Xavier, *J. Magn. Reson.* **11**, 172 (1973).

Cross-references

Continuous-wave spectroscopy
Difference spectroscopy
Fourier transformation
Resolution enhancement
Sensitivity enhancement

Cooley–Tukey Algorithm

The time-domain free induction decay* can always be represented by a superposition of several different sine and cosine waves of appropriate amplitudes, which mutually interfere. This is the principle of Fourier analysis. In the section on Fourier transformation* a spectrum analyser was employed to seek out these components one at a time. It accomplishes this by choosing a monochromatic frequency (the reference frequency f_r), multiplying it by the free induction signal and integrating the result from time zero to infinity. Only the components at or near the reference frequency give a finite integral, the rest quickly get out of phase with the reference frequency and give near zero integrals. The procedure is then repeated for different values of the reference frequency until all the appropriate frequencies have been explored.

Transformation in the computer requires that the free induction signal be sampled in the time domain at discrete intervals, and hence the transformation process is necessarily discrete, and the integration is replaced by a summation. Suppose that the free induction decay is represented by an array of values X, with a typical value X_k, where k runs from 0 to N − 1. The frequency-domain spectrum is an array A with typical value A_r, where r runs from 0 to N − 1. The elements of the array X are simply multiplied by the appropriate values of a sine wave and a cosine wave, and the products summed over all values of the index k. This may be written

$$A_r = \sum_{k=0}^{N-1} X_k W^{rk} \quad r = 0, 1, \ldots, N-1 \qquad k = 0, 1, \ldots, N-1, \qquad [1]$$

where $W^{rk} = \exp(-2\pi irk/j)$. In some computer programs a table of values of sines is calculated beforehand and used as a look-up table instead of computing the sine or cosine (via a Taylor series) at each stage of the calculation. Because of the periodicity of the expression for W, there are some useful relationships between certain powers of W. For example

$$W^{rk+N} = \exp(-2\pi irk/N)\exp(-2\pi i) = \exp(-2\pi irk/N) = W^{rk} \quad [2]$$

$$W^{rk+\frac{1}{2}N} = \exp(-2\pi irk/N)\exp(-\pi i) = -\exp(-2\pi irk/N) = -W^{rk}. \quad [3]$$

Thus for $N = 4$, $W^6 = W^2$ and $W^3 = -W^1$. Note also that $W^2 \times W^3 = W^5$, etc.

The first computer programs for Fourier transformation operated on exactly this principle, multiplying the elements as in eqn [1] and summing the result. Unfortunately, multiplication is a very time-consuming operation in a mini-computer and the transformation times by this method were measured in tens of minutes. The operation of addition is much faster, and in what follows this time will be neglected, only the slow multiplication steps being considered. There are N^2 complex multiplication steps required to evaluate all the N values.

A breakthrough in the speed of computation was achieved by the *fast Fourier transform* algorithm of Cooley and Tukey (1) as modified by Bergland (2). It happened at just the opportune moment for the development of Fourier transform NMR (1965), for it would be hard for today's spectroscopist to have to wait 20 minutes for his high-resolution spectrum to be computed.

The way in which this time-saving comes about has been presented in some detail by Oran Brigham (3). His treatment is followed here by considering the simple case of $N = 4$ and then generalizing the result. For $N = 4$ the summation of eqn [1] can be written

$$\begin{aligned} A_0 &= X_0W^0 + X_1W^0 + X_2W^0 + X_3W^0 \\ A_1 &= X_0W^0 + X_1W^1 + X_2W^2 + X_3W^3 \\ A_2 &= X_0W^0 + X_1W^2 + X_2W^4 + X_3W^6 \\ A_3 &= X_0W^0 + X_1W^3 + X_2W^6 + X_3W^9. \end{aligned} \quad [4]$$

Using the relationship between the different powers of W,

$$\begin{bmatrix} A_0 \\ A_1 \\ A_2 \\ A_3 \end{bmatrix} = \begin{bmatrix} W^0 & W^0 & W^0 & W^0 \\ W^0 & W^1 & W^2 & W^3 \\ W^0 & W^2 & W^0 & W^2 \\ W^0 & W^3 & W^2 & W^1 \end{bmatrix} \begin{bmatrix} X_0 \\ X_1 \\ X_2 \\ X_3 \end{bmatrix}. \quad [5]$$

Now suppose that we reorder the column vector A_r and the matrix W^{rk}

$$\begin{bmatrix} A_0 \\ A_2 \\ A_1 \\ A_3 \end{bmatrix} = \begin{bmatrix} 1 & 1 & 1 & 1 \\ 1 & W^2 & W^0 & W^2 \\ 1 & W^1 & W^2 & W^3 \\ 1 & W^3 & W^2 & W^1 \end{bmatrix} \begin{bmatrix} X_0 \\ X_1 \\ X_2 \\ X_3 \end{bmatrix} \quad [6]$$

Here we have interchanged the second and third rows and replaced W^0 by 1 except in one location where it is retained in order to be consistent with the more general results. This matrix can be factorized into two

$$M = \begin{bmatrix} 1 & W^0 & 0 & 0 \\ 1 & W^2 & 0 & 0 \\ 0 & 0 & 1 & W^1 \\ 0 & 0 & 1 & W^3 \end{bmatrix} \begin{bmatrix} 1 & 0 & W^0 & 0 \\ 0 & 1 & 0 & W^0 \\ 1 & 0 & W^2 & 0 \\ 0 & 1 & 0 & W^2 \end{bmatrix}. \quad [7]$$

The frequency domain values A_r may now be evaluated by taking the two matrix multiplications separately. First define an intermediate column vector

$$\begin{bmatrix} x_0 \\ x_1 \\ x_2 \\ x_3 \end{bmatrix} = \begin{bmatrix} 1 & 0 & W^0 & 0 \\ 0 & 1 & 0 & W^0 \\ 1 & 0 & W^2 & 0 \\ 0 & 1 & 0 & W^2 \end{bmatrix} \begin{bmatrix} X_0 \\ X_1 \\ X_2 \\ X_3 \end{bmatrix}. \qquad [8]$$

Suppose we choose to calculate x_0 and x_2, remembering that $W^0 = -W^2$

$$x_0 = X_0 + W^0X_2 \qquad x_2 = X_0 - W^0X_2.$$

Thus in calculating this pair of values only one complex multiplication is involved instead of two. Moreover, X_0 and X_2 are never needed again in the calculation so that x_0 and x_2 may be stored in their locations. Exactly the same reasoning applies to the pair of values x_1 and x_3, and these may be stored in the locations previously used for X_1 and X_3. Hence the calculation can be performed 'in place' without increasing the size of the data table. This first 'pass' through the calculation is usually represented diagrammatically thus

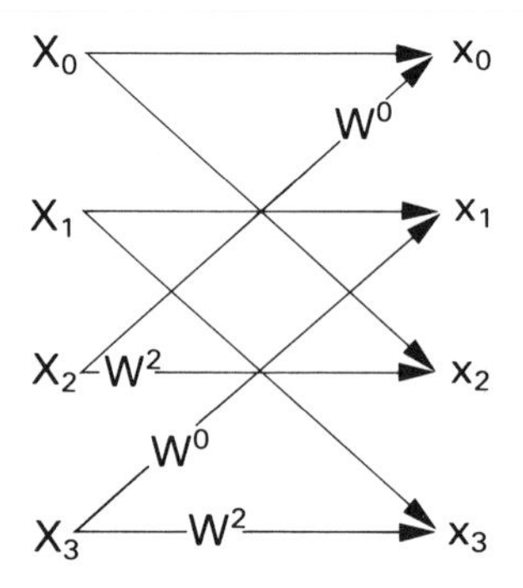

The connecting lines with a W attached involve multiplication, but only two independent multiplications occur.

The next 'pass' multiplies the intermediate column vector x_k by the left-hand matrix of eqn [7]

$$\begin{bmatrix} A_0 \\ A_2 \\ A_1 \\ A_3 \end{bmatrix} = \begin{bmatrix} 1 & W^0 & 0 & 0 \\ 1 & W^2 & 0 & 0 \\ 0 & 0 & 1 & W^1 \\ 0 & 0 & 1 & W^3 \end{bmatrix} \begin{bmatrix} x_0 \\ x_1 \\ x_2 \\ x_3 \end{bmatrix}. \qquad [9]$$

An exactly analogous reasoning indicates that again only two complex multiplications are involved, provided the calculations are carried out on the appropriate pairs of intermediate values (x_0, x_1) and (x_2, x_3). The 'flow diagram' can thus be extended

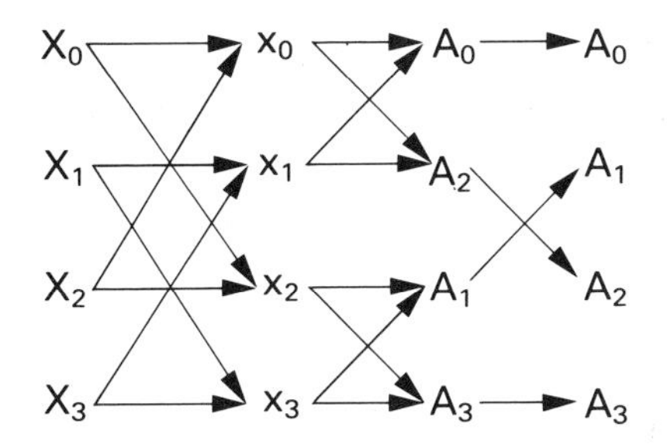

This factorization procedure and the pairing of the calculations has thus reduced the total of complex multiplications from sixteen to four. If the argument is generalized for $N = 2^p$ the fast Fourier transform algorithm factorizes the $N \times N$ matrix into p matrices of size $N \times N$, each containing many zeros, and p passes are required in the calculation. Instead of N^2 complex multiplications, only $\frac{1}{2}Np$ complex multiplications are required, giving a reduction in overall computing time of approximately 2N/p, which for N = 8K is roughly an advantage of 1300.

Because of the reordering that was employed in eqn [6] the values for the frequency-domain points A_r are in a scrambled order and must be unscrambled before the spectrum is displayed. In the case considered, where N = 4, this simply involves interchanging A_1 with A_2, as indicated in the last step of the flow diagram. The recipe for reordering the general case is quite interesting. It requires that the index r be expressed in binary digits and then be reversed. This 'bit-reversed order' is illustrated for the slightly less trivial case of N = 8.

A_0	000	→	000	A_0
A_1	001	→	100	A_4
A_2	010	→	010	A_2
A_3	011	→	110	A_6
A_4	100	→	001	A_1
A_5	101	→	101	A_5
A_6	110	→	011	A_3
A_7	111	→	111	A_7

Reordering could equally well be accomplished before the transformation by an operation on the matrix X_k.

Note that the flow diagrams shown above suggest that Fourier transformation could be further speeded up by parallel processing of the data in contrast to the serial processing discussed here. Array processors are now being used increasingly in high-resolution spectrometers as the data tables become larger and larger, and particularly in the growing field of two-dimensional spectroscopy*, where two-dimensional arrays are involved, and transformation times for serial processing can be several minutes.

REFERENCES

1. J. W. Cooley and J. W. Tukey, *Math. Comput.* **19**, 297 (1965).

2. G. D. Bergland, *Commun. A. C. M.* **11**, 703 (1968).

3. E. Oran Brigham, *The Fast Fourier Transform*. Prentice-Hall: New Jersey, 1974.

Cross-references

Fourier transformation
Free induction decay
Two-dimensional spectroscopy

Difference Spectroscopy

Overcrowded spectra are the bane of the NMR spectroscopist's life, and considerable efforts have been devoted to methods of spectral simplification. One general method is to impose some kind of selective perturbation and then to record the difference between the perturbed and the unperturbed spectrum (1). Although difference spectra can be obtained by chemical modification of the sample, for example by adding a relaxation reagent, the best results are obtained by physical perturbations such as double irradiation. In principle, the milder the perturbation, the cleaner the difference spectrum. A good example is a proton spin echo* experiment with and without a 180° spin flip pulse applied to a coupled carbon-13 or nitrogen-15 site. If perturbations of the field/frequency regulation* can be avoided, nothing changes in the proton spectrum except that the echoes become modulated. If the delay between excitation and acquisition of the echo is chosen to be a half cycle of the modulation, a clean difference spectrum is obtained showing the carbon-13 or nitrogen-15 satellites.

It should be remembered that difference spectroscopy merely alters the display mode without introducing any new information. It *highlights* the features that change in the spectrum and forces them upon our attention but does not in itself improve the accuracy of the measurement. For example, it has become common to record proton–proton Overhauser spectra in a difference mode on the grounds that the intensity variations are usually very small, as little as 1% of the unperturbed proton line intensities. Difference spectroscopy ensures that we do not overlook any of these small changes, but the accuracy of the measurement of the Overhauser enhancement is not improved.

One source of frequency displacement between the original spectrum and the double resonance spectrum is the Bloch–Siegert effect.* For determinations of the nuclear Overhauser enhancement this can be avoided by switching off the second irradiation field B_2 just before acquisition of the free induction decay, since coherence effects (the Bloch–Siegert shift) are removed instantaneously, whereas population disturbances (the Overhauser effect) persist for times comparable with the spin–lattice relaxation* time. Otherwise, the Bloch–Siegert shifts interfere with the subtraction process and spurious difference-mode responses appear, looking rather like dispersion-mode signals. Similar artifacts appear in the difference spectrum if there is a shift caused by improper operation of the field/frequency control. Carried over into two-dimensional spectroscopy*, this becomes the problem of 't_1 noise', spurious baseline fluctuations wherever there is a strong signal in the two-dimensional spectrum (2).

Since Fourier transformation is a linear operation, a difference spectrum can be obtained by subtraction either in the time domain or in the frequency domain. Subtraction of free induction decays before transformation is sometimes preferable since it reduces the dynamic range of the signal to be transformed. The unperturbed free induction decay should not, however, be discarded since it contains potentially useful information not present in the difference mode.

APPLICATIONS OF DIFFERENCE SPECTROSCOPY

Difference spectroscopy is often used in inversion-recovery measurements of spin–lattice relaxation and in saturation transfer* experiments. For this purpose, it is important to *interleave* the 'control' and 'perturbed' measurements rather than time-average a large number of the former and then a similar number of the latter. This tends to reduce the effects of slow drifts in the gain or resolution of the spectrometer.

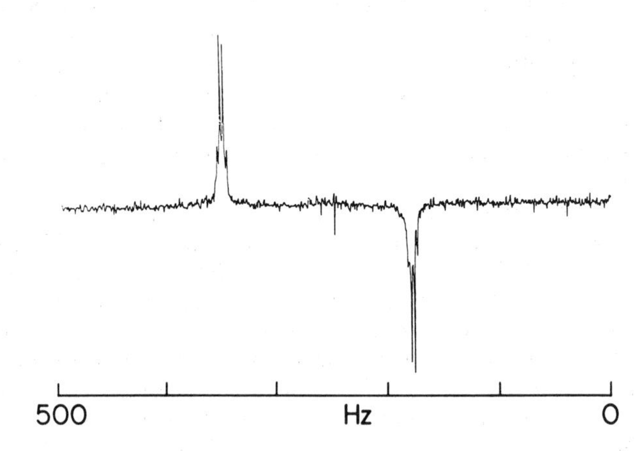

Fig. 1. The carbon-13 satellites in the proton spectrum of the methyl group of acetaldehyde with the parent proton line (centre) suppressed by difference spectroscopy. Polarization transfer from carbon-13 to protons was carried out by the 'reverse INEPT' experiment first to give 'up–down' multiplets and then 'down–up' multiplets. When the difference spectrum is calculated the strong parent proton signal is eliminated.

Networks of spin–spin coupling are nowadays best investigated by homonuclear chemical shift correlation* (COSY). An alternative is decoupling difference spectroscopy. This allows a well-separated multiplet to be irradiated by the B_2 field and all coupled groups to be identified as responses in the difference spectrum even though they are normally hidden in a crowded region of the spectrum. Indeed, the experiment can be performed 'blind' when even the irradiated group is obscured by other overlapping resonances, and it may be advantageous to conduct a computer-controlled search where the frequency of B_2 is incremented in small steps through the appropriate chemical shift range. Similar heteronuclear double-resonance difference experiments can be performed. In this context, when high decoupling fields are employed, the possibility of a heteronuclear Bloch–Siegert shift cannot be dismissed. A decoupling field strength $\gamma B_2/2\pi = 10$ kHz applied to carbon-13 resonance would, for example, shift a deuterium internal field/frequency signal 20 MHz distant by 2.5 Hz.

One of the most demanding applications of difference spectroscopy is the 'isotope filter' experiment mentioned above, involving the detection of the very weak satellite resonances which appear in proton spectra due to a low-abundance nucleus such as carbon-13, nitrogen-15 or silicon-29 (3). It puts severe demands on the spectrometer stability since it involves the observation of very small differences in a very intense proton signal. It may be advisable to employ a polarization transfer* technique here, rather than spin-echo difference spectroscopy, since this permits presaturation of the proton signal, the detected signal being excited by transfer from the low-abundance nuclei. Figure 1 illustrates the application of the reverse 'INEPT' experiment (3) to the detection of the carbon-13 satellites in the proton spectrum of the methyl group of acetaldehyde. Two experiments are performed; in the second, the sense of the polarization transfer is inverted. When the difference signal is displayed, the strong parent proton peak is almost entirely suppressed.

The full power of difference spectroscopy is demonstrated when it is used to eliminate spurious instrumental artifacts. For example, if the genuine magnetic resonance response can be reversed in sense by changing the phase of a radiofrequency pulse, while an undesirable instrumental effect persists in the same sense, then difference spectroscopy eliminates the latter but retains the 'true' magnetic resonance signal. This is the basis of several phase-cycling techniques used in two-dimensional spectroscopy and in studies of multiple-quantum coherence*.

REFERENCES

1. J. K. M. Sanders and J. D. Mersh, *Progress in NMR Spectroscopy* **15**, 353 (1982).
2. A. F. Mehlkopf, D. Korbee, T. A. Tiggelman and R. Freeman, *J. Magn. Reson.* **58**, 315 (1984).
3. T. H. Mareci and R. Freeman, *J. Magn. Reson.* **44**, 572 (1981).

Cross-references

Bloch–Siegert effect
Field/frequency regulation
Multiple-quantum coherence
Phase-cycling
Polarization transfer
Saturation transfer
Shift correlation
Spin echoes
Spin–lattice relaxation
t_1 Noise
Two-dimensional spectroscopy

Digitization

Fourier transformation* is most conveniently carried out on a digital computer, and consequently an essential step in the data processing chain is to convert the analogue free induction signal from the spectrometer into digital form. There are two aspects of this process, sampling in the time domain and conversion of each analogue ordinate into digital form.

SAMPLING IN THE TIME DOMAIN

Suppose we need to represent a pure sine wave by a set of discrete samples at regular intervals in time. What is the minimum number of samples required to define this frequency unambiguously? The answer is provided by the *sampling theorem*, which states that there must be at least two samples per cycle. Applied to the problem of the free induction decay* in NMR, this means that the sampling rate should be high enough to give at least two data points per cycle of the highest frequency component. If this *Nyquist condition* is violated, the corresponding frequency is decreased by an integral multiple of the sampling rate; the frequency is said to have been *aliased*. This phenomenon is not so surprising. It is related to the stroboscopic effect well known from Western movies where stage coach wheels appear to slow down and even turn backwards as a subharmonic of the rotation frequency (related to the number of spokes) interacts with the frame speed of the film. Clearly, if this subharmonic exactly matches the frame speed, there is no evidence on the screen that the wheels are moving at all; any slight difference between these two frequencies translates into a slow apparent rotation, forwards or backwards.

The phenomenon of aliasing may be appreciated diagrammatically by considering the sampling operation on a frequency F which has exactly two samples per cycle and on a slightly higher frequency $F + \Delta F$. From Fig. 1 we see that for a cosine wave the sampling points for $F + \Delta F$ are identical with those for $F - \Delta F$, whereas for a sine wave the same is true except that the sine wave is inverted in phase, equivalent to changing the sign of the frequency. By sampling too slowly, we have converted $F + \Delta F$ into $-F + \Delta F$, that is to say *we have subtracted the sampling rate from the true frequency*.

Consider a practical example appropriate to carbon-13 spectroscopy performed in the usual quadrature detection* mode where positive and negative frequencies are discriminated and the transmitter is positioned near the centre of the spectral range under investigation. Suppose that the chemical shifts fall within a range of 20 kHz. The spectral width would therefore run from +10 kHz to −10 kHz, and the sampling rate would be 20 kHz, each sampling operation involving one 'real' and one 'imaginary' data point, acquired simultaneously. If there happened to be a chemical species with a resonance frequency outside this range, say at +11 kHz, its frequency appears at −9 kHz in the transformed spectrum (Fig. 2). In Fourier spectroscopy, resonances outside the chosen 'window' are always aliased inside this window unless they are removed by a suitable low-pass filter. In an analogous way, the filming of a stage coach at a relatively low framing speed always makes the wheels appear to rotate slowly enough to be within the frequency window of the camera.

Since the acquisition time is determined by the desired resolving power, a high sampling rate means that a large data storage space must be set aside for the free induction decay. Because aliasing can lead to considerable confusion, precautions must be taken to avoid it in all normal experimental situations by

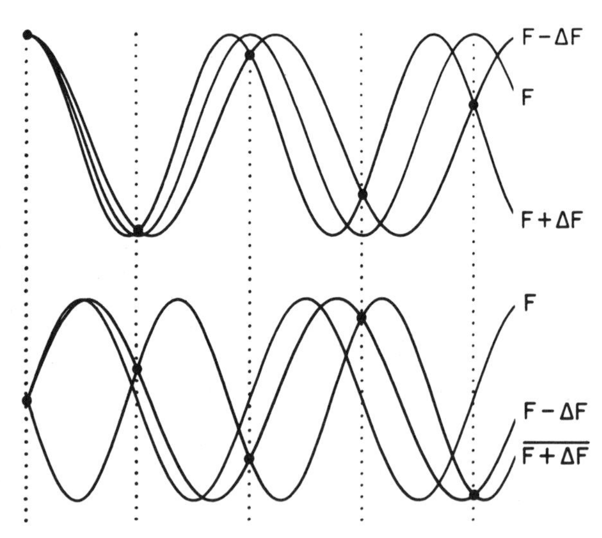

Fig. 1. Aliasing of frequencies above the Nyquist frequency (F). For a cosine wave (top) the sampled ordinates are the same for a frequency $(F - \Delta F)$ as for a frequency $(F + \Delta F)$. For a sine wave (bottom) the sampled ordinates are the same for a frequency $(F - \Delta F)$ as for an *inverted* sine wave of frequency $(F + \Delta F)$, shown as $\overline{F + \Delta F}$. The result is to subtract the sampling rate 2F Hz from the actual frequency.

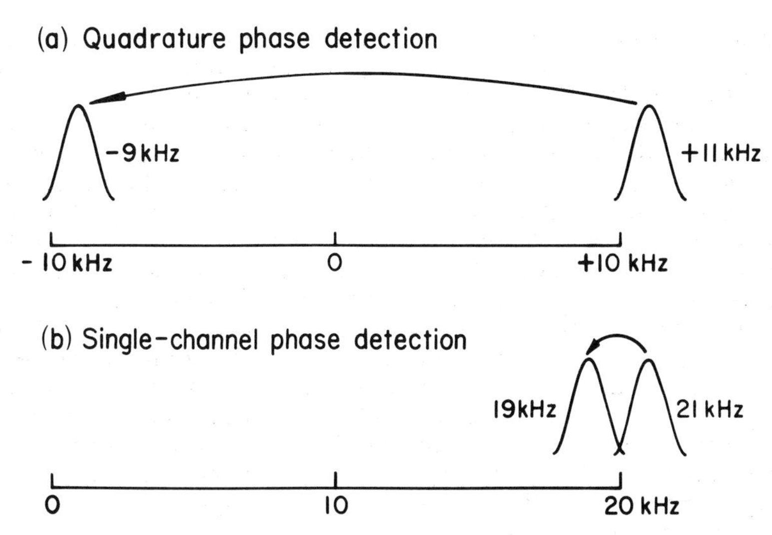

Fig. 2. Aliasing of frequencies above the Nyquist frequency (F). (a) With quadrature phase detection the apparent frequency is the actual frequency reduced by an integral multiple of the sampling rate (20 kHz). (b) With single-channel phase detection the frequency is reduced by a multiple of the sampling rate (40 kHz) but the sign information is lost, giving the effect of 'folding' about the Nyquist frequency.

filtering out frequencies above the Nyquist frequency. It is also important to filter out high-frequency noise which would otherwise be aliased into the spectral window, reducing the sensitivity.

In a spectrometer without quadrature detection the same general considerations apply. Now, the transmitter is set at one edge of the spectral window because there is no discrimination between positive and negative frequencies. In our example, the spectral width would run from 0 to 20 kHz and the sampling rate would be 40 kHz, but there would be only one data point taken for each sampling operation. Now, a chemical species with a resonance outside the spectral window (say at 21 kHz) appears at 19 kHz, giving rise to the term 'folding'. This is seen to be equivalent to aliasing (reducing the frequency by 40 kHz) and concomitant loss of the sign information (Fig. 2).

Some Fourier spectrometers employ the 'pseudo-quadrature' detection scheme suggested by Redfield and Kunz (1), where sampling in the 'real' and 'imaginary' channels is not performed simultaneously but interleaved in time. In this mode, violation of the Nyquist condition leads to folding of the high-frequency signals, as in single-channel phase detection (2).

ANALOGUE-TO-DIGITAL CONVERSION

Representation of the experimental free induction decay in digital form also entails conversion of the analogue voltage into a digital signal. A typical analogue-to-digital converter (ADC) works by successive approximations and requires a finite time (microseconds) to accomplish a conversion (3). In order to avoid errors due to variation of the input signal during this conversion period, the first step is to store the incoming signal on a sample-and-hold device which 'freezes' the voltage. Suppose the ADC has N binary digits available and has been designed for a 10 volt full-scale signal. The first thing is to set the spectrometer gain so that the excursions of the free induction signal almost fill this 10 volt range. The ADC employs a digital-to-analogue converter (DAC) to generate an analogue level of 5 volts, and compares this with the incoming signal on the sample-and-hold device. If the latter is higher than 5 volts, the most significant bit of the ADC is set. For the next stage of the approximation the DAC supplies a reference signal of 7.5 volts and determines whether the incoming signal is higher or lower than this, setting the next-most-significant bit appropriately. This cycle is repeated until the least significant bit has been set. Clearly, this leaves the final digitized signal in error by as much as half a bit; we have reached the limit of *resolution* of the ADC. It follows as a general rule that the higher the resolution the lower the maximum sampling rate, but of course special analogue-to-digital converters can be obtained that have both high resolution and high conversion rates.

The free induction decay is now made up of a sequence of digitized ordinates, each of which can be thought of as a true signal plus a *digitization error* due to the finite resolution of the ADC. Consequently, the resulting frequency-domain spectrum is also the sum of two parts, the Fourier transform of the true free induction decay superimposed on the Fourier transform of the digitization errors, the *digitization noise* (Fig. 3).

This is the origin of the serious *dynamic range* problem which occurs when a solute is present at low concentration in a solvent, such as water, that has a very intense NMR line. The spectrometer gain cannot be increased without running into overflow in the ADC, while the solute peaks are so weak that they are swamped by the digitization noise. What we need is an ADC that has a

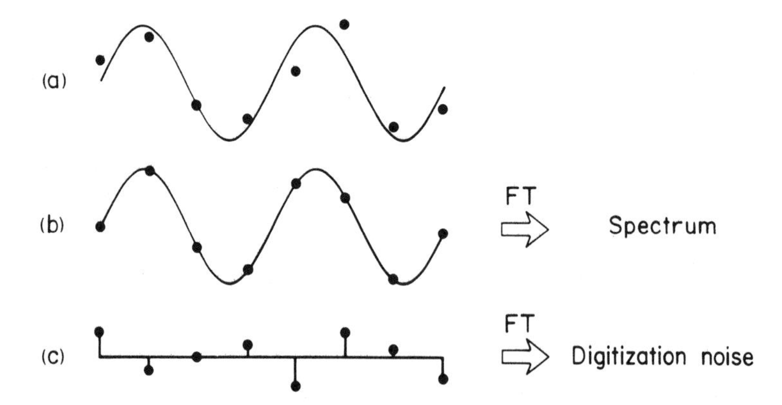

Fig. 3. Digitization errors in the analogue-to-digital converter (grossly exaggerated). Since the transformation is a linear process, (a) may be represented as the sum of a perfectly digitized signal (b) and digitization errors (c). The Fourier transform of (c) is the 'digitization noise' in the spectrum.

resolution higher than the dynamic range of the signals in the free induction decay, but we have seen above that this normally conflicts with the requirement for high sampling rate. In order to counter this difficulty, a whole armoury of NMR techniques has been developed to suppress the strong water resonance before detection. (See Solvent suppression*.)

We might have been tempted to visualize this problem in a much simpler form, arguing that a signal weaker than the least significant bit of the ADC would be lost forever and could not appear in the transformed spectrum at all. In practice, this is not the case, for a variety of reasons (3). Noise can carry over weak signal components so that they influence the least significant bit on some fraction of the conversions. In the absence of an appreciable level of noise, spectrometer instabilities achieve a similar effect if time averaging is used. Even in the case of a single-shot experiment, the presence of a strong signal component permits the very weak components to trigger the least significant bit on some of the conversions, and these weak signals appear in the spectrum.

There have been attempts to circumvent the limitations imposed by the restricted ADC resolution by operating in a differential mode with the strong water signal near zero frequency (4) or by deliberately sampling much faster than the Nyquist frequency with a view to averaging the digitization noise.

REFERENCES

1. A. G. Redfield and S. D. Kunz, *J. Magn. Reson.* **19**, 250 (1975).
2. C. J. Turner and H. D. W. Hill, *J. Magn. Reson.* **66**, 410 (1986).
3. J. C. Lindon and A. G. Ferrige, *Progress NMR Spectr.* **14**, 27 (1980).
4. S. J. Davies, C. J. Bauer, P. B. Barker and R. Freeman, *J. Magn. Reson.* **64**, 155 (1985).

Cross-references

Fourier transformation
Free induction decay
Quadrature detection
Solvent suppression

Field/Frequency Regulation

NMR enjoys an advantage over many forms of spectroscopy as a result of the narrow width of the resonance lines, holding out the promise of very high resolution. In order to achieve this resolution the applied static field must be homogeneous in space to about 1 part in 10^9 (taking into account the effect of sample spinning*). We might also anticipate that it should be stable in time to the same degree. For experiments which involve accumulation of many transient signals, this stability may need to be maintained for several hours. Fortunately, the critical requirement is rather less rigorous – it is sufficient to maintain the *ratio* of applied field to frequency within these stability limits, since a proportional change in both field and frequency, if small enough, has a negligible effect on the spectra (it involves a small change in the very small chemical shielding term σ).

Short-term fluctuations in magnetic field can be attenuated with the aid of a feedback arrangement incorporating a *pick-up* and a *buck-out* coil with their axes along the field direction; variations in field induce a voltage in the pick-up coil proportional to the rate of change of flux. This voltage is integrated, amplified and applied to the buck-out coil in order to oppose the initial change in magnetic field. This *flux stabilizer* is more effective the more rapid the change; it fails to cope with long-term drifts of the magnetic field. Electromagnets, in particular, need this type of field regulation. Permanent magnets and superconducting solenoids have better short-term stability, and hence can often dispense with a flux stabilizer.

It was realized at an early date that this drawback of flux stabilization schemes could be countered by providing a fixed reference point for the magnetic field, derived from the nuclear resonance condition itself. This involves the concept of two simultaneous nuclear resonance experiments, giving a 'reference' and an 'analytical' signal. If both share the same excitation frequency, or if the two excitation frequencies are derived from the same source, then the magnetic field can be said to be 'locked' to the frequency used to observe the spectrum. For many years, two separate NMR samples were employed, and since they were necessarily in different regions of the magnetic field, drifts would occur whenever the field distribution changed. Primas (1) was the first to suggest that these *external lock* schemes should be replaced by *internal lock* systems, where the reference signal is derived from the sample under investigation. For example, the proton resonance of tetramethysilane may be used for field/frequency regulation when proton spectra are being recorded (1, 2). More recently, the widespread use of deuterated solvents for NMR has permitted the use of deuterium resonances as the reference signal. Fluorine reference signals are also used.

The key to this method of regulation is the dispersion-mode signal of the reference material. At the desired field/frequency ratio, a voltage corresponding to the centre of the dispersion-mode line is detected, essentially zero volts; if the field begins to drift away from this chosen condition a finite *error signal* is generated, and the sign of this signal depends on whether the field has increased or decreased. If this signal is integrated it constitutes an ideal discriminator curve, pulling back the field in proportion to the extent of the deviation from exact resonance. This error signal may then be used to supplement the correction signal in a flux stabilization system, providing it with an absolute reference point, so that no long-term drift can occur. Very large fluctuations in magnetic field can take this regulator outside its operating range, unlocking the system. The resonance condition for the reference material must then be found by searching manually, or by a suitable electronic 'sweep and lock' device. When samples are changed, the disturbance is usually small enough that the regulator can recapture the lock condition.

Most field/frequency regulation systems now employ pulse excitation, the repetition rates being usually in the region of 10 Hz to 10 kHz, and the signal being monitored in the intervals between pulses. Where the signal of another nuclear species is used as reference, the probe is equipped either with a separate coil, tuned, for example, to deuterium resonance, or the existing NMR coil is double-tuned, and suitable radiofrequency traps included to separate the two radiofrequencies. One area where particular care is needed in the design of a field/frequency regulator is the choice of the characteristics of the amplifiers in the feedback loop, in particular the manner in which the gain varies as a function of the frequency of the disturbance.

While the dispersion-mode signal is being used for regulation, it is common to display the absorption mode signal on a meter or similar device. The peak height of the absorption-mode signal provides a very convenient method of monitoring the resolution. If, for example, the linewidth is narrowed by adjustment of the field homogeneity coils, the peak signal increases accordingly. This proves to be a surprisingly effective procedure, even though in principle a reduction in width at half height does not necessarily correspond to an increase in peak height. Certain adjustments may be made with the sample spinning; others are made with a non-spinning sample. Since many spectroscopists use this as the sole criterion for optimization of resolution, it is particularly important that the effective sample volume for the reference nucleus corresponds reasonably closely with the effective volume for the observed nucleus; hence the tendency to use double-tuned receiver coils.

As may be appreciated from Fig. 1, noise on a field/frequency reference signal is converted into instability of the field/frequency ratio. This is particularly undesirable if the reference nucleus is one of low gyromagnetic ratio (such as deuterium) while the observed nucleus is of high gyromagnetic ratio (protons) because there is a 'leverage' of a factor 6.5 in the resulting disturbances at the proton resonance condition. Where there is some freedom of choice of deuterated solvent, it is therefore advisable to choose one with a strong signal – for example acetone-d_6 is better than chloroform-d in this respect. It is also important that the reference material has a narrow natural linewidth since it is to be used in optimizing the resolution. (Heavy water can be undesirable when very high resolution is required, because the signal is so strong that it may be significantly broadened by radiation damping*.) Fortunately, the spin–spin relaxation* times of the commonly used deuterated solvents are sufficiently long for this purpose even though deuterium possesses a quadrupole moment and is therefore relaxed by interaction with electric field gradients.

Occasionally, some spectrometer operation perturbs the field/frequency lock condition. This may be particularly severe when a field gradient pulse is applied in order to disperse transverse components of magnetization. Under these circumstances it may be advisable to inhibit the regulator loop momentarily while the gradient pulse is applied, using a 'sample and hold' circuit. During certain heteronuclear double resonance experiments with intense B_2 fields, the Bloch–Siegert effect* may be large enough to disturb the deuterium resonance condition and upset the field/frequency regulation.

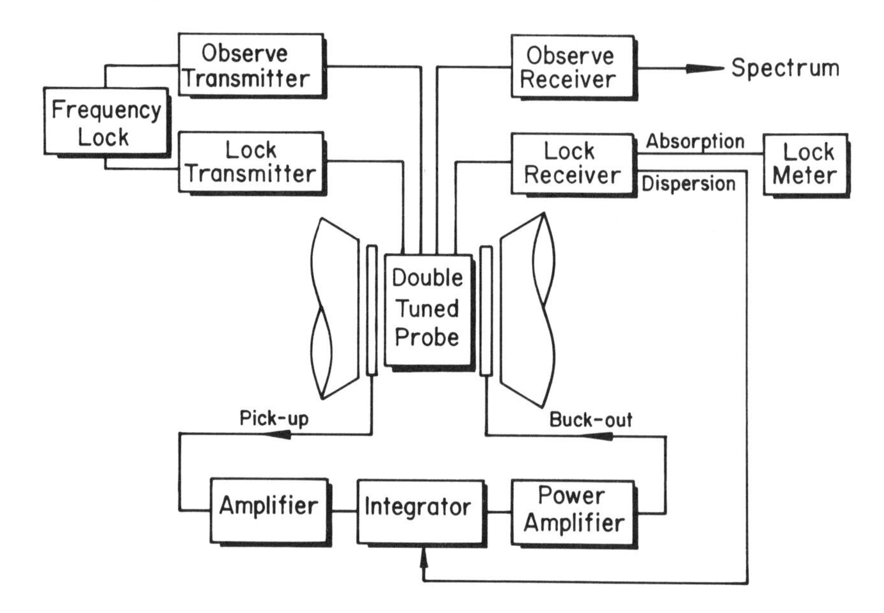

Fig. 1. Block schematic of a typical internal lock field/frequency regulation scheme. In many cases, the lock signal is from the deuterium resonance of the deuterated solvent. Monitoring the peak of the absorption-mode lock signal provides a convenient method of adjusting the homogeneity coils.

REFERENCES

1. H. Primas, *Fifth European Congress on Molecular Spectroscopy*. Amsterdam, 1961.
2. R. Freeman and D. H. Whiffen, *Proc. Phys. Soc. (Lond.)* **79**, 794 (1962).

Cross-references

Bloch–Siegert effect
Radiation damping
Sample spinning
Spin–spin relaxation

Fourier Transformation

Some years ago it was considered something of an art-form to demolish a grand piano with a sledgehammer. In the vocabulary of NMR this would be known as the impulse response, while the more usual note-by-note excitation would be the equivalent of slow-passage (digitized) frequency sweep. Unlike the piano, the nuclear spins recover from a pulse in a few seconds, the spin–lattice relaxation time, and may then be excited again. Ernst and Anderson (1) showed that this gives an enormous improvement in the sensitivity of the NMR

method, and opens up a wealth of new transient experiments, so that now essentially all new research-type NMR spectrometers operate in the pulse-excited mode.

The sensitivity advantage arises because the sweep method excites only one resonance line at a time, whereas the pulse method excites all resonances simultaneously. This can be appreciated by consideration of the motion of the nuclear magnetization vectors in the presence of a strong radiofrequency pulse acting in the rotating frame*. The resulting sensitivity gain, known to infra-red spectroscopists as the 'Fellgett advantage' (2), can be predicted approximately using the principle that signal-to-noise increases as the square root of the number of independent signals that are added together.

Suppose that a typical line in the spectrum has a width Δ given by $1/(\pi T_2^*)$ Hz, where T_2^* is the decay time constant due to instrumental effects. Consider first the case where the slow passage condition is satisfied, the sweep rate being adjusted so that a time nT_2^* is taken to sweep through a single resonance line (n being a small number, say 2 or 3). If the total sweep width to be covered is F Hz then this experiment requires a total time of nT_2^*F/Δ seconds. In order to achieve comparable resolution in the pulse mode, it is necessary to sample each transient for a time long compared with T_2^*, say nT_2^* s, where n is again 2 or 3. However, in the long time taken by the sweep experiment it is possible to repeat the pulse experiment N times and to add the N transient signals together in registration. Thus, by equating the total times required by the two experiments,

$$nT_2^*F/\Delta = NnT_2^*$$

$$N = F/\Delta.$$

The repeated pulse experiment consequently shows a sensitivity improvement of $N^{1/2} = (F/\Delta)^{1/2}$. For a typical carbon-13 spectrum F could be as large as 10 kHz and Δ might be 1 Hz, giving two orders of magnitude sensitivity advantage for the pulse mode.

In practice, the proper digitization of such a spectrum would require well over 10 000 words of data storage, which is not always available. Moreover, there is a significant improvement in sensitivity to be gained in the sweep experiment by violating the slow passage condition to some extent, thus sacrificing some resolving power. These factors make the 'Fellgett advantage' rather lower than calculated above, although the order of magnitude is still correct.

A piano certainly sounds more pleasant played in the 'continuous-wave' mode and NMR spectroscopists find that the interpretation of a frequency-domain spectrum is far easier than attempting to disentangle all the component frequencies of the transient free induction decay* after a radiofrequency pulse; yet the information content is the same. Indeed, the time-domain signal is related to the slow-passage frequency-domain spectrum by the process of Fourier transformation.

High-resolution pulse spectrometers are therefore known as Fourier transform spectrometers, this transformation facility being essential. This necessarily involves temporary storage of the time-averaged free induction signal, hence the process of digitization*. Two methods are available for Fourier transformation: the *spectrum analyser* and the discrete Fourier transform.

THE SPECTRUM ANALYSER

While this method is now little used in practice, it is important in that it provides some insight into the process of Fourier transformation. The spectrum analyser (or wave analyser) is a device which accepts a complex wave form S(t), examines the intensities of the various sine waves that may be said to make up the input waveform and presents the result in the form of a graph of intensity against frequency S(f). S(f) is the Fourier transform of S(t).

One implementation of this method is illustrated in Fig. 1. The waveform S(t) is stored digitally and read repetitively into a lock-in detector*, basically an amplifier with significant gain only near the reference frequency f_r. This device measures the content of S(t) carried at or near the frequency f_r, and represents the result as a point on the frequency-domain spectrum S(f). Then f_r is incremented and the waveform S(t) is again read into the synchronous detector, giving the next point in the spectrum S(f). The method is thus intrinsically discrete, but carries the advantage that the frequency steps may be made arbitrarily small, giving very fine digitization. The first computer programs used for Fourier transformation used a similar step-by-step examination of the waveform S(t) to find one frequency component at a time. These programs could take as long as 20 minutes to perform the Fourier transformation and were soon superseded by the Cooley-Tukey algorithm* (3).

Since the spectrum analyser method of Fourier transformation requires a multichannel storage device essentially equivalent to a minicomputer, it turns out to be quicker and more convenient to calculate the transforms in a small dedicated computer; essentially all the commercial Fourier transform spectrometers have adopted this method.

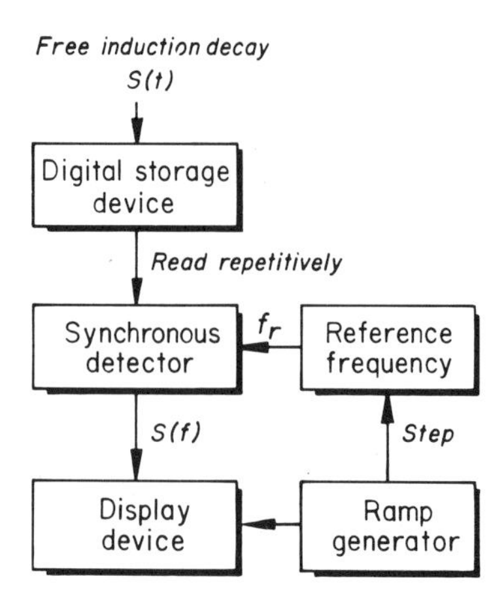

Fig. 1. Discrete Fourier transformation as implemented in a spectrum analyser or wave analyser. Digitization in the frequency domain may be as fine as desired, but each point involves a separate operation of reading the free induction decay into the synchronous detector.

ARTIFACTS IN THE TRANSFORMED SPECTRUM

It is important for the NMR user to be aware of the possible artifacts which may appear in the final frequency-domain spectrum. Most of these may be attributed to improper handling of the free induction signal before the transformation, and some of these are treated in more detail under the heading 'Digitization'.

Both the analogue-to-digital converter and the computer storage word have finite dynamic range, leading to the possibility of clipping the free induction signal. The principal result of this non-linear process is to introduce harmonics and sums and differences of NMR frequencies into the spectrum, together with some baseline distortion (see below). If the incoming free induction decay is scaled down to avoid clipping, then the digitization errors of the analogue-to-digital converter introduce *digitization noise* into the spectrum and this may obscure the very weak signals.

The sampling theorem requires that there should be at least two sampling points per period of the highest frequency component of the free induction signal. If this condition is violated, then some high-frequency resonances are aliased. The sampling rate may be said to define a spectral window which should encompass all the NMR lines of the spectrum. Continuous-wave spectroscopy* with an insufficiently wide sweep runs the risk of losing certain resonances; in Fourier transform spectroscopy these resonances are aliased into the spectral window and appear at the wrong frequencies. This is an ever-present danger in the interpretation of spectra from unknown compounds. High-frequency noise may also be aliased in this way and it is important to remove it with a suitable low-pass filter *before* the free induction signal is sampled, otherwise the sensitivity is unnecessarily impaired.

Because of instrumental shortcomings, the free induction signal entering the analogue-to-digital converter is not always exactly balanced about zero volts. After Fourier transformation, this d.c. offset gives rise to a spurious spike in the frequency-domain spectrum at zero frequency, a particularly unfortunate artifact when quadrature detection* is employed. If such an unbalanced free induction decay is multiplied by a sensitivity enhancement weighting function, the spike is broadened and affects several data points near zero frequency. In order to remedy this effect, many computer programs calculate the mean value of the free induction signal (or the last section of it) and subtract it from all the data points before any weighting function is applied.

A sensitive NMR receiver requires a finite time to recover from the effects of a strong radiofrequency pulse, although, like a radar receiver, it is gated off at the time of the pulse. The practical result is that the receiver gain may vary somewhat during the acquisition of the first few points of the free induction decay. This has come to be called *pulse breakthrough*. If only the very first data point is falsified, this displaces the baseline of the frequency-domain spectrum, a relatively trivial matter, even when the spectrum is to be integrated. However, if the next data points are significantly affected, this distortion transforms into a rolling baseline corresponding to the introduction of spurious low-frequency sine waves. One solution is to delay acquisition until the receiver has recovered, but this introduces a frequency-dependent phase shift into the spectrum which must be corrected by the phasing routine. An alternative remedy is to apply a baseline correction to the frequency-domain spectrum. The third possible approach is to apply a 180° pulse in order to generate a spin echo*, all frequency components being brought back into phase at a time when the receiver has recovered from the two pulses. This method has serious complications for proton spectroscopy where there are homonuclear spin–spin couplings which modulate the echo, but is satisfactory for carbon-13 spectroscopy where homonuclear coupling can be neglected. Note that none of these solutions is really suitable when there are very broad NMR lines in the spectrum.

If there is a discontinuity in the time-domain function, the Fourier transform program requires that it be rounded off before the transformation takes place. The free induction decay is treated as a periodically repeating signal so there is normally a discontinuity between the tail of one free induction decay and the first (intense) data point of the next. This should be replaced by the mean value, usually half the height of the first data point (4). Some spectrometer programs neglect this precaution, with the consequence that the Fourier transformation algorithm interprets the first data point as being too intense, giving a positive offset to the baseline of the resulting spectrum. In one-dimensional spectroscopy this effect is mitigated by the action of the bandpass filter used to prevent aliasing of noise, since this rounds off the discontinuity before the free induction signal reaches the transformation stage. However, in two-dimensional spectroscopy there is no equivalent filter acting in the t_1 dimension. Whenever there is a strong NMR signal therefore, there is a discontinuity at the point $t_1 = 0$, leading to ridges in the spectrum parallel to the F_1 axis.

NMR pulse experiments should be repeated as rapidly as possible if high sensitivity is required. This often entails a compromise with resolution requirements, such that the free induction signal is truncated before the decay is complete. When the technique of zero filling* is used, this gives a step function discontinuity in the envelope of the free induction decay. Resolution enhancement* shaping functions may exaggerate this discontinuity. Since the Fourier transform of a step function is a sin x/x function, this introduces some sin x/x character into the frequency-domain lineshape, the degree depending upon the severity of the time-domain discontinuity. It is therefore important to 'round off' this step in the envelope by means of an apodization* function which affects only a section of the data table just before the discontinuity, or more effectively, by a sensitivity enhancement* weighting function modifying the entire data table.

A digital computer necessarily introduces round-off errors in a calculation, which is aggravated when there is a cascade of operations as in Fourier transformation. Normally, these errors are outweighed by the noise present on the NMR signals and no harm is done. However, if time averaging is continued to the point where dynamic range is beginning to be limited by the computer word length (the noise level being reduced to the level of one bit) then these round-off errors in the transformation program become significant. Furthermore, these errors cannot be treated as noise and reduced by time-averaging several frequency-domain spectra, since under these conditions the round-off errors are coherent from one Fourier transform to the next. This is most probably the reason why very protracted time-averaging experiments sometimes fail to yield the expected $N^{1/2}$ improvement in signal-to-noise ratio. The remedy is to use double-precision arithmetic or a floating-point Fourier transform program.

REFERENCES

1. R. R. Ernst and W. A. Anderson, *Rev. Sci. Instr.* **37**, 93 (1966).
2. P. Fellgett, Ph.D. Thesis, Cambridge University (1951).
3. J. W. Cooley and J. W. Tukey, *Math. Comput.* **19**, 297 (1965).
4. G. Otting, H. Widmer, G. Wagner and K. Wüthrich, *J. Magn. Reson.* **66**, 187 (1986).

Cross-references

Apodization
Digitization
Continuous-wave spectroscopy
Cooley–Tukey algorithm
Free induction decay
Modulation and lock-in detection
Quadrature detection
Resolution enhancement
Rotating frame
Sensitivity enhancement
Spin echoes
Zero filling

Free Induction Decay

The response of a nuclear spin system to a radiofrequency impulse is known as the 'free precession signal' or the 'free induction decay' and is of fundamental importance because its Fourier transform is the frequency-domain high resolution spectrum. It is here that the Bloch vector model* is useful. Instead of considering individual nuclear spins obeying the laws of quantum mechanics, Bloch (1) showed that it is sufficient to focus attention on the net macroscopic nuclear magnetization obtained by taking the ensemble average over all the spins. This macroscopic magnetization obeys the laws of classical mechanics in its interaction with the applied static and radiofrequency fields, while the effects of spin–spin and spin–lattice relaxation may be accounted for phenomenologically by the introduction of simple damping terms into the Bloch equations (1). For the pulse excitation used in Fourier transform spectrometers, the transient solutions of the Bloch equations are used. For the older continuous-wave spectrometers, the more complicated steady-state solutions are required. The Bloch equations take on a particularly simple form when described in a reference frame* rotating synchronism with the radiofrequency field B_1.

The high resolution NMR spectroscopist tends to generalize this picture to the case of systems with many different chemical shifts, and with spin–spin couplings. Most of the observed NMR phenomena can be accounted for on the assumption that each individual resonance in the high-resolution spectrum can be represented by a vector M with a characteristic intensity and a characteristic precession frequency. Each such vector M is assumed to obey the Bloch equations. Difficulties can arise when the spins are subjected to more than one radiofrequency pulse, for example when multiple-quantum coherence is involved.

For a spin system that has been allowed to reach Boltzmann equilibrium, all these vectors are aligned along the Z axis of the rotating frame, and a 90° radiofrequency pulse rotates them about the X axis, leaving them along the Y axis. For the purposes of this section it is convenient to assume that the pulse is 'perfect' in that it affects all parts of the effective sample volume uniformly, and that its action is independent of the offset of a given line from the transmitter frequency. The reference phase of the receiver is usually taken to be such as to detect the Y magnetization in phase. Thus the signal intensity immediately after the pulse is at its maximum value. In fact, this first ordinate of the free induction decay represents the integral of the total intensity in the absorption spectrum.

The individual magnetization vectors are then free to precess at their characteristic frequencies, generating an interference pattern. (This has led some spectroscopists to refer to free induction decays as interferograms by analogy with the infra-red experiment.) In a very simple case, such as the familiar 1 : 3 : 3 : 1 quartet, the beat pattern itself may be recognized (see Fig. 1) and the frequency difference measured directly, but in the general case the pattern is not decipherable by inspection and a Fourier transform must be computed. When there are both strong and weak components present in the free induction decay the dynamic range of the detection equipment must be sufficiently high to handle both adequately. This contrasts with continuous-wave spectroscopy where the strong and weak signals are not usually riding on top of each other, so the weak signal can be examined while the strong signal is being truncated.

The envelope of the free precession signal decays with time, in many cases falling to a negligible level before acquisition is stopped. This is a result of spin–spin relaxation* and the mutual interference between macroscopic signals from different regions of the sample due to the inhomogeneity of the field B_0. Residual instabilities in the field/frequency regulation may also contribute to this decay, particularly if several sequential free induction signals are being

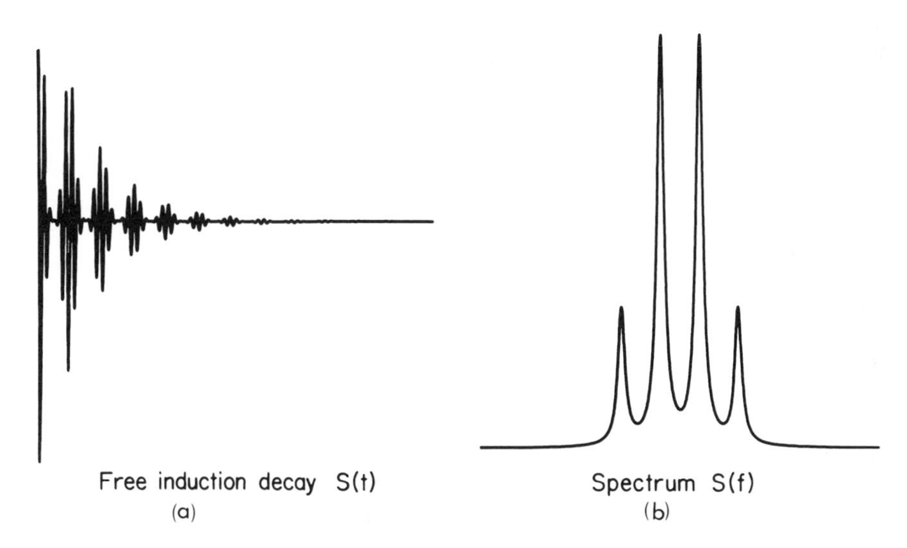

Fig. 1. (a) A free induction decay S(t) and (b) its Fourier transform S(f). In practice, S(t) was extended to four times its length by zero-filling in order to improve the digitization of S(f).

summed together. For a single line, the Fourier transform of this envelope determines the frequency-domain lineshape. For example, an exponentially decaying envelope gives a Lorentzian lineshape. The common practice of imposing an exponential envelope on the free induction signal for the purposes of sensitivity enhancement* introduces a strong Lorentzian character on the spectral lines. For the purposes of high resolution in spectra of many overlapping lines, a Gaussian shape might be preferable since it falls off more rapidly in the skirts of the line. Because of this Fourier transform relationship between the envelope of the free induction decay and the lineshape in the frequency domain, a slowly decaying free precession signal gives a narrow line. High resolution therefore requires a long acquisition time; early termination of the free induction decay will broaden and distort the lines. Thus by setting a lower limit on the interval between excitation pulses, high resolution competes with the requirements of high sensitivity. Artificial techniques for slowing down the decay come under the heading of resolution enhancement*.

Most modern NMR spectrometers acquire two free induction signals simultaneously by having two synchronous detectors operating in quadrature. Thus the real and the imaginary parts of the complex susceptibility are being detected, and the input to the Fourier transform program is an array of complex numbers. This makes it possible to determine the signs of the precession frequencies in the rotating frame, thus removing any ambiguity about whether a given resonance is at a higher or lower frequency than the transmitter. (See Quadrature detection*.) Sensitivity is increased by a factor $2^{1/2}$ by this technique, and it allows the transmitter frequency to be set in the centre of the spectrum, thus using the pulse intensity more effectively.

ABNORMALITIES IN THE FREE INDUCTION SIGNAL

Some distortions of the free induction decay may be ascribed to shortcomings of the detection scheme, for example receiver dead time or improper sampling; these are treated under Fourier transformation* and Digitization*. Quite another type of problem can arise because of the physics of the excitation process, and this is treated below.

Under ideal operating conditions, the interval between successive pulse excitations would be long compared with all relaxation times. In practice, sensitivity considerations rule out such an extravagant mode of operation, the normal practice being to excite the spins again as soon as a free induction decay of sufficient length has been acquired. This gives rise to the possibility of steady-state effects* which may be divided into those concerned with longitudinal magnetization (pulse interval comparable with T_1), and those concerned with transverse magnetization (pulse interval comparable with T_2). Needless to say, both types of steady-state effects tend to occur together since T_1 and T_2 tend to be closely related in liquid samples used for high-resolution work.

Repetition of the pulse excitation at a rate comparable with the rate of spin–lattice relaxation sets up a steady-state Z magnetization which is weaker than the equilibrium magnetization M_0; this is the equivalent of saturation in continuous-wave spectroscopy. It causes both loss of overall intensity and variations in relative intensities within a given spectrum if the individual spin–lattice relaxation times differ. In fact, this technique can be made the basis of a method for determining the spin–lattice relaxation times (2). As a consequence, considerable care should be exercised when using relative intensities for quantitative measurements if the free induction decays were obtained under steady-state conditions. Usually, a second experiment at a different repetition rate will resolve any doubts about the reliability of the relative intensity measurements.

Steady-state effects on transverse magnetization are more complex. A free precession signal which has started to decay due to field inhomogeneity may be seen to refocus again into a spin echo*. This can be seen in terms of spin *isochromats* (signals from small volume elements of the sample), each isochromat having a slightly different precession frequency because that volume element sits in a slightly different applied field B_0. In the steady-state regime, a vector representing a given isochromat is not aligned along the Z axis at the time of the pulse but has a significant X–Y component. The pulse rotates this vector about the X axis, maintaining the same X component and changing the Y component in the same way as for all other isochromats. Since the individual vectors have a particular phase relationship immediately after the pulse, they must also have a very similar phase relationship just before the pulse. Free precession trajectories are thus constrained by the steady-state requirements so that refocusing occurs. Since the actual trajectory depends quite strongly on offset from the transmitter, the lines in the final spectrum exhibit anomalies in phase and intensity which themselves depend on offset (3). These anomalies can be avoided by introducing a small random variation in the interpulse interval, or by a suitable phase cycling* technique.

WHAT IS A FREE INDUCTION DECAY?

There are difficulties in attempting to apply the spectroscopic concepts of stimulated absorption, stimulated emission or spontaneous emission to the free precession signal following a radiofrequency pulse. Spontaneous emission probabilities are known to depend on the cube of the radiation frequency, and for NMR are known to be quite negligibly small (one event in 10^{23} seconds for a typical example). Continuous-wave spectroscopy may be discussed in terms of stimulated absorption of energy by the nuclei, but during a free induction decay there is no applied radiation field to stimulate either absorption or emission. Furthermore, this signal is detected in many cases while the populations of upper and lower energy levels are equal. Hence the importance of Bloch's concept of *nuclear induction* – a concerted motion of the individual nuclear spins, generating a current in the receiver coil according to the well-known laws of electromagnetic induction. Immediately before the excitation pulse there is a very small excess spin population Δ in the lower energy level, but no phase coherence between the precession of individual spins. There is thus a very small longitudinal magnetization M_0 which can induce no signal in the receiver coil. Now a radiofrequency pulse acts on each spin independently, rotating it about the X axis of the rotating frame. When we consider the ensemble of spins in the sample, the net effect is to rotate the macroscopic magnetization M_0 away from the Z axis toward the Y axis. The existence of a macroscopic magnetization M_Y can be rationalized by saying that the pulse has induced a small degree of phase coherence into the previously random motion of the individual precessing spins.

Viewed in the laboratory frame, M_Y is precessing at the Larmor frequency and cuts the turns of the receiver coil, inducing a voltage at this frequency. Under most high-resolution conditions the induced signal is very weak, but at high fields and with strong samples, this induced signal generates a rotating magnetic field strong enough to affect the motion of the nuclei, as in a continuous-wave experiment. This radiation damping* effect broadens the lines involved.

It is important to bear in mind that in many high-resolution situations, the concerted precession of the nuclear spins usually persists for considerably longer than the observed free induction decay. This is because the 'phase memory time' or spin–spin relaxation time may well be of the order of 10 seconds, while the decay time constant due to field inhomogeneity T_2^* would normally be of the order of one second. Certain NMR techniques involve the application of B_0 field gradients which are often said to 'destroy' transverse nuclear magnetization. Actually, they only disperse the isochromats from different sample regions, and the possibility always remains that these isochromats may be refocused into a spin echo, if the spin–spin relaxation time is long. This is the origin of one of the steady-state effects described above.

REFERENCES

1. F. Bloch, *Phys. Rev.* **102**, 104 (1956).
2. R. Freeman and H. D. W. Hill, *J. Chem. Phys.* **54**, 3367 (1971).
3. R. Freeman and H. D. W. Hill, *J. Magn. Reson.* **4**, 366 (1971).

Cross-references

Digitization
Fourier transformation

Phase cycling
Quadrature detection
Radiation damping
Resolution enhancement
Rotating frame
Sensitivity enhancement
Spin echoes
Spin–spin relaxation
Steady-state effects
Vector model

Gated Decoupling

Double resonance effects in NMR can be divided into two main categories. In the first, only intensity changes are observed and these can be described in terms of the disturbance of the spin populations (with or without cross-relaxation effects). In the second category, lines are observed to split into submultiplets, or multiplets are seen to coalesce into simpler structures, often into a single line. The second mechanism relies on the coherence of the second radiofrequency field B_2, while the first type of effect would also occur even if a degree of incoherence were present.

In many situations, both effects occur together, but they may be easily separated if necessary, since the population disturbances persist after the B_2 field is extinguished, for a period of the order of the spin–lattice relaxation* time, whereas the coherence effects disappear at once. Experiments designed to exploit this difference are usually known as *gated decoupling*. One of the more important population effects is the intensity enhancement created by the nuclear Overhauser effect*, particularly in spectra from nuclei of low isotopic abundance such as carbon-13 (1). In the conventional experiment all proton resonances are irradiated continuously and the carbon-13 spectrum is decoupled and enhanced in intensity. If the proton-coupled carbon-13 spectrum is to be studied, the nuclear Overhauser enhancement can be retained by a gated decoupling technique. The protons are saturated by a broadband B_2 irradiation for some time which is long compared with the cross-relaxation time, building up non-Boltzmann populations on the spin energy levels. The decoupler is then switched off and the carbon-13 free induction signal excited by a radiofrequency pulse; since the carbon-13 spins precess in the absence of proton irradiation, the coupled spectrum is obtained, but the nuclear Overhauser enhancement is retained. Similar experiments have been used in proton spectra to reveal homonuclear Overhauser effects (2) untrammelled by decoupling or Bloch–Siegert effects*.

Gated heteronuclear decoupling can be very useful in the reverse mode – the generation of a decoupled carbon-13 spectrum with no nuclear Overhauser enhancement, the proton irradiation being switched on at the time of the carbon-13 excitation pulse. There is thus no time for the cross-relaxation to take place before the pulse creates transverse carbon-13 magnetization, and although a population rearrangement begins to build up during the acquisition of the carbon free induction decay, this only affects longitudinal carbon-13 magnetization and *not* the intensities in the resulting spectrum. In fact, the rate at which the nuclear Overhauser effect is established may be measured by performing a series of experiments as a function of the period of proton pre-irradiation just before the carbon excitation pulse (3). If the gated decoupler results are compared with those from the conventional spectrum where protons are continuously irradiated, then the intensity ratios observed for a given carbon-13 resonance allow the nuclear Overhauser enhancement to be determined.

The expression for the nuclear Overhauser enhancement may be written

$$\eta = (\gamma_H/\gamma_X)\,\frac{(W_2 - W_0)}{(W_2 + 2W_1 + W_0)}. \qquad [1]$$

The last factor, involving the transition probabilities W_2, etc., reflects the degree to which the relaxation mechanism is dipolar. In the extreme narrowing limit and with predominantly dipolar relaxation, this factor takes on its maximum value, $\frac{1}{2}$. For nuclei such as ^{15}N or ^{29}Si, which have negative gyromagnetic ratios γ_X, the enhanced signal is reversed in sign (emission rather than absorption). An inverted signal poses no particular problem in itself, but if the relaxation is only partly dipolar, the observed enhancement may happen to lie in the range 0 to −2. This subtracts from the natural (unenhanced) signal to give a loss rather than a gain in sensitivity, and then it is better to suppress the Overhauser effect by means of the gated decoupler experiment or by quenching it with a relaxation reagent (4).

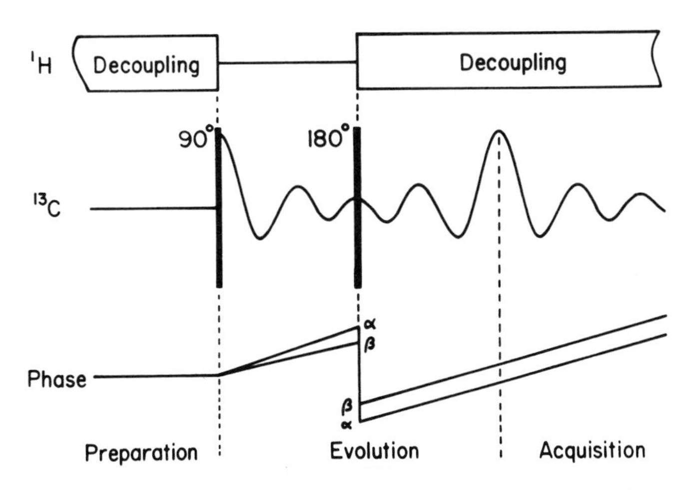

Fig. 1. Gated decoupling employed to introduce J-modulation of carbon-13 spin echoes. Only the second half of the spin echo is acquired. The phase evolution diagram is shown for the carbon-13 spins of a CH group.

Broadband heteronuclear decoupling* can induce serious heating effects in the sample, particularly if there are ions present in solution. Since the criteria for efficient decoupling are more demanding than the requirements for the nuclear Overhauser effect (proton saturation), there are situations where two-level irradiation is indicated. The proton irradiation level would be kept relatively low prior to the carbon-13 excitation pulse and then switched to maximum during the acquisition of the free induction decay. In this way, the mean radiofrequency power dissipated in the sample can be reduced. In extreme cases the nuclear Overhauser enhancement might have to be sacrificed altogether in order to minimize heating effects.

In homonuclear decoupling experiments, for example proton–proton decoupling, the intense irradiation field B_2 can cause overloading and non-linear effects in the preamplifier and receiver. This problem may be circumvented by a time-share modulation* of the irradiation field B_2, with the receiver gated off while the decoupler is gated on. The modulation frequency is high (typically of the order of 10 kHz), and under these conditions the spins behave as though they were being continuously irradiated at a correspondingly lower level of B_2. This apparent paradox may be rationalized by saying that the sidebands generated by the modulation are so far separated from the centreband that

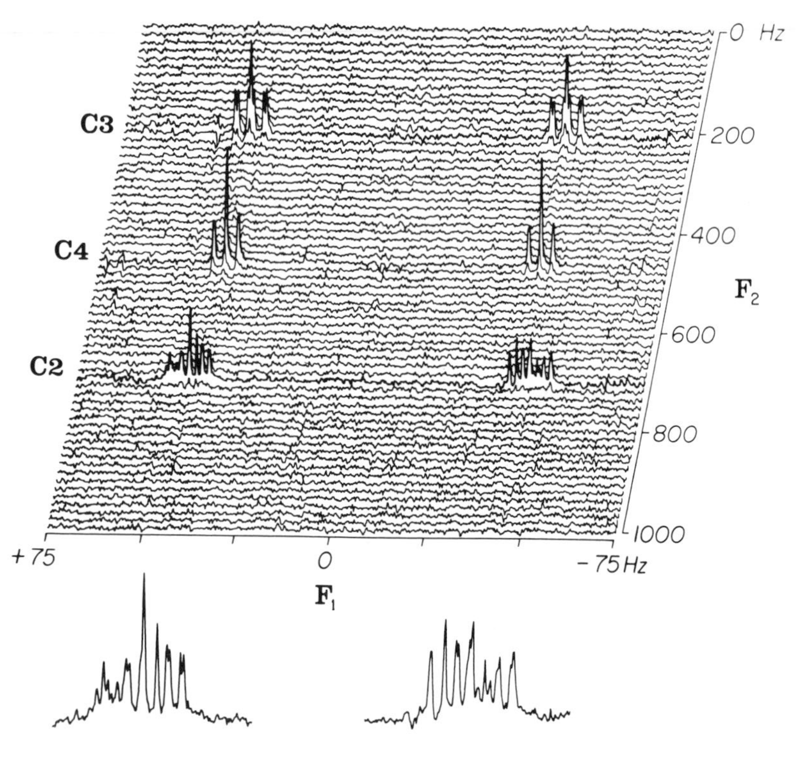

Fig. 2. Two-dimensional J-spectrum of the three carbon-13 sites of pyridine obtained by the gated decoupling technique. The C2 site shows a marked asymmetry attributable to strong coupling between protons (inset). The spectrum is in the pure absorption mode.

throughout the entire proton chemical shift range only the centreband has any significant effect.

Double resonance provides an effective method of distinguishing coupled spin systems from isolated spins. For example, there is considerable interest in the weak satellite lines which appear in the flanks of proton resonances due to coupling to low-abundance nuclei such as carbon-13 or nitrogen-15. One method of suppressing the strong 'parent' proton lines to leave the pure satellite spectrum is double resonance difference spectroscopy*. In its simple form, this experiment records the normal proton spectrum and stores it in a computer, and then subtracts the corresponding spectrum with the low-abundance nucleus strongly irradiated. If care is taken to prevent the irradiation field from interfering with the field-frequency regulation, the parent signal cancels quite effectively, leaving a satellite spectrum together with an inverted decoupled satellite spectrum. Subtraction is usually performed on the free induction decays before Fourier transformation.

Spin echo experiments on a spin S coupled to a heteronuclear spin I normally show no modulation due to spin–spin coupling J_{IS}. One method of introducing echo modulation is to decouple I from S during one half of the interval between the initial pulse and the centre of the spin echo. For example, I and S might be coupled during the time that the S spin isochromats were defocusing but be decoupled during the refocusing period. Consideration of the appropriate phase evolution diagram (Fig. 1) shows that this introduces a phase modulation of the echoes as a function of the time at which the echo appears. This phenomenon has been exploited in two-dimensional spectroscopy* for observation of the proton multiplet structure on carbon-13 resonances. It is particularly useful in cases where the proton spins form a strongly coupled system, as at the C2 site of pyridine (Fig. 2). A related experiment can be performed by selective excitation* of a *single* carbon-13 resonance line while the proton decoupler is switched on, allowing the carbon-13 free precession to occur with the decoupler switched off. The result is a proton-coupled multiplet from each carbon site in the molecule (5).

REFERENCES

1. J. H. Noggle and R. E. Schirmer, *The Nuclear Overhauser Effect: Chemical Applications.* Academic Press: New York, 1971.
2. J. K. M. Sanders and J. D. Mersh, *Progress in NMR Spectroscopy* **15**, 353 (1982).
3. R. Freeman, H. D. W. Hill and R. Kaptein, *J. Magn. Reson.* **7**, 327 (1972).
4. R. Freeman, K. G. R. Pachler and G. N. La Mar, *J. Chem. Phys.* **55**, 4586 (1971).
5. G. A. Morris and R. Freeman, *J. Magn. Reson.* **29**, 433 (1978).

Cross-references

Bloch–Siegert effect
Broadband decoupling
Difference spectroscopy
Nuclear Overhauser effect
Selective excitation
Spin–lattice relaxation
Time-share modulation
Two-dimensional spectroscopy

Hartmann–Hahn Experiment

The original experiment performed by Hartmann and Hahn (1) was a solid state technique and is therefore outside the scope of this book. However, it has more recently been extended to liquid-phase high-resolution work as a method of polarization transfer* and affects some experiments in two-dimensional spectroscopy*.

Consider then the solid-state Hartmann–Hahn experiment. The aim is to detect the NMR response of a nuclear species S which is of low natural abundance or low magnetogyric ratio, surrounded by abundant spins I which have a strong NMR response. The first step is to apply to the I spins a 90° pulse followed by a phase-shifted continuous radiofrequency field B_1. The I spin resonance is spin locked (see Spin locking*) for a relatively long period, comparable with the spin–lattice relaxation time T_1. During this period the I spins have a large nuclear magnetization but in the rotating frame* they experience only a weak effective field B_1 and they can thus be regarded as having an extremely low *spin temperature*, equal to the lattice temperature multiplied by the ratio B_1/B_0. It is permissible to speak of spin temperatures of the I spin system since in the solid state the spins are strongly coupled together (short T_2) but are relatively isolated from the lattice (long T_1). A second radiofrequency field B_2 is applied to the rare S spins, saturating them. The idea is to couple these 'hot' spins to the 'cold' I spins by a cross-relaxation experiment. In this way, the loss of I spin magnetization (monitored as a free induction decay at the end of the spin-locking period) is used to detect the resonance condition for the rare S spins. The longer the 'contact' between the two spin systems, the greater the loss of observed I spin signal.

The trick is to find a mechanism for establishing thermal contact between the two spin systems. Coupling takes place by way of the dipole–dipole interaction. The S spins are irradiated at resonance with a radiofrequency field B_2. In the reference frame rotating in synchronism with this radiofrequency field (the S spin frame) these spins precess at a frequency $\gamma_S B_2/2\pi$ (Hz) generating a Z component of magnetization which oscillates at this frequency. Because of the dipole–dipole interaction with the neighbouring I spins, the latter see an oscillating magnetic field at this frequency with an amplitude which depends on the strength of the dipole–dipole interaction. This behaves like an excitation field, inducing I spin transitions if it matches the precession frequency of the I spins in their reference frame. This Hartmann–Hahn matching is achieved by applying a radiofrequency field B_1 to the I spins chosen so that

$$\gamma_S B_2/2\pi = \gamma_I B_1/2\pi. \quad [1]$$

This is called the Hartmann–Hahn condition. The I spins can then absorb energy by 'resonance in the rotating frame' analogous to resonance absorption at the Larmor condition in the laboratory frame. The I and S spin systems then move towards a common spin temperature. In practice, it is important to optimize the rate of energy transfer so as to minimize the length of the 'contact time'; one factor which inhibits the energy transfer is a mismatch of the Hartmann–Hahn condition (1).

Cross polarization by Hartmann–Hahn rotating frame resonance has become almost routine for solid-state studies of the low-abundance carbon-13 nucleus. This technique is often combined with 'magic angle spinning' to remove static dipolar interactions. A practical difficulty with these heteronuclear experiments is that unless the same radiofrequency coil is used to generate both B_1 and B_2, the spatial inhomogeneities of these two fields will probably differ and there will be a mismatch of the Hartmann–Hahn condition in some regions of the sample.

HIGH RESOLUTION SPECTRA OF LIQUIDS

In the liquid phase there are several important differences in the physics. The static dipole–dipole interaction averages to zero in an isotropic liquid so the two spin systems are only coupled by the scalar spin–spin interaction J_{IS} and polarization transfer can only occur within the molecule. This means that there is no I-spin thermal reservoir. Furthermore, it is no longer permissible to use the concept of spin temperature since the spin–spin and spin–lattice relaxation times are normally comparable. In liquid-phase Hartmann–Hahn experiments magnetization is transferred back and forth between the I and the S spins at a rate determined by J_{IS}.

Nevertheless, useful experiments can be performed to transfer proton polarization to carbon-13 nuclei by Hartmann–Hahn contact in the rotating frame (2). The maximum enhancement is governed by the population advantage $\gamma_H/\gamma_C = 4$. Effective transfer requires that the mismatch of the Hartmann–Hahn condition be less than the interaction between the spins $^1J_{CH}$ of the order of 200 Hz. Unfortunately, this is difficult to achieve in practice and this has inhibited the adoption of this method. The same maximum enhancement of 4 can be achieved with the 'INEPT' polarization transfer sequence which is easier to implement. Often, the enhancement provided by the nuclear Overhauser effect* is preferred over polarization transfer since it comes as a by-product of broadband decoupling*.

HOMONUCLEAR SYSTEMS

Hartmann–Hahn effects can also occur in homonuclear systems. Obviously, a transfer of energy between one proton and another is not a method of sensitivity enhancement, but it can be used for shift correlation*, in order to determine which protons are spin coupled, either directly or through a chain of spin–spin couplings. Spectral assignment, particularly for large molecules, often requires an examination of these long-range or relayed correlations and the appropriate two-dimensional experiment (3) has been called 'TOCSY' (totally correlated spectroscopy), to be compared with COSY (correlated spectroscopy) which only detects single-stage transfers.

For a homonuclear system there is only a single radiofrequency field (B_1) and a single rotating frame, but the I and S spins in general have different offsets from exact resonance and hence different effective fields in the rotating frame. A spin-locking sequence is employed, consisting of a 90°(X) pulse followed by a continuous B_1 field applied along the Y axis. If resonance offsets could be neglected, the Hartmann–Hahn condition would be exactly satisfied (for all regions of the sample) and magnetization transfer between I and S would occur at a rate $2J_{IS}$ (twice as fast as in the heteronuclear case). Finite resonance offsets that differ for the I and S spins cause a mismatch of the Hartmann–Hahn condition, interfering with magnetization transfer and reducing the intensity of the corresponding cross-peak in the two-dimensional correlation spectrum. Efforts are now directed towards methods for broadening the range over which the Hartmann–Hahn condition can be satisfied – one method involves the use of proton broadband decoupling sequences (4).

A practical advantage of this method over the COSY experiment is that it generates phase-sensitive spectra without intensity alternation, giving better sensitivity for poorly-resolved multiplets. It may also be used for a version of the nuclear Overhauser experiment where transverse magnetization is involved; these cross-relaxation effects have the advantage that they always have the same sign, independent of the motional correlation time, and since they give negative cross peaks they are easily distinguished from the Hartmann–Hahn transfer through spin–spin coupling. Finally, the technique can be used to generate a subspectrum characteristic of part of a molecule that has no scalar couplings to other spins in the molecule.

BROADBAND DECOUPLING

Rotating-frame resonance can also produce undesirable effects in carbon-13 spectroscopy when proton broadband decoupling is employed. For example, spin-echo experiments on carbon-13 are adversely affected by Hartmann–Hahn contact with the irradiated protons (5). If coherent proton irradiation is used, a series of rotating-frame resonance conditions is observed where the echo decay rate is enormously accelerated. For noise decoupling, the interference effect is spread over the whole range of values of B_2, making spin echo experiments quite unsuitable for measuring long spin–spin relaxation times of carbon-13 in the liquid phase.

REFERENCES

1. S. R. Hartmann and E. L. Hahn, *Phys. Rev.* **128**, 2042 (1962).
2. G. C. Chingas, A. N. Garroway, R. D. Bertrand and W. B. Moniz, *J. Chem. Phys.* **74**, 127 (1981).
3. L. Braunschweiler and R. R. Ernst, *J. Magn. Reson.* **53**, 521 (1983).
4. A. Bax and D. G. Davis, *J. Magn. Reson.* **65**, 335 (1985).
5. R. Freeman and H. D. W. Hill, *Dynamic Nuclear Magnetic Resonance Spectroscopy*, Eds.: L. M. Jackman and F. A. Cotton. Academic Press: San Francisco, 1975, Ch. 5.

Cross-references

Broadband decoupling
Nuclear Overhauser effect
Polarization transfer
Rotating frame
Shift correlation
Spin-locking
Two-dimensional spectroscopy

Intensities

High resolution NMR possesses the very valuable property that the intensity of a given line is proportional to the number of nuclei contributing to that line, provided that certain instrumental precautions are taken. This allows a quantitative determination of the relative amounts of two substances present in the sample, and the measurement can be made effectively absolute by adding a carefully measured concentration of one of these substances followed by a second recording of the spectrum. This property may also be used as an assignment technique by counting the number of equivalent protons giving rise to a given resonance line and thereby identifying, for example, a methyl group.

Intensity is measured by the area under the line. In situations where all the lines in the spectrum have the same width and shape, peak heights may be used as a simpler measure of intensity, but in general it is safer to measure the area by means of integration (see below). In the case of only partially resolved lines, some method of decomposition into the individual constituents is required, which may involve curve-fitting through a computer program.

The precautions necessary in order to ensure reliable intensity measurements are listed below.

(1) A slow-passage spectrum is recorded, as otherwise the lines are distorted and the intensities become functions of the relaxation times. Fortunately, Fourier transform spectra are equivalent to slow-passage spectra except in very unusual circumstances.

(2) No significant saturation is allowed to occur. For slow-passage continuous-wave spectrometers (see Continuous-wave spectroscopy*) this means ensuring that $\gamma^2 B_1^2 T_1 T_2 \ll 1$ for all lines in the spectrum; for Fourier spectrometers this requires either a long waiting time between 90° pulses (T of the order of $5T_1$) or a sufficiently small flip angle that the steady-state magnetization M_Z is not significantly different from M_0. (See Steady-state effects*.)

(3) In pulse Fourier transform spectrometers, no appreciable steady-state transverse magnetization is set up, since this leads to anomalies in phase and intensity which vary in a cyclic fashion across the spectrum. This can be a problem if the interval between pulses is shorter than, or comparable with, T_2, the spin–spin relaxation* time. One remedy is to introduce a short random timing delay between excitation pulses (1).

(4) If pulse excitation is used, then the excitation is uniform across the entire spectrum. This requires that the intensity of the pulse be high enough that $B_1 \gg |\Delta B|$, the offset of the resonance line furthest from the transmitter frequency. Quadrature phase detection* helps somewhat in that it allows the transmitter frequency to be set near the centre of the spectrum.

(5) The low-pass filters used in Fourier spectrometers in order to avoid aliasing of noise should have a flat response over the frequency range of interest. These filters sometimes encroach at the ends of the spectrum, reducing the intensities there.

(6) For the observation of nuclei such as carbon-13 where there may be differential nuclear Overhauser enhancements, the effect is quenched by adding suitable relaxation reagents (2), or is suppressed by gating the proton decoupler off in the period immediately before the carbon excitation pulse. (See Gated decoupling*.)

(7) The entire spectrum is correctly phased for the absorption-mode signal. This is a routine adjustment in a Fourier transform spectrometer, but it is particularly important when measuring integrals.

(8) Nothing must be allowed to falsify the early part of the free induction decay. Thus pulse breakthrough must be prevented. Only well-behaved weighting functions should be used – those which do not alter the initial value of the free induction decay. *Convolution-difference* or *pseudo-echo* weighting functions (3) should be avoided in this context.

(9) The spectrometer response must be linear. Problems arise when the signals span a large dynamic range, giving difficulties with amplification and analogue-to-digital conversion.

(10) Sampling in the frequency domain should be sufficiently fine in comparison with the linewidths and there should be adequate digitization* of the intensity ordinates. Sometimes an interpolation routine may be required.

(11) For carbon-13 spectroscopy or for any nuclear species that requires broadband heteronuclear decoupling, the decoupling efficiency should be essentially uniform over the appropriate frequency range. This suggests the use of the new broadband decoupling* methods.

It should be emphasized that these precautions are suggested for applications which demand accurate intensity comparisons; they are difficult to reconcile with sensitivity* considerations and some are therefore ignored in routine high-resolution spectroscopy. For example, carbon-13 spectra are often run under conditions where lines with long spin–lattice relaxation times are partially saturated. Indeed, the resulting low intensity is often used to recognize a quaternary carbon site. Similarly, it is not usual to sacrifice the advantage of the nuclear Overhauser effect in carbon-13 spectroscopy. The remaining precautions, however, incur very little penalty in routine operations.

Figure 1 shows the carbon-13 spectrum of p-ethoxybenzaldehyde obtained under conditions where the intensities are essentially uniform across the entire spectrum, allowing for the fact that there are two pairs of equivalent carbon sites (4).

INTEGRATION

There are several methods of measuring the area under a resonance line. The more primitive methods of counting squares on graph paper or electronic integration based on the charging of a capacitor have given way to summation of the digitized signal in the spectrometer computer. This is essentially adding together the ordinates of a histogram representing the spectrum.

The practical difficulties arise not from the process of integration itself but in eliminating sources of systematic error. The most serious of these is the difficulty in defining the baseline level of the spectrum. The integral is extremely sensitive to small changes in the baseline. The total integral of the entire spectrum is determined by the value of the first data point of the free induction decay; if this is falsified by 'pulse breakthrough' this introduces a spurious contribution to the mean baseline level of the resulting spectrum. The spectrometer computer can be instructed to examine regions of signal-free baseline near the ends of the spectrum and thus perform a baseline correction* to

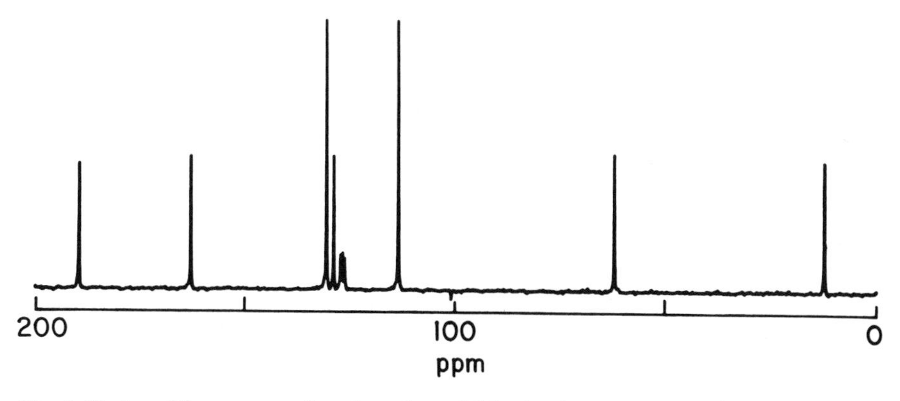

Fig. 1. Carbon-13 spectrum of p-ethoxybenzaldehyde after precautions have been taken to avoid all factors which perturb the intensities. Apart from the signal from the solvent $CDCl_3$, all the resonance lines have essentially equal intensities (two lines are doubly degenerate).

remove offset and tilt. More sophisticated programs can correct the low-frequency 'baseline roll' which sometimes occurs when the first few points of the free induction signal are in error. Integration errors arising from such baseline problems are reduced if the frequency range of the integration is carefully limited to the resonance line of interest. Accurate integration presupposes that the resonance lines are adequately sampled and digitized.

Some spectrometer computer programs incorporate these features and automatically print out integrals along with the line frequencies. Others give intensities in the form of peak heights, using some form of peak-finding routine (with a minimum threshold); this is clearly less satisfactory unless all lines have been broadened to the same linewidth. For measurements that are sensitive to systematic error, in particular weak nuclear Overhauser enhancements or spin–lattice relaxation* studies based on a small number of measurements, a check should be made on exactly how the program handles intensity measurements.

Integration of peaks in two-dimensional spectroscopy* presents special difficulties. Data storage limitations for the two-dimensional array may impose quite coarse digitization in the two frequency dimensions and this reduces the accuracy of the integrals. Perhaps more serious is the problem of deciding just where to truncate the integration, since a relatively high proportion of the total intensity resides in the skirts of the line. In one-dimensional spectra the truncated portion is a small *area*, but in two-dimensional spectroscopy it is a *volume* encircling the peak, falling off only quite slowly with offset. Measurements of the nuclear Overhauser effect* (5) would appear to be particularly susceptible to this problem.

REFERENCES

1. R. Freeman and H. D. W. Hill, *J. Magn. Reson.* **4**, 366 (1971).
2. R. Freeman, K. G. R. Pachler and G. N. La Mar, *J. Chem. Phys.* **55**, 4586 (1971).
3. J. C. Lindon and A. G. Ferrige, *Progress in NMR Spectroscopy* **14**, 27 (1980).
4. M. H. Levitt, R. Freeman and T. Frenkiel, *Advances in Magnetic Resonance* **11**, 47 (1983).
5. J. Jeener, B. H. Meier, P. Bachmann and R. R. Ernst, *J. Chem. Phys.* **71**, 4546 (1979).

Cross-references

Baseline correction
Broadband decoupling

Continuous-wave spectroscopy
Digitization
Gated decoupling
Nuclear Overhauser effect
Quadrature detection
Sensitivity
Spin–lattice relaxation
Spin–spin relaxation
Steady-state effects
Two-dimensional spectroscopy

J-Spectroscopy

It would be very useful to be able to separate chemical shifts from spin coupling constants. For the special case of carbon-13 spectroscopy where homonuclear coupling is usually ignored, this goal is partially achieved by broadband decoupling* of the protons, leaving only carbon-13 chemical shifts. Two-dimensional spectroscopy* allows us to go further and to obtain the spin multiplet structure for each chemically shifted site, even for proton spectra. This is achieved by exploiting the fact that in a spin-echo experiment, the chemical shift effect is eliminated by refocusing, whereas spin–spin coupling information is retained.

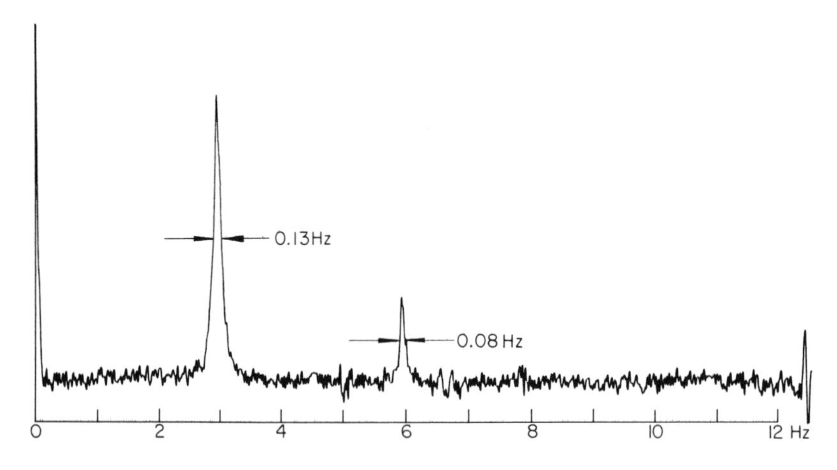

Fig. 1. J-spectrum of protons in 1,1,2-trichloroethane obtained by Fourier transformation of the spin echo modulation. The spectrum is the superposition of a 1 : 1 doublet and a 1 : 2 : 1 triplet but only positive frequencies have been calculated. The coupling constant is 5.95 Hz.

Consider first a system of two weakly coupled protons. If the time evolution of the peak of the spin echo* is monitored it is found to carry an amplitude modulation due to the spin–spin coupling J_{IS}. In the general case of many coupled protons, there is a superposition of many such modulations, and the corresponding waveform is usually called an *interferogram* in order to distinguish it from a free induction decay. The Fourier transform of this interferogram is a new kind of spectrum which contains no chemical shift information but only J-splittings. We may visualize it as a hypothetical single resonance line at zero frequency split successively by all the couplings to the nucleus in question. Each chemically distinct site would produce one such multiplet, and they all appear superimposed in the resulting *J-spectrum*. A J-spectrum exhibits responses at frequencies determined by sums and differences of coupling constants, and it is symmetrical about zero frequency. It may be thought of as being derived from the conventional high-resolution spectrum by eliminating all chemical shifts (but nevertheless maintaining first-order coupling); so that the centres of all spin multiplets converge at the same point. This point is a true zero frequency, which is not related to the frequency of any chemical shift reference. If instrumental instabilities and spin diffusion can be neglected, the lines in a J-spectrum approach the natural linewidths $(\pi T_2)^{-1}$ rather than the field inhomogeneity width $(\pi T_2^*)^{-1}$. Spin–spin coupling constants may therefore be measured with high accuracy. Figure 1 shows such a proton J-spectrum obtained from 1,1,2-trichloroethane (1). Note that the highest frequency line (from the CH group) is only 0.08 Hz wide; this is much narrower than the normal field inhomogeneity linewidth.

With the advent of two-dimensional Fourier transformation (2), this concept was extended to include experiments where the refocusing to form an echo occurs at the end of a variable 'evolution period' t_1, the second half of the spin

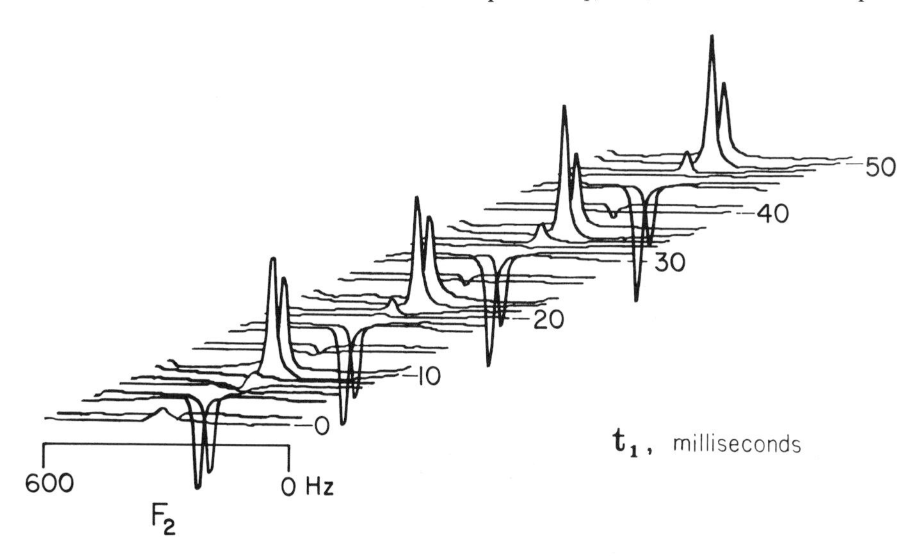

Fig. 2. J-modulation observed when the carbon-13 spin echoes from the methyl group are Fourier transformed. The spectra are followed as a function of the evolution period t_1.

echo being acquired as a function of a running time variable t_2. We consider first of all the heteronuclear case (protons coupled to a carbon-13 nucleus). The acquisition of the half-echo is repeated several times, incrementing the evolution time t_1 until a suitable time-domain data matrix $S(t_1, t_2)$ is built up. Fourier transformation of this data matrix as a function of t_2 generates a series of specta $S(F_2)$ as a function of the evolution time t_1 as illustrated by the carbon-13 spectra of Fig. 2. They are modulated by the coupling J_{CH} to the three protons of a methyl group. Corresponding points are then examined from each spectrum. If there is a signal present it will be seen as an interferogram carrying the modulation due to J_{CH}. We may therefore perform a second Fourier transformation (as a function of t_1) and repeat this for every point on the F_2 spectrum. The result is a spectrum in two orthogonal frequency dimensions $S(F_1, F_2)$. A typical result is illustrated in Fig. 3 for the carbon-13 J-spectrum of 2-(1-methylcyclohexyl)4,6-dimethylphenol. Arrayed in the F_1 dimension are the CH spin coupling constants, while the projection on to the F_2 axis gives the carbon-13 chemical shifts. This type of J-spectrum can be used to examine the fine structure due to long-range CH coupling, or simply to determine the multiplicities of the carbon-13 resonances. (See Multiplicity determination*.)

J-spectra from homonuclear systems are similar, except for the fact that there is no instrumental method of homonuclear broadband decoupling, so the F_2

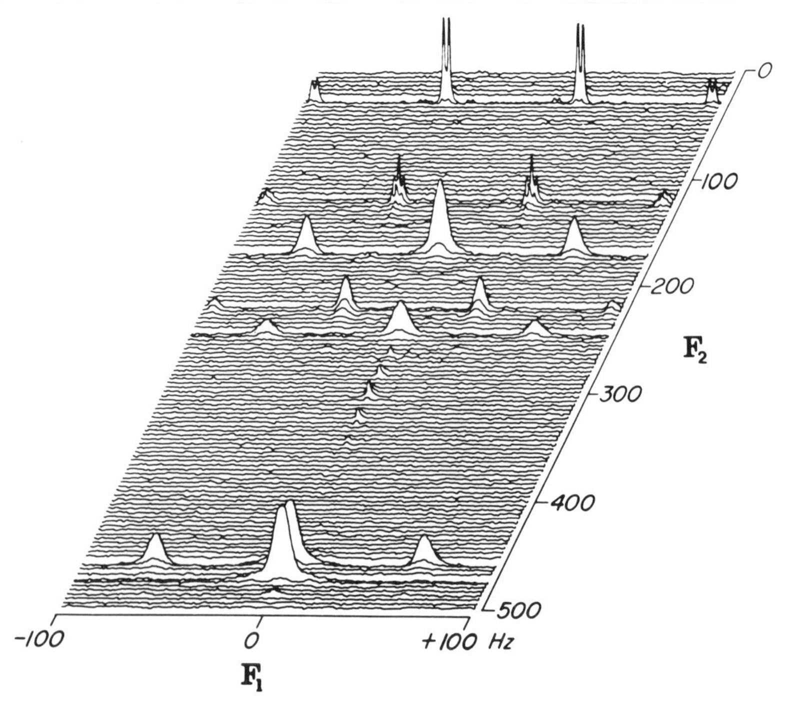

Fig. 3. Two-dimensional J-spectroscopy of carbon-13 in 2 (1-methylcyclohexyl) 4,6-dimethylphenol. The three methyl resonances are clearly distinguished by their different fine structure (long-range CH coupling).

dimension contains both shifts *and* couplings. As an example, we can take the two-dimensional proton J-spectrum of a hydroxytricyclodecanone derivative (illustrated in Fig. 4). Note that each spin multiplet lies along a diagonal; when both frequency axes are on the same scale, these are 45° diagonals. Each multiplet is separated from its neighbours by the appropriate chemical shift in the F_2 dimension. This provides a powerful method (3) for distinguishing the effects of chemical shifts and spin–spin couplings in systems which show complicated overlapping multiplets in the conventional spectrum. By projecting the two-dimensional spectrum (absolute-value mode) at 45° to the F_1 and F_2

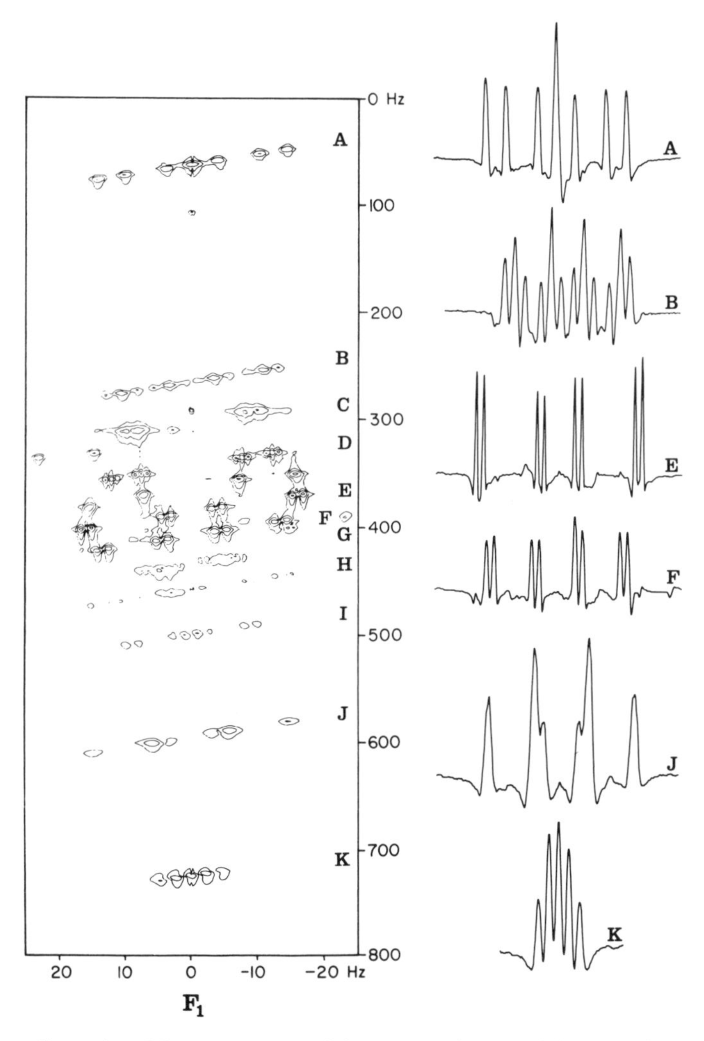

Fig. 4. Two-dimensional J-spectroscopy of the protons in one of the stereoisomers of 9-hydroxytricyclodecan-2,5-dione, shown as a contour plot. Each spin multiplet lies on a separate diagonal and may be retrieved by taking the appropriate section through the data matrix.

axes, catching the spin multiplets in enfilade, a spectrum is obtained which has resonances only at the chemical shift frequencies, *as if* the protons had been subjected to broadband decoupling. By taking sections through the two-dimensional spectrum at the appropriate proton chemical shift frequencies, we may obtain the spin multiplet structure from each proton site individually .

REFERENCES

1. R. Freeman and H. D. W. Hill, *Dynamic Nuclear Magnetic Resonance Spectroscopy*, Eds.: L. M. Jackman and F. A. Cotton. Academic Press: New York, 1975, Chapter 5.
2. J. Jeener, *Ampère International Summer School.* Basko Polje, Yugoslavia (1971).
3. W. P. Aue, J. Karhan and R. R. Ernst, *J. Chem. Phys.* **64**, 4226 (1976).

Cross-references

Broadband decoupling
Multiplicity determination
Spin-echoes
Two-dimensional spectroscopy

Lineshapes in Two-Dimensional Spectra

Many two-dimensional experiments, for example correlation spectroscopy (COSY) and nuclear Overhauser enhancement spectroscopy (NOESY) (see Nuclear Overhauser effect*), involve polarization transfer* from one group of spins to another. It is characteristic of these experiments that the observed signal is *amplitude* modulated as a function of the evolution time t_1; this can be represented as two equal counter-rotating phase modulations. Consequently, the sense of the precession during t_1 remains ambiguous; if there is a response in the two-dimensional spectrum at $(+\Omega_1, \Omega_2)$ there is always a mirror image response at $(-\Omega_1, \Omega_2)$. The section of the spectrum lying above the transmitter frequency is folded over into the remainder of the spectrum (and vice versa) causing considerable confusion. In early two-dimensional experiments (see Two-dimensional spectroscopy*), this could only be avoided by shifting the transmitter frequency to one extreme end of the spectrum so that all resonances had offsets of the same sign. Only the positive frequencies are then displayed, but a great deal of data storage is wasted.

One way out of this dilemma involves relating the sense of precession during t_1 with the sense of precession during t_2. For opposite senses of rotation, the observed response is called a *coherence transfer echo* (1, 2) because static field inhomogeneity effects are refocused. When the senses of precession are the same, the response can be called an antiecho – field inhomogeneity effects are not refocused. Selection of the coherence transfer echo is achieved by a phase cycling* scheme; on successive transients the receiver reference phase is shifted in 90° steps while the phases of all radiofrequency pulses prior to the evolution period are shifted by 90° in the opposite sense (for multiple-quantum coherence of order n, the phase shifts of the pulses should be 90°/n). This is equivalent to imposing a phase modulation during the evolution period rather than an amplitude modulation. There are some other forms of two-dimensional spectroscopy where phase modulation is inherent; these include J-spectroscopy*.

Spectra obtained by Fourier transformation of a phase-modulated signal exhibit a very characteristic two-dimensional line-shape known as the *phase-twist*, illustrated in Fig. 1. Sections parallel to the F_1 or F_2 axes through the exact centre of this response show a pure absorption lineshape, but all other sections have dispersion-mode contributions which increase as the offset from exact resonance increases. If several parallel sections are examined sequentially, the phase of the NMR signal appears to rotate from pure dispersion through pure absorption to dispersion in the other sense, hence the term phase-twist.

The phase-twist lineshape has several undesirable properties. The tails of the line (mainly dispersion-mode) extend out a long way from the centre and interfere with adjacent lines in a complicated manner. A more serious problem occurs with projections of a phase-twist line. In two-dimensional proton J-spectroscopy, where the F_1 axis displays J-couplings and the F_2 axis displays chemical shifts and J-couplings, it would be possible to obtain a spectrum devoid of all H–H couplings if a 45° projection could be made, since this would catch all the spin multiplets in *enfilade*. Unfortunately, a projection in this direction vanishes because the negative lobes of this lineshape exactly cancel the positive central peak.

In many early two-dimensional experiments, the awkward nature of the phase-twist lineshape was sidestepped by plotting the absolute-value mode, the square root of the sum of the squares of the absorptive and dispersive

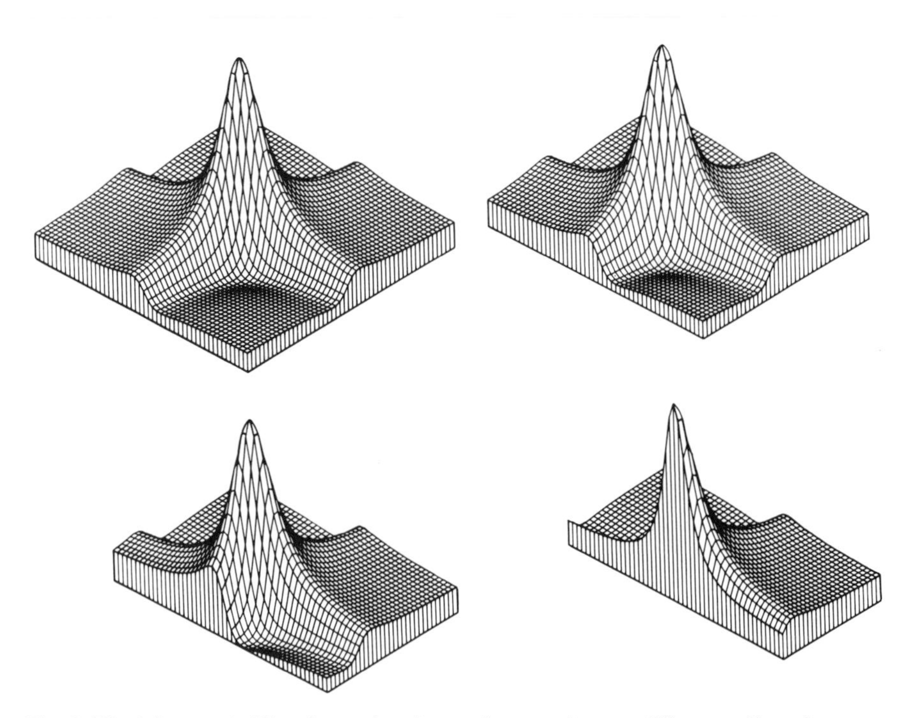

Fig. 1. The 'phase twist' lineshape showing various sections at different offsets from exact resonance. Far from resonance the section is predominantly dispersion-mode, but closer to resonance there is an increasing absorption-mode character and at exact resonance it is pure absorption.

components. This allows projections of two-dimensional spectra at any angle and incidentally removes the necessity of spectrometer phase adjustment in two dimensions. The problem of overlapping lines is still serious of course. It can be alleviated by multiplying the time-domain signals (in both dimensions) by suitable weighting functions, such as the *sine-bell* and *pseudo-echo*. These reshape the time-domain envelopes so that they peak at their midpoints and decay approximately symmetrically on either side. Now, the Fourier transform of a symmetric function is itself symmetric and therefore contains no dispersive part. The absolute-value mode must still be used but now it contains only the absorption-mode component. Unfortunately, this procedure carries a severe sensitivity penalty since these are quite drastic resolution enhancement* functions.

For these reasons it is important to be able to achieve sign discrimination in the F_1 dimension while retaining the pure two-dimensional absorption-mode lineshape. Procedures for accomplishing this were developed by States *et al.* (3) and Marion and Wüthrich (4). The treatment outlined here is based on the analysis of Keeler and Neuhaus (5). Since these techniques offer an important advantage in resolving power and a small increase in sensitivity, they seem likely to be widely adopted in two-dimensional spectroscopy.

Consider a signal which is amplitude-modulated as a function of t_1, represented by a cosine wave

$$C(t_1, t_2) = \cos(\Omega_1 t_1)[\cos(\Omega_2 t_2) + i \sin(\Omega_2 t_2)]E_1E_2 \qquad [1]$$

where $E_1 = \exp(-t_1/T_2)$ and $E_2 = \exp(-t_2/T_2)$ are the decay constants in the two time dimensions. The form of this equation indicates that the sense of rotation during t_2 is determined, but not the sense during t_1. The amplitude modulation can be viewed as two counter-rotating phase modulations $\cos(\Omega_1 t_1) + i \sin(\Omega_1 t_1)$ and $\cos(\Omega_1 t_1) - i \sin(\Omega_1 t_1)$.

The usual complex Fourier transformation is performed with respect to t_2 (quadrature detection*). The absorption mode spectrum is

$$C_a(t_1, F_2) = \cos(\Omega_1 t_1)A_2E_1. \qquad [2]$$

Here A_2 represents a Lorentzian absorption-mode signal

$$A_2 = T_2/(1 + \Delta\omega^2 T_2^2), \qquad [3]$$

where $\Delta\omega$ is the offset from the centre of the line. (The Lorentzian shape is a result of transforming an exponentially-decaying time-domain function). The dispersion mode spectrum is

$$C_d(t_1, F_2) = \cos(\Omega_1 t_1)D_2E_1, \qquad [4]$$

where D_2 represents a Lorentzian dispersion-mode signal

$$D_2 = \Delta\omega T_2^2/(1 + \Delta\omega^2 T_2^2). \qquad [5]$$

These are the normal results obtained by quadrature detection; the problems arise when the second transformation is performed. We can now define a cosine transform (for a continuous function) by

$$C^{\cos}(F_1, F_2) = \int C(t_1, F_2) \cos(2\pi F_1 t_1)\, dt_1. \qquad [6]$$

We assume for simplicity that the spectrometer phase has been properly adjusted in both frequency dimensions. The cosine transform of eqn [2] represents a response that has no sign discrimination of its frequency in the F_1 dimension. We choose to write this as

$$C_a^{cos}(F_1, F_2) = A_1^+A_2 + A_1^-A_2, \qquad [7]$$

which corresponds to a peak A_1^+ in the $+F_1$ dimension and a symmetrically related peak A_1^- in the $-F_1$ dimension. This is equivalent to regarding the amplitude modulation of eqn [2] as two counter-rotating phase modulations. Both peaks have the two-dimensional absorption-mode lineshape illustrated in Fig. 2(d).

We may also calculate the sine transform defined by

$$C_a^{sin}(F_1, F_2) = \int C(t_1, F_2) \sin(2\pi F_1 t_1)\, dt_1. \qquad [8]$$

This gives

$$C_a^{sin}(F_1, F_2) = D_1^+A_2 + D_1^-A_2. \qquad [9]$$

This represents two lines that have dispersion-mode profiles in the F_1 dimension but absorption-mode profiles in the F_2 dimension (Fig. 2(b)). Alternatively, we could transform the dispersion-mode spectrum, eqn [4], to give

$$C_d^{cos}(F_1, F_2) = A_1^+D_2 + A_1^-D_2. \qquad [10]$$

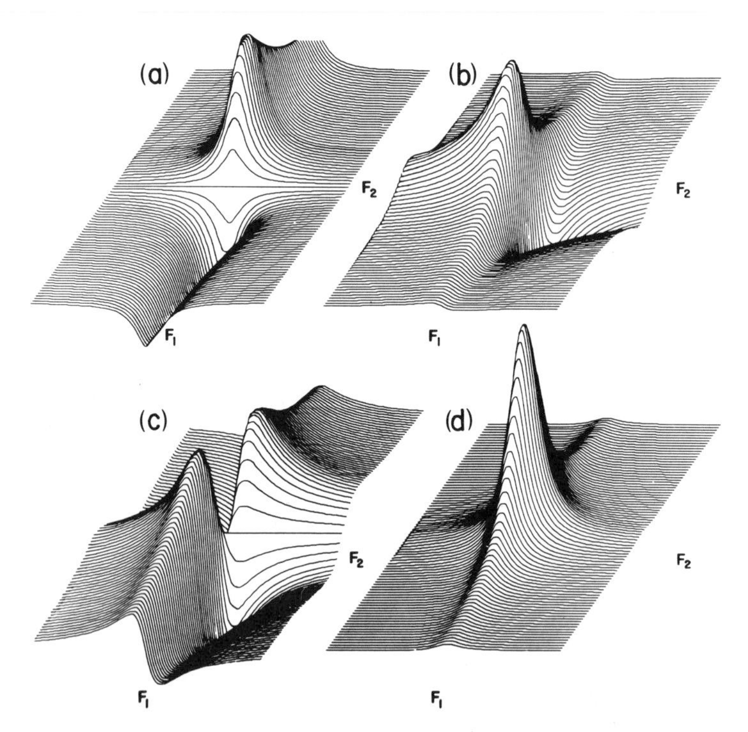

Fig. 2. Four types of two-dimensional lineshape: (a) absorption-mode in F_1 and dispersion-mode in F_2; (b) dispersion-mode in F_1 and absorption-mode in F_2; (c) dispersion-mode in both dimensions; and (d) absorption-mode in both dimensions.

These two lines have absorption-mode profiles in F_1 and dispersion profiles in F_2 (Fig. 2(a)). While the sine transform of eqn [4] gives

$$C_d^{sin}(F_1, F_2) = D_1^+D_2 + D_1^-D_2. \quad [11]$$

These two lines have dispersion-mode profiles in both dimensions (Fig. 2(c)).

Clearly, additional information is required to determine the sense of precession during t_1. The basic problem is to determine, without ambiguity, the phase angle β of a vector evolving in t_1. If we knew that $x = \sin\beta$, then we could say that $\beta = \sin^{-1} x$ or $\beta = \sin^{-1} x - \pi$. If we knew that $y = \cos\beta$, then we could say that $\beta = \pm\cos^{-1} y$. Only if we know $\sin\beta$ and $\cos\beta$ simultaneously can we be sure about the angle β. This additional information is provided by a second experiment in which the modulation during t_1 follows a sine rather than a cosine wave.

$$S(t_1, t_2) = \sin(\Omega_1 t_1)[\cos(\Omega_2 t_2) + i\sin(\Omega_2 t_2)]E_1E_2 \quad [12]$$

For example, in a homonuclear chemical shift correlation experiment (COSY) this would involve a 90° phase shift in the coherence that evolves during t_1, which is achieved by shifting the phase of the initial pulse by 90°. For multiple-quantum coherence of order n, all the pulses which precede the evolution period would need to be shifted in phase by 90°/n. This function $S(t_1, t_2)$ is always kept separate from the function $C(t_1, t_2)$.

As above we perform quadrature detection in the t_2 dimension and compute the absorption spectrum

$$S_a(t_1, F_2) = \sin(\Omega_1 t_1)A_2E_1, \quad [13]$$

and the dispersion spectrum

$$S_d(t_1, F_2) = \sin(\Omega_1 t_1)D_2E_1. \quad [14]$$

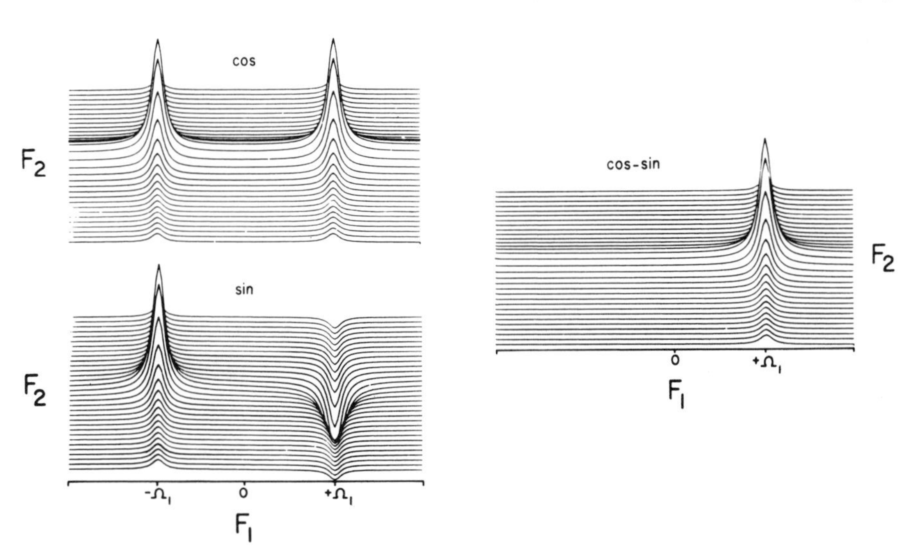

Fig. 3. Discrimination of the sign of the frequency for pure absorption mode spectra. By subtraction of the spectra obtained by transformation of the cosine-modulated signal and the sine-modulated signal, a single peak is obtained at a positive frequency.

If we were to proceed as with the first experiment, the result would again be a pair of lines at $+\Omega_1, \Omega_2$ and $-\Omega_1, \Omega_2$. The method of States *et al.* (3) avoids this by zeroing the dispersion-mode spectrum eqn [14], before the next stage of complex Fourier transformation. The absorption part corresponds to a line with no sign discrimination of its F_1 frequency, which we choose to write

$$S_a^{sin}(F_1, F_2) = A_1^+A_2 - A_1^-A_2. \qquad [15]$$

Only the sine transform is retained, S_a^{cos} being discarded. The negative sign in eqn [15] is important; it reflects the fact that sin $(\Omega_1 t_1)$ reverses its sign for negative frequencies (A_1^-). Combination with the results from the first experiment, eqn [7], gives

$$C_a^{cos}(F_1, F_2) + S_a^{sin}(F_1, F_2) = 2A_1^+A_2. \qquad [16]$$

This is a two-dimensional absorption-mode line at a positive frequency in the F_1 dimension. The component at a negative frequency has been cancelled as indicated schematically in Fig. 3.

The alternative method proposed by Marion and Wüthrich was developed for spectrometers which do not have a complex Fourier transform program. It is based on Redfield's method (6) for obtaining the equivalent of quadrature detection with only a single phase-sensitive detector. This involves doubling the sampling rate and shifting the receiver phase in 90° steps at each new sampling operation. Translated into the t_1 dimension this method is usually known as TPPI (time-proportional phase incrementation) (7). It has the effect of increasing the apparent frequencies in the F_1 dimension by a frequency Ω_c equal to half the spectral width in this dimension, just as if the reference frame were rotating at $\Omega_o - \Omega_c$ rather than at the Larmor frequency Ω_o. This keeps all precession frequencies positive and avoids aliasing problems. Real Fourier transforms are used in this method rather than complex transforms, but otherwise the principles are the same as in the method of States *et al.* Both techniques use the same amount of data storage, but the method of Marion and Wüthrich is particularly easy to program. The concept of *coherence transfer pathways* (8) provides a clear description of both methods for obtaining pure absorption-mode lineshapes.

There are several practical advantages of the pure absorption-mode display and it seems likely to be widely adopted. It avoids the difficulties of the *phase twist* or absolute-value lineshapes without the complications associated with

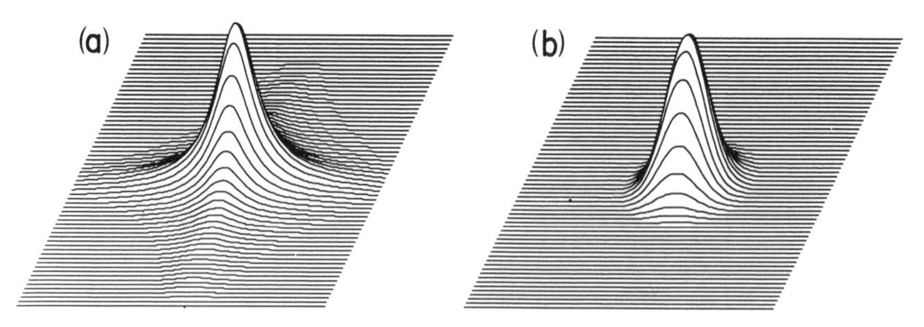

Fig. 4. (a) A two-dimensional Lorentzian lineshape. (b) A two-dimensional Gaussian lineshape.

the *pseudo-echo* or *sine-bell* weighting functions. It achieves high resolution of the multiplet components within a given cross-peak and does not obscure the 'up–down' pattern of their intensities. These considerations are important when pattern recognition techniques (9) are employed to characterize cross-peaks in two-dimensional spectra.

LORENTZIANS AND GAUSSIANS

When exponential weighting functions are imposed in both time dimensions of a two-dimensional experiment, pure absorption-mode responses have lineshapes that are two-dimensional Lorentzians (Fig. 4(a)). This can be an undesirable shape because it has ridges that run parallel to the F_1 and F_2 axes, and they can cause misleading effects in the case of overlap with adjacent lines. Looked at in another way, the line has intensity contours that are not circular (or elliptical) but the shape of a starfish with only four legs. A more suitable time-domain weighting function is the Gaussian, for this gives a Gaussian lineshape in the frequency domain (Fig. 4(b)), with circular (or elliptical) intensity contours.

REFERENCES

1. K. Nagayama, K. Wüthrich and R. R. Ernst, *Biochem. Biophys. Res. Commun.* **90**, 305 (1979).
2. A. Bax, R. Freeman and G. A. Morris, *J. Magn. Reson.* **42**, 164 (1981).
3. D. J. States, R. A. Haberkorn and D. J. Ruben, *J. Magn. Reson.* **48**, 286 (1982).
4. D. Marion and K. Wüthrich, *Biochem. Biophys. Res. Commun.* **113**, 967 (1983).
5. J. Keeler and D. Neuhaus, *J. Magn. Reson.* **63**, 454 (1985).
6. A. G. Redfield and S. D. Kunz, *J. Magn. Reson.* **19**, 250 (1975).
7. G. Drobny, A. Pines, S. Sinton, D. Weitekamp and D. Wemmer, *Faraday Div. Chem. Soc. Symp.* **13**, 49 (1979).
8. G. Bodenhausen, H. Kogler and R. R. Ernst, *J. Magn. Reson.* **58**, 370 (1984).
9. B. U. Meier, G. Bodenhausen and R. R. Ernst, *J. Magn. Reson.* **60**, 161 (1984).

Cross-references

J-spectroscopy
Nuclear Overhauser effect
Phase cycling
Polarization transfer
Quadrature detection
Resolution enhancement
Two-dimensional spectroscopy

Maximum Entropy Method

Imagine the following scenario. An eyewitness to a serious crime is describing the suspect so that a police artist can make an identification sketch of the wanted man, a notoriously difficult operation. Influenced by his revulsion at the details of the crime, the witness lets his own impressions influence his recollections, for example by imagining the villain to be scowling and unshaven. This situation would be improved if half a dozen eyewitnesses could be found.

'How did they catch you then?'
'Dunno, something to do with maximum entropy'

Assuming that they were all reliable observers and that the artist was competent, this should produce a set of sketches, all of which are *compatible* with the raw data, although they will obviously differ in detail. The detective in charge of the case is presented with six drawings – which one should he choose? To be of the most use for identification purposes, the chosen sketch should obviously contain the least amount of *extraneous* information, but since there is no criterion for judging what is extraneous the only safe solution is to choose the drawing with the lowest information content. In the (inelegant) language of information theory this is said to be the 'maximally non-committal' solution. On the grounds that entropy is the negative of information content, it is also called the *maximum entropy solution*. This data processing technique (1–7) is now being applied to problems of magnetic resonance where it has been successful in suppressing undesirable artifacts in the spectra.

Present-day NMR spectrometers yield their information in the form of a free induction decay* and the conversion into the frequency domain is almost always carried out by Fourier transformation*. There are some possible alternatives. The free induction decay could be fitted to a superposition of decaying oscillations with suitable frequencies, phases and intensities, using a least-squares procedure. The result could then be displayed in the form of a frequency-domain spectrum with the appropriate linewidths and lineshapes, without any need for Fourier transformation. Computational time would be saved by terminating the iterations before the program began to fit the noise components in the raw data. This is where the danger lies, since the final spectrum would then appear to be noise-free, but since the search program would also miss signal components comparable in intensity to the noise, no sensitivity advantage would ensue.

The real advantages of this method are more subtle. Often an experimental free induction signal has to be truncated before it has decayed to a negligible level, in order to minimize the time spent on acquisition. After zero filling* and Fourier transformation, this step function introduces a degree of sinc function character to the spectral lineshapes, which is undesirable. A direct computer fitting of the free induction decay can take into account the fact that the experimental data set had been truncated in this manner, thus avoiding sinc function artifacts. Similarly, the distortions which often appear on the first one or two sample points of the free induction decay due to receiver dead-time need not be carried over into the final spectrum.

The maximum entropy method represents a rather more sophisticated approach to this type of problem, and one that need not incorporate a model for the spectrum. In the application of this method to NMR spectroscopy, we might imagine a number of possible 'trial spectra' which, by inverse Fourier transformation, give free induction decays compatible with the experimental one. This inability to settle on a single 'correct' spectrum arises out of the uncertainty associated with the presence of noise in the experimental data. None of the free-induction decays obtained by back-transformation of the trial spectra corresponds exactly with the experimental time-domain signal because each sample point carries a component of noise.

The choice between several 'acceptable' spectra is made on the basis of the maximum entropy criterion. This chosen spectrum is no more likely to be 'correct' than any of the other acceptable spectra, since there is just no hard evidence available to make a meaningful distinction. It nevertheless represents the safest choice simply because it has the lowest information content.

In this context, entropy represents the negative of the information content

$$S = -\sum_{k}^{N} p_k \log p_k, \qquad [1]$$

where the p_k values represent the intensity ordinates which define the trial spectrum and N is the total number of ordinates. These intensities must be suitably normalized in order to prevent the algorithm from simply raising the level of the baseline in order to increase S. The aim is to choose the trial spectrum that has the largest value of S subject to the constraint that it is consistent with the experimental data. It can be shown that there is normally a unique maximum entropy solution to a given problem. The spectrum chosen according to the maximum entropy principle has the important property that it contains no more 'structure' than is strictly justified by the experimental evidence. No unnecessary artifacts are introduced, in contrast to the much more common Fourier transform method.

Since it would be incorrect to attempt to fit the experimental free induction decay exactly (because of the presence of noise) we only use those trial spectra that have a suitable chi-squared statistic

$$\chi^2 = \sum_{a}^{M} (I_a - I'_a)^2/\sigma_a^2, \qquad [2]$$

where I represents an actual signal intensity in the time-domain free-induction decay and I′ the corresponding time-domain intensity derived from the trial spectrum, while σ is the r.m.s. noise. The parameter χ^2 should be of the order of M, the number of measurements made on the free induction signal (M is less than or equal to N). This means that the error in fitting a given point on the free induction decay is of the order of the r.m.s. noise.

The entropy function is non-linear and its maximization entails an iterative procedure, but the starting point for this iteration is not critical. It is possible to assume an initial trial spectrum that is merely a flat baseline, and this has the advantage of avoiding prejudice on the part of the spectroscopist. In other circumstances it can be advantageous to incorporate prior knowledge about the spectrum, measuring the entropy with respect to that of an initial model spectrum m rather than with respect to a completely featureless flat baseline.

$$S = -\sum_{k}^{N} p_k (\log p_k - \log m_k)/m_k. \qquad [3]$$

Of course, if some aspect of this model spectrum is erroneous, the maximum entropy algorithm can do nothing to correct it, so the procedure must be used with caution.

The maximum entropy method is used in a wide variety of applications including radioastronomy, tomographic reconstruction, deconvolution of blurred photographic images and X-ray crystallography. Sometimes the experimental data may be incomplete; some pixels may be missing from an image or a free induction decay may be prematurely truncated. The maximum entropy method handles this kind of data corruption very effectively, rejecting trial images or spectra that contain artifacts attributable to defective input data. Thus sinc function artifacts are suppressed very efficiently. The maximum entropy method can also be used for resolution enhancement* as an alternative to manipulating the decay rate of the free induction signal. As an example, Fig. 1 shows one of the proton resonances (H_5) of 3-bromonitrobenzene with and without resolution enhancement by the maximum entropy method. The central triplet (doublet of doublets) now appears well-resolved, whereas in the unprocessed spectrum it had only a hint of shoulders on the line profile.

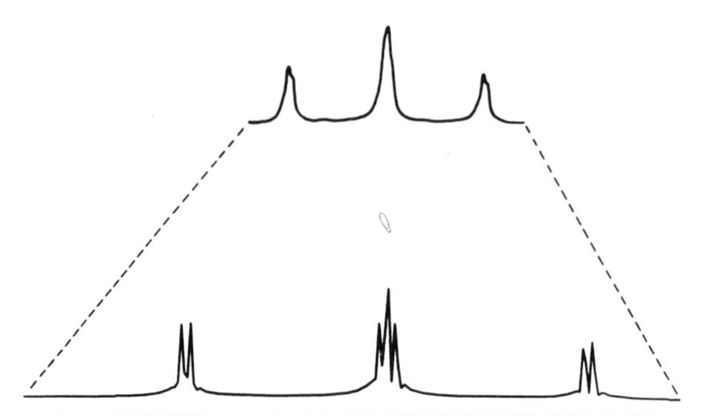

Fig. 1. Resolution enhancement by the maximum entropy method. The multiplet from H_5 in 3-bromonitrobenzene is shown before and after processing by the maximum entropy method. The width of the spectrum is 22.4 Hz.

SENSITIVITY ENHANCEMENT*?

NMR spectroscopists are accustomed to working with linear systems and to processing the data with linear operations such as Fourier transformation. They may therefore be forgiven for using the terms *sensitivity* and *signal-to-noise ratio* essentially interchangeably. If a test sample with a concentration C mol dm^{-3} gives a signal-to-noise ratio of, say, 10, then in a linear system, we can estimate the minimum concentration that would just give a detectable signal, something near C/4 or C/5 mol dm^{-3}. This is a measure of the sensitivity of the spectrometer.

With non-linear operations (maximum entropy is one of these) we must be careful to distinguish the two terms, since an improvement in signal-to-noise ratio is no longer equivalent to an improvement in sensitivity. For signals significantly more intense than the noise, the signal-to-noise ratio can be increased by very large factors by non-linear amplification, by plotting on a recorder with an appreciable dead-band, or by many possible data processing tricks. None of these serves any really useful purpose, because for the case that matters – signals comparable with noise – they fail to enhance the signal-to-noise ratio and may even degrade it. They *do not* improve sensitivity.

Noise should not be thought of as an 'artifact' that can be separated from true signals and discarded. High-frequency components of the noise can be filtered out, and it is common practice to do this with the optimum matched filter when sensitivity is crucial. Since noise is uncorrelated between one experiment and the next, it can be reduced by time averaging*. Beyond that, there are no obvious properties that distinguish signal from noise and we should not expect any miraculous new data processing technique to be able to enhance sensitivity. Indeed, the very crux of the maximum entropy principle is not to introduce any feature into the final spectrum for which there is no clear evidence in the experimental data. Weak signals hidden in the noise remain unobservable. There is then no strong incentive to use these non-linear operations to 'clean up' the spectrum by reducing baseline noise; if we do so, it is important to remember that some very weak signals have been similarly suppressed.

Noise in a spectrum actually serves a useful purpose provided that we are dealing with a linear system. If the baseline noise has an amplitude equal to (say) 10% of a given NMR signal, then the operator recognizes that the peak height of this signal is unreliable to the same degree. On the other hand, if the noise is suppressed by a non-linear data processing method (such as maximum entropy) the intensity of this line is clearly no more reliable than before, although the operator might be misled into thinking that it is.

REFERENCES

1. C. E. Shannon, *Bell Syst. Tech. J.* **27**, 623 (1948).
2. S. F. Gull and G. J. Daniell, *Nature (Lond.)* **272**, 686 (1978).
3. S. Sibisi, *Nature (Lond.)* **301**, 134 (1983).
4. J. Skilling *Nature (Lond.)* **309**, 748 (1984).
5. S. Sibisi, J. Skilling, R. G. Brereton, E. D. Laue, and J. Staunton, *Nature (Lond.)* **311**, 446 (1984).
6. E. D. Laue, J. Skilling, J. Staunton, S. Sibisi, and R. G. Brereton, *J. Magn. Reson.* **62**, 437 (1985).
7. P. J. Hore, *J. Magn. Reson.* **62**, 561 (1985).

Cross-references

Fourier transformation
Free induction decay
Resolution enhancement
Sensitivity enhancement
Time averaging
Zero filling

Modulation and Lock-in Detection

It is not too difficult to set up a lecture demonstration with an infra-red detector, a high-gain amplifier and a display meter, so that the meter gives a strong deflection due to the heat radiated by a single member of the audience. Of course in a lecture theatre containing many people the meter needle fluctuates wildly as the different infra-red sources move about, fidget, fall asleep, and do all the usual things expected of a scientific audience. Now it is quite feasible to make this detection scheme sensitive to only one chosen member of the audience by suitably coding the radiation from this particular source and then decoding the signal in the detection system. A simple modulation suffices for the encoding process, this is easily done with a mechanical chopper operating at some low audiofrequency. Decoding is performed by a device variously known as a *synchronous detector*, *lock-in detector* or *phase-sensitive detector*. This is simply an amplifier that has appreciable gain only within a small bandwidth around a particular 'reference' frequency derived from the same audiofrequency source as was used to chop the infra-red radiation. With this modulation and lock-in detection scheme all radiation falling on the infra-red sensor is filtered out unless it is modulated at the chosen reference frequency.

This same principle is widely used in magnetic resonance to record the desired signals while suppressing unwanted 'interference'. It is particularly important in continuous-wave spectroscopy*, where small changes in transmitter level, probe balance or amplifier gain give rise to baseline drifts which make integration inaccurate. (See Baseline correction*.) The more modern pulsed spectrometers avoid these problems by gating the receiver off when the transmitter is on.

Modulation is the systematic perturbation of a pure sinewave, usually as amplitude modulation or frequency modulation. The effect of modifying the original signal is to produce a waveform which contains a number of frequencies: the original 'centreband' frequency and a series of new 'sideband' frequencies on each side of the centreband.

Consider a sinewave of frequency f_0 multiplied by a second sinewave of much lower frequency f_m (an extreme case of amplitude modulation). The resultant signal is

$$S(t) = \sin(2\pi f_0 t) \sin(2\pi f_m t). \quad [1]$$

Simple trigonometry decomposes this into the superposition of two cosine waves

$$S(t) = \tfrac{1}{2}\{\cos[2\pi(f_0 - f_m)t] - \cos[2\pi(f_0 + f_m)t]\}. \quad [2]$$

Thus the resultant waveform contains just two frequencies, the upper and lower sidebands $f_0 + f_m$ and $f_0 - f_m$, the centreband f_0 having been suppressed. The multiplication of two signals to obtain sum and difference frequencies is readily achieved using a 'double balanced mixer', and is used throughout NMR spectrometers for the generation of new frequencies.

Frequency-modulation gives a more complicated Fourier spectrum, with an array of modulation sidebands, but in the simple case where the depth of modulation is small in comparison with its frequency, only the first sidebands are observed and they are quite weak. A modulation of the applied static field B_0 gives rise to a frequency modulation of the nuclear precession. A very common example is the modulation which results from spinning the sample in an inhomogeneous static field. The resulting frequency-modulation gives rise to the well-known spinning sidebands.

There is a real-life analogy of frequency modulation in the case of an ice skater performing a rapid spin, let us say at a frequency f_0. For interesting physiological reasons associated with the balance mechanism in the ear, skaters are taught to 'spot' during this exercise to avoid becoming dizzy. This means fixing the eyes on some suitable prominent object as long as possible during the revolution by turning the head counter to the rotation of the body. There then must be a rapid compensating rotation of the head in the same sense as the body so as to be able to 'spot' the same object during the next revolution. In our 'laboratory frame' the motion of the skater's head is frequency-modulated at a frequency f_m which is automatically adjusted to equal f_0. The Fourier spectrum thus contains frequencies at 0, f_m and $2f_m$. To a member of the audience it is the zero-frequency component that makes the most impact since it appears that the skater is looking directly at him (yet spinning rapidly). A process of integration is being carried out, selecting the d.c. component of the total signal.

Frequency-modulation (or amplitude modulation) causes the NMR signal at the detector to be carried at the modulation frequency f_m and it may then be separated from all other signal components by feeding it to a lock-in detector supplied with a reference frequency f_m. This lock-in detector can be thought of as a square-wave switching circuit, passing the input signal only on positive half-cycles (Fig. 1) and feeding it to a circuit with a long RC time constant. If the input signal is of the right phase and has the same frequency as the reference (or very close to it) a large charge builds up on the capacitor; if the two frequencies differ significantly, the capacitor is alternately charged and discharged and the net output is zero. This RC time constant determines the bandwidth of the lock-in detector. Note that the phase relationship between the input signal and the reference is important; if these are 90° out of phase, the output is zero even when the frequencies are exactly matched. This is why the device is also known as a phase-sensitive detector.

Modulation and lock-in detection improve the quality of magnetic resonance spectra by suppressing spurious 'interference' while retaining the true NMR signal. The process does not improve the signal-to-noise ratio except under certain special circumstances. If the noise is 'white', that is to say it has a flat Fourier spectrum as a function of frequency, then an ordinary d.c. detector and a lock-in detection scheme both pass the same amount of noise for the same final bandwidth. However, for non-white noise, the modulation scheme has a better signal-to-noise ratio, because the noise level tends to rise at very low frequencies. This is an important factor in electron spin resonance spectrometers where the dominant noise source is the crystal detector which has a noise spectrum that increases as 1/f at very low frequencies. Consequently, most ESR spectrometers employ a modulation scheme at a reasonably high audiofrequency, say 10 kHz.

More sophisticated uses of this principle can be devised by making the coding more specific. For example, carbon-13 spectra can be examined indirectly by applying a modulated decoupling field B_2 which is swept through the spectral region of interest. The proton resonance response is examined by way of a lock-in detector with the modulation frequency as reference. Only protons coupled to carbon-13 give modulated signals; the much more intense signals from protons attached to carbon-12 are rejected by the lock-in detector (1).

Investigation of any weak perturbation of the nuclear spin system is probably better carried out by a modulation scheme than a simple 'on–off' experiment. For example, strong electrostatic fields are believed to cause a slight net

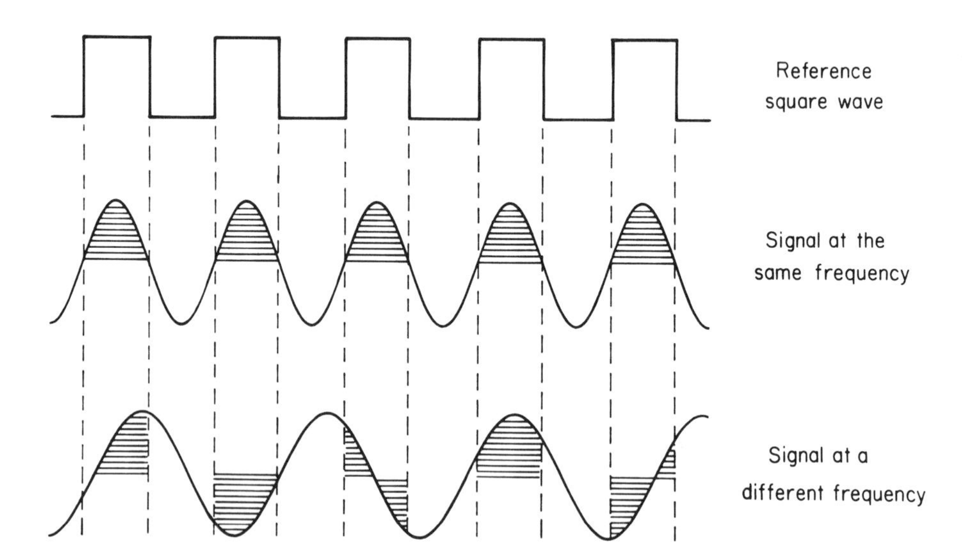

Fig. 1. The reference square wave accepts incoming signals only during the positive half-cycles. If the incoming signal is of the same frequency and the correct phase, the shaded areas add constructively to give a large integral. If the incoming signal is at a different frequency the shaded areas tend to cancel and the integral tends to zero.

orientation of the molecules in a polar liquid, introducing a small splitting due to dipole–dipole interactions that are no longer averaged to zero by isotropic motion (2). Rather than comparing the NMR spectrum with and without an applied d.c. electrostatic field, it might be better to apply a sinusoidally modulated electrostatic field and search for the weak modulation sidebands that would be generated, assuming that the alignment is established rapidly compared with the modulation frequency.

Nowadays, coding of individual chemically shifted sites with a characteristic frequency can be carried out by two-dimensional spectroscopy*, thus reducing the need for the modulation and lock-in detection method. Time-share modulation* schemes also employ lock-in detectors.

REFERENCES

1. R. Freeman and W. A. Anderson, *J. Chem. Phys.* **42**, 1199 (1965).
2. A. D. Buckingham and K. A. McLauchlan, *Proc. Chem. Soc.* (*Lond.*) 144 (1963).

Cross-references

Baseline correction
Continuous-wave spectroscopy
Time-share modulation
Two-dimensional spectroscopy

Multiple-Quantum Coherence

The selection rule for NMR transitions is $\Delta m = \pm 1$. A transition which obeys this selection rule can be excited by weak continuous-wave radiation or a single strong radiofrequency pulse. It generates a macroscopic transverse nuclear magnetization which induces a signal in the receiver coil. All other transitions are formally forbidden; and yet double-quantum transitions can be observed quite simply in a continuous-wave spectrometer if the radiofrequency level B_1 is deliberately increased beyond the usual setting. (A similar breakdown of the selection rule can be brought about by the imposition of a strong second radiofrequency field B_2 as in the spin tickling* experiment.) Excitation of a double-quantum transition by a strong B_1 field is most easily visualized as the simultaneous excitation of two progressively connected single-quantum transitions. The frequency of irradiation is set at the mean frequency of these two single-quantum transitions and they are simultaneously excited coherently in their tails far from resonance. The observation of this double-quantum line may be used to establish the connectivity of the two single-quantum transitions and also to determine relative signs of spin coupling constants. Since the same information can be obtained by double-resonance methods, interest in these experiments waned for over a decade, until quite recently new methods of excitation and detection were devised (1).

One way of considering the effect of a radiofrequency pulse is to say that it creates a *coherence* between two states of the spin system differing in quantum number by one unit. If two or more pulses are used, they may act in cascade and thereby excite multiple-quantum coherence. The interesting difference is that this type of coherence does not induce any voltage in the receiver coil and is thus not detectable in the NMR spectrometer: it must be detected indirectly by reconversion into single-quantum coherence. Consider the simple case of a homonuclear two-spin system with the energy-level diagram shown in Fig. 1(a). Suppose that transition 1–2 is excited by a selective 90° pulse and the resulting signal recorded for a short period. The selective 180° pulse applied to transition 2–4 transfers the coherence to the levels 1–4 (Fig. 1(b)). A double-quantum coherence has thus been created, but now no signal is observed in the spectrometer. It requires another selective 180° pulse at the frequency of transition 2–4 to restore the 1–2 coherence and the observable NMR signal (Fig. 1(c)). The relative phase of this signal is a function of the double-quantum frequency and the time for which it was allowed to evolve. This provides the clue to the method used to measure multiple-quantum frequencies by introducing a variable evolution period t_1 which is incremented in a series of experiments. This is the principle of two-dimensional spectroscopy*.

There are many methods for exciting multiple-quantum coherence. One of the most useful employs a $90° - \tau - 180° - \tau - 90°$ sequence with all pulses non-selective and the delays τ adjusted to the condition $\tau = 1/(4J)$. The importance of this method stems from the refocusing effect of the 180° pulse which makes the excitation independent of offset from the transmitter frequency (2). Only even orders of multiple-quantum coherence are excited by this sequence. Recently, Pines (3) has suggested pulse sequences where the excitation is specific for one chosen order of coherence; this is important for coherences involving many quanta in dipolar coupled systems.

There are some interesting special properties of multiple-quantum coherences, of which three are of particular importance. The problem must be considered in a reference frame rotating at the frequency f_1 of the radiofrequency transmitter. (See Rotating frame*.) If we transform the energy level

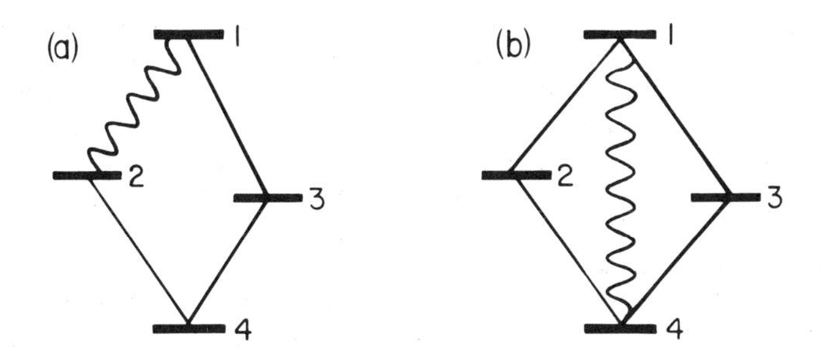

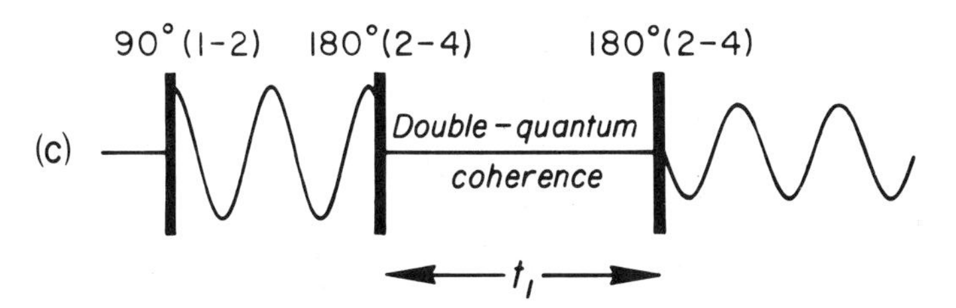

Fig. 1. (a) A selective 90° pulse excites a single-quantum coherence between levels 1 and 2. If this is followed by a selective 180° pulse at the 2–4 frequency, there is a conversion into non-observable double-quantum coherence (b) which may precess for an evolution period t_1. (c) It may be reconverted into observable single-quantum coherence by a second selective 180° pulse at the same frequency.

diagram of Fig. 1(b) into this frame, level 1 is lowered by f_1 Hz and level 4 is raised by f_1 Hz, thus reducing all the allowed single-quantum transitions by f_1 Hz. By contrast, the double-quantum frequency (1–4) is reduced by $2f_1$, while the zero-quantum frequency (2–3) is unchanged. Consequently, double-quantum frequencies are twice as sensitive to changes in radiofrequency offset as single-quantum transitions. This may be generalized to state that the frequency of a multiple-quantum coherence of order p exhibits a p-fold dependence on offset from the transmitter frequency. Since the excitation of multiple-quantum coherence can be considered as a cascade of p single-quantum steps, the phase of a multiple-quantum coherence of order p is changed by $p\phi$ for a change of ϕ radians in the excitation pulses. If the *detection* pulse is phase shifted by ϕ radians this has the effect of a $(p \pm 1)\phi$ phase shift of the multiple-quantum coherence. This characteristic sensitivity to radiofrequency phase has been made the basis of methods of separating the various orders. Figure 2 shows the separation of zero, single, double and triple quantum coherences by the method of 'time-proportional phase increments' (4, 5) where the phases of the multiple-quantum coherences are incremented by a small angle for each increment of the evolution period t_1.

The idea that a multiple-quantum jump can be considered as a cascade of single-quantum jumps (not necessarily in the same sense) provides a rationale for describing the special effects of spatial inhomogeneity of the applied field B_0. Homonuclear zero-quantum coherence is thus almost completely insensitive to the broadening effect of field inhomogeneity (compare regressively connected transitions in a spin tickling experiment), while double-quantum coherences are broadened to twice the extent of single-quantum coherences. In general, the influence of inhomogeneity is amplified by a factor p, where p is the order of the coherence. This property can be used to separate the orders by the application of suitably matched pairs of field gradient pulses (6).

The detection of multiple-quantum coherence is always indirect, by reconversion into observable transverse magnetization. If the frequency of the multiple-quantum coherence is to be measured, it is necessary to introduce a variable evolution period (t_1) which is incremented in a series of experiments, and a two-dimensional Fourier transformation is performed as a function of both time variables t_1 and t_2 (the running parameter for acquisition of the free induction signal). There is a general rule that the reconversion into single-quantum coherence gives no *net* signal: there is always an equal number of positive

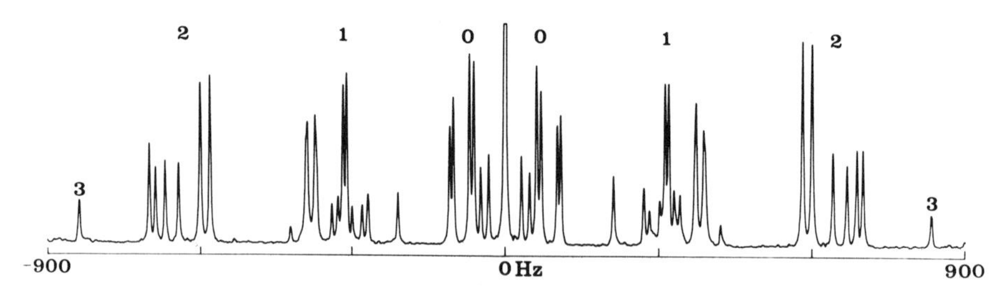

Fig. 2. Multiple-quantum spectrum of the three-spin system of protons in acrylic acid. The zero- and double-quantum spectra consist of six lines each. The single-quantum spectrum contains 15 lines, of which three are 'combination lines' involving simultaneous flipping of all three spins. There is only one triple-quantum line.

and negative signals which cancel in pairs at $t_2 = 0$. For a single coupled spin system, each trace in the F_1 dimension of this spectrum contains the same information, and it is usual to project the entire spectrum (in the absolute-value mode) on to the F_1 axis. Figure 2 was obtained by this method.

APPLICATIONS OF MULTIPLE-QUANTUM EXPERIMENTS

The insensitivity of zero-quantum coherence to spatial inhomogeneity of the B_0 field suggests that it might be used to extract high resolution chemical shift or spin–spin relaxation information from samples that are in quite inhomogeneous magnetic fields for which the conventional NMR spectrum would be hopelessly broad. Some biological and medical applications of NMR are forced to deal with samples that are magnetically inhomogeneous, and it is possible that zero-quantum experiments could be useful here. Only the *relative* chemical shifts between nuclei in a coupled spin system would be accessible by this method. In principle, linewidths in zero-quantum spectra may be used as a measure of spin–spin relaxation rates. It is also feasible to extract spin–spin relaxation times from multiple-quantum spectra if a 180° refocusing pulse is applied at the midpoint of the evolution period, generating a coherence transfer echo (5).

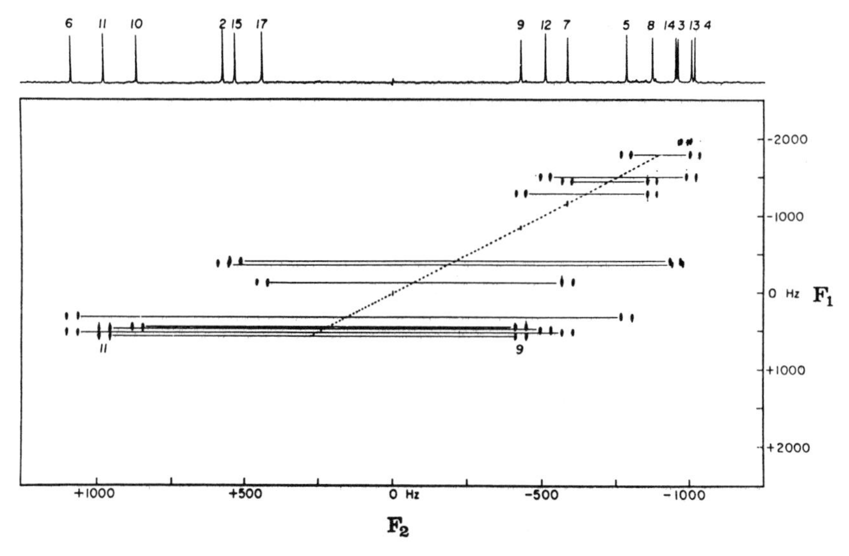

Fig. 3. Two-dimensional 'INADEQUATE' spectrum of carbon-13 in the molecule of sparteine. Each pair of coupled carbon-13 spins gives rise to a four-line AX (or AB) spectrum at an ordinate in the F_1 dimension equal to the double-quantum frequency $(\delta_A + \delta_X)$. By indicating the connectivity of the carbon atoms in the molecule, this permits the molecular framework to be established.

Another application is to the simplification of spectra. In a complicated system of coupled spins where the conventional NMR spectrum has too many lines for a straightforward assignment, the zero-quantum or multiple-quantum spectra contain fewer lines, particularly the high-order multiple-quantum spectra. This simplification technique has been used successfully for spectra derived from liquid-state samples and partially oriented samples in liquid crystal matrices (4).

Perhaps the most practically useful application to date has been the separation of spectra from different spin systems by exploiting the special properties of multiple-quantum coherences. An example is the detection of the weak carbon-13 satellite lines which appear in the flanks of the 'parent' carbon-13 resonances. The satellites carry the interesting carbon–carbon coupling information while the parent lines are an embarrassment since they tend to obscure the satellites. The satellites arise from coupled AB or AX spin systems in which one can generate double-quantum coherence, whereas the parent lines are from isolated carbon-13 spins and cannot sustain multiple-quantum coherence under any circumstances. By the momentary conversion of the satellite signals into double-quantum coherence and their reconversion into observable precessing magnetization in a manner which exploits the unique phase properties of double-quantum coherence, a pure satellite spectrum can be obtained and the parent signals suppressed. Figure 3 shows a two-dimensional spectrum derived in this way from naturally abundant carbon-13 in a sample of sparteine. Each four-line subspectrum represents a pair of directly bonded carbon-13 nuclei (AB or AX spin system) and each such subspectrum is dispersed in the F_1 dimension according to its characteristic double-quantum frequency, equal to the sum of the chemical shifts of the two sites. Spectra such as this can be used directly to establish the connectivity of the carbon skeleton of the molecule (7). The technique has been called 'INADEQUATE' (Incredible Natural Abundance Double-Quantum Transfer Experiment).

REFERENCES

1. W. P. Aue, E. Bartholdi and R. R. Ernst, *J. Chem. Phys.* **64**, 2229 (1976).
2. D. P. Burum and R. R. Ernst, *J. Magn. Reson.* **39**, 163 (1980).
3. W. S. Warren, S. Sinton, D. P. Weitekamp and A. Pines, *Phys. Rev. Lett.* **43**, 1791 (1979).
4. A. Pines, D. E. Wemmer, J. Tang and S. Sinton, *Bull. Am. Phys. Soc.* **23**, 21 (1978).
5. G. Bodenhausen, R. L. Vold and R. R. Vold, *J. Magn. Reson.* **37**, 93 (1980).
6. A. Bax, P. G. de Jong, A. F. Mehlkopf and J. Smidt, *Chem. Phys. Lett.* **69**, 567 (1980).
7. A. Bax, R. Freeman, T. A. Frenkiel and M. H. Levitt, *J. Magn. Reson.* **43**, 478 (1981).

Cross-references

Product operator formalism
Rotating frame
Spin tickling
Two-dimensional spectroscopy

Multiplicity Determination

Nowadays, the recording of a decoupled high-resolution carbon-13 spectrum is a straightforward matter, but the assignment of the resonances to individual carbon sites presents a considerable challenge. Fortunately there is now quite an armoury of techniques available to assist in assignment. Apart from deductions based on chemical shifts, relaxation properties, double resonance or chemical shift correlation*, there are several methods designed to identify the number of directly attached protons, relying on the fact that $^{1}J_{CH}$ is very much larger than the long-range couplings $^{n}J_{CH}$. This is called *multiplicity determination*.

OFF-RESONANCE DECOUPLING

The earliest such method for determining multiplicity was that of off-resonance decoupling* using a coherent monochromatic irradiation field B_2 applied at some suitable frequency in the proton spectrum (1). The idea is to shrink the multiplet structure in the frequency dimension so as to avoid overlap between adjacent multiplets, while still leaving the multiplet structure well enough resolved that quartets, triplets, doublets and singlets can be distinguished. Long-range couplings disappear within the linewidth. The disadvantages of the method are well-known; the scaling factors depend on the proton resonance offsets, so that it may be necessary to try several different decoupler frequencies before a suitable carbon-13 spectrum is obtained. The multiplet structure can be distorted by 'second order' effects when the protons are strongly coupled, particularly for non-equivalent methylene protons. These effects may take the operator by surprise since there may be no evidence of strong coupling in the normal proton spectrum, although the carbon-13 satellite spectrum *is* strongly coupled, and the presence of the decoupler field can aggravate the effect. Spatial inhomogeneity in the decoupling field B_2 may also distort the carbon-13 multiplets by differential broadening of the outer lines (2).

J-SCALING

This is a method designed to preserve a constant scaling factor for all carbon-13 spin multiplets independent of the relevant proton chemical shifts. Figure 1 illustrates the application of this method to the carbon-13 spectrum of camphor. The ability to change the scaling factor at wil! turns out to be a powerful method of resolving ambiguities presented by overlapping multiplets. In this spectrum there is a doublet and triplet at almost the same carbon-13 shift; they are readily recognized by running the J-scaled spectrum twice, with different scaling factors. These spectra were obtained by a time-consuming 'interferogram' method (3), but now there is a 'real time' method which is much simpler and faster. It employs a broadband decoupling* scheme such as WALTZ-16 in which the proton spins execute a periodic spin-inversion which is insensitive to proton resonance offset and pulse imperfections because it uses a self-compensating 'magic cycle' of composite pulses (4). At the end of the decoupler cycle t_w, the decoupler is switched off for a short period t_p in order to allow the

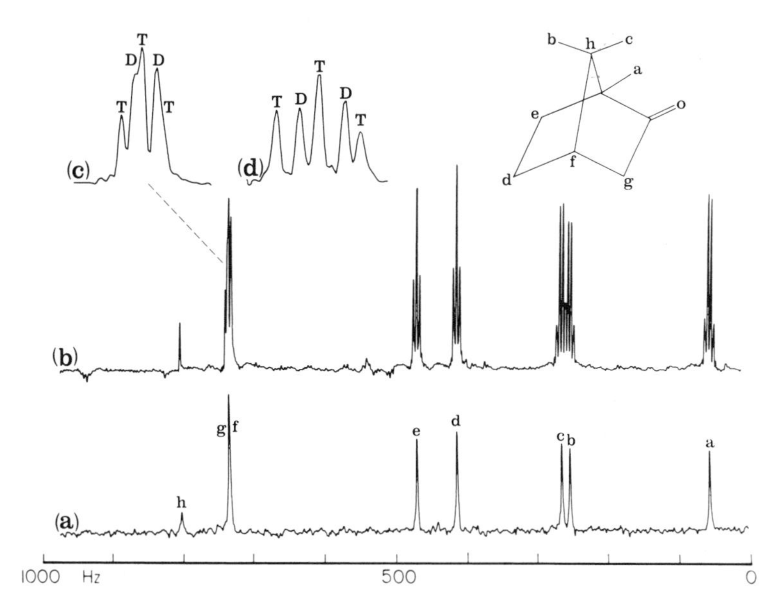

Fig. 1. A conventional decoupled carbon-13 spectrum (a) and the same spectrum with J-scaling (b). The multiplicity of each resonance in spectrum (b) indicates the number of directly attached protons. One multiplet has been recorded with two different scaling factors (c) and (d).

carbon-13 spins to precess in the proton-coupled mode. The apparent multiplet splitting is then scaled down from $^1J_{CH}$ by a factor $t_p/(t_p + t_w)$ independent of proton offset over a considerable frequency range. This scaling factor is normally adjusted to be some small fraction (for example 1/20) in order to display well-resolved multiplets that are nevertheless well-separated from their nearest neighbours (5).

TWO-DIMENSIONAL SPECTROSCOPY*

Another approach is to record the two-dimensional carbon-13 J-spectrum, obtained by Fourier transformation of the modulation which appears on carbon-13 spin echoes due to CH coupling. There are basically two variants of this experiment – one employs a proton spin inversion pulse at the midpoint of the evolution period, the other uses broadband proton decoupling for half of the evolution period. The advantage of this technique is that the carbon-13 spin multiplets appear in a new frequency dimension (F_1), whereas the carbon-13 chemical shifts remain in the old frequency dimension (F_2), thus avoiding overlap between multiplets. Figure 2 shows the two-dimensional spectrum of

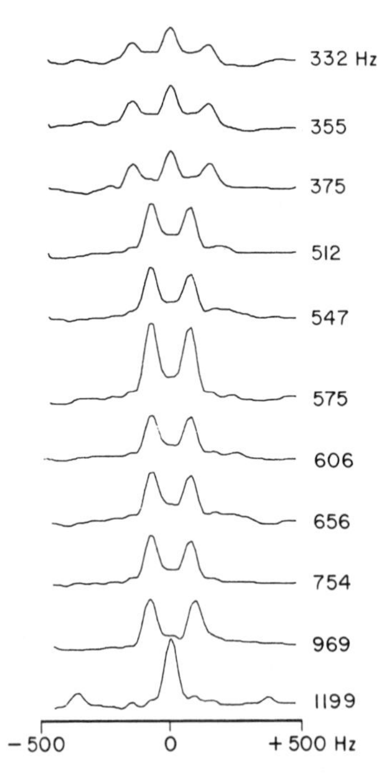

Fig. 2. Multiplicity of the carbon-13 sites of sucrose obtained by taking sections through the two-dimensional J-spectrum at the appropriate chemical shifts (indicated on the right).

sucrose obtained by this method. The principal disadvantage of two-dimensional methods is the time required to perform several measurements at different increments of the evolution time t_1 in order to achieve fine digital resolution in the F_1 dimension. Recently, it has been demonstrated that quartets, triplets, doublets and singlets can still be distinguished even when the number of measurements in the t_1 domain is drastically reduced, provided that precautions are taken to eliminate the sinc function wiggles caused by premature truncation of the signal in this domain. Good discrimination has been achieved with as few as five measurements (6).

EDITING OF CARBON-13 SPECTRA

Polarization transfer* from protons to carbon-13 by the INEPT pulse sequence (7), followed by observation of the carbon-13 signal under decoupled conditions, gives a time evolution of the signal intensity which depends on the multiplicity (see Fig. 3, page 160)

$$\begin{aligned} &\text{CH groups: } S = \sin\theta \\ &\text{CH}_2 \text{ groups: } S = \sin 2\theta \\ &\text{CH}_3 \text{ groups: } S = 0.75[\sin\theta + \sin 3\theta], \end{aligned} \qquad [1]$$

where $\theta = 360\ J_{CH}\tau_2$. Polarization transfer to quaternary carbon sites is negligible on this time scale since the long-range CH couplings are small in comparison with $^1J_{CH}$. In these experiments it is customary to cancel the natural carbon-13 signal, leaving only signals generated by magnetization transfer from protons. By suitable choice of the timing interval $2\tau_2$, for example by setting $\theta = 90°$, the CH groups can be separately displayed since the CH_2 and CH_3 intensities are at a null condition. This has been made the basis of some useful methods of multiplicity determination. A straightforward extension of these techniques permits a virtually complete separation of the decoupled carbon-13 spectrum into four subspectra – one for methyl sites, one for methylene, one for methine and one for quaternaries. This is known as *editing* the carbon-13 spectrum.

Another approach makes use of the modulation which appears on carbon-13 spin echoes* due to $^1J_{CH}$ coupling. Here the time evolution follows a different pattern

$$\begin{aligned} &\text{CH groups: } S = \cos\theta \\ &\text{CH}_2 \text{ groups: } S = \cos^2\theta \\ &\text{CH}_3 \text{ groups: } S = \cos^3\theta, \end{aligned} \qquad [2]$$

where $\theta = 360\ J_{CH}\tau$. Quaternary carbons remain essentially unmodulated on this time scale since the long-range CH couplings are small in comparison with the direct couplings. There are several different strategies for spectral editing through spin-echo modulation. Figure 3 shows the carbon-13 spectrum of estrone methyl ether in a form where the CH and CH_3 resonances have been inverted while the CH_2 and quaternary carbon resonances remain upright (8).

The choice of technique for spectral editing hinges on practical aspects of the methods, in particular the amount of 'cross-talk' between the subspectra. Cross-talk arises because of the range of ${}^1J_{CH}$ values; there is no single 'correct' setting of the timing parameter 2τ which gives a complete separation. For example, there will be weak anomalous responses in the methine subspectrum arising from the methyl groups. However, the DEPT experiment (9) is relatively insensitive to this problem and has been widely adopted in practice for this reason.

A clear visualization of the spin physics of the DEPT experiment has been slow to emerge. It is easier to approach it through relationships to other, more straightforward pulse sequences. For example, it is possible to establish a relationship to the early heteronuclear multiple-quantum coherence* experiment first described by Müller (10) for magnetization transfer from protons to carbon-13. The following exercise derives DEPT from the INEPT experiment (which has the advantage that it is easily couched in terms of the vector model*).

Consider a simple IS spin system and an INEPT sequence stripped of both sets of 180° refocusing pulses. The result of this simplification is to make the amount of magnetization transfer a function of the proton chemical shifts (measured with respect to the proton transmitter frequency) and to introduce a frequency-dependent phase shift into the observed carbon-13 spectrum due to the evolution of the carbon-13 shifts. These chemical shift effects will in fact be eliminated later. Now it is customary to write the INEPT sequence (Fig. 4(a)) with the 90°(Y) proton pulse just before the 90° carbon-13 read pulse, for this makes the vector diagrams clear and simple. We know in practice that these pulses could be simultaneous or even in reversed order – the same spectra are obtained. Suppose then that the 90° carbon pulse actually *precedes* the 90°(Y) proton pulse by a short interval δ (Fig. 4(b)). In this case a quite different language is required to describe the spin gymnastics. The 90° carbon pulse excites heteronuclear zero- and double-quantum coherence during the

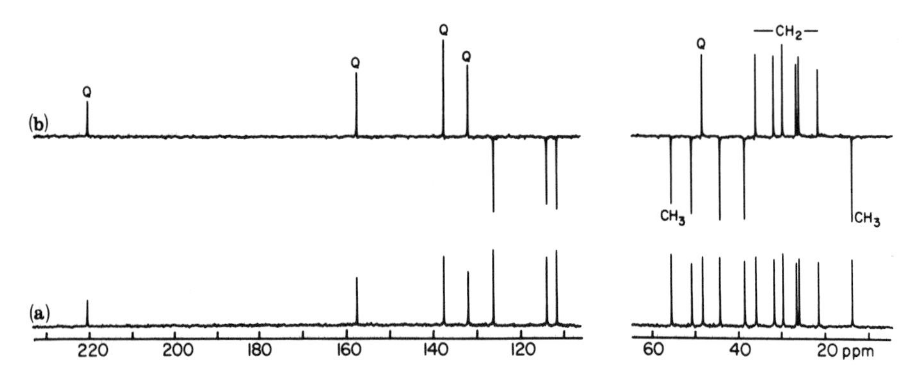

Fig. 3. (a) Conventional carbon-13 spectrum of estrone methyl ether. (b) Spectrum obtained by Fourier transformation of the spin echoes recorded with a delay (7 ms) calculated to allow inversion of the methyl and methine signals, while the quaternary and methylene signals remain upright.

interval δ. The relative proportions depend on the proton chemical shifts, but this can be disregarded because these effects will later be cancelled. These multiple-quantum coherences 'belong' to both protons and carbon-13; in terms of the IS energy level diagram they correspond to coherences between levels 2 and 3 and between levels 1 and 4. The 90°(Y) proton pulse reconverts these multiple-quantum coherences into observable carbon-13 transverse magnetization in the form of two antiphase absorption lines. In the last interval τ the corresponding vectors precess into parallel orientations, giving a net carbon-13 signal. Thus the overall result of this modified INEPT experiment is identical to that of the experiment with the original ordering of pulses.

Now there is no reason why δ should not be extended to some finite interval, giving time for appreciable precession of the zero- and double-quantum coherences at the frequencies $(\nu_H - \nu_C)$ and $(\nu_H + \nu_C)$ respectively. Note that there are no terms involving $^1J_{CH}$ in these expressions. It turns out to be convenient to set $\delta = \tau$, for this facilitates a simple scheme for removing the chemical shift effects. Proton shifts are effective during the first and second τ intervals and are therefore cancelled by a 180° pulse at the midpoint (coincident with the carbon 90° pulse). Carbon-13 chemical shifts are effective during the second and third τ intervals and are therefore cancelled by a 180° pulse at the corresponding midpoint (coincident with the proton 90°(Y) pulse). The conver-

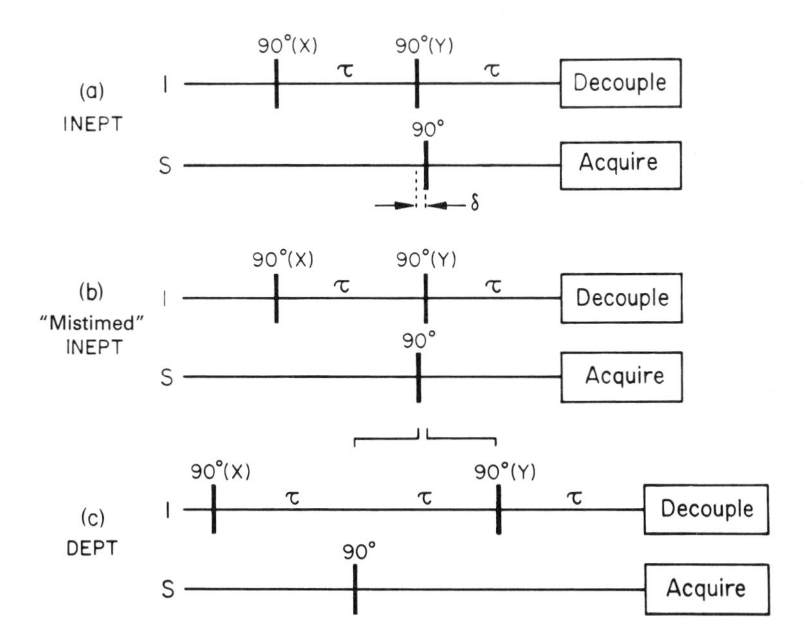

Fig. 4. Derivation of the 'DEPT' experiment from the 'INEPT' experiment. Whereas the INEPT experiment (a) can be described in terms of the vector model, the mistimed INEPT experiment (b) involves the creation of multiple-quantum coherence, although it gives the same end-result. DEPT (c) is obtained by expanding the δ delay to equal the τ delay. In practice, the last pulse of the DEPT experiment is allowed to have a variable flip angle θ.

sion into multiple-quantum coherence is then independent of proton resonance offsets, and the relative phases of the carbon-13 resonances are independent of carbon-13 resonance offsets. This is the DEPT sequence (Fig. 4(c)).

The three different spin systems, methine (IS), methylene (I_2S), and methyl (I_3S) respond differently to the last proton pulse, the 'reconversion' pulse. Thus when used for editing, the DEPT experiment employs a variable pulse $\beta(Y)$. This is the key to the entire process of editing and the clue to how it works has been provided by Levitt (11). Suppose we replace $\beta(Y)$ by a 'virtual composite pulse' $90°(X)$ $\beta(Z)$ $90°(-X)$ since these are known to be equivalent (see Composite pulses*). The first pulse $90°(X)$ creates I-spin coherences; single-quantum coherence for IS systems, double-quantum coherence for I_2S systems and a mixture of single and triple-quantum coherence for I_3S systems. The second pulse $\beta(Z)$ simply acts as a phase shift but is n times as effective for coherence of order n. Thus when the third pulse $90°(-X)$ reconverts the multiple-quantum coherence into observable carbon-13 magnetization, the IS system gives a maximum at $\beta = 90°$, the I_2S system a maximum at 45° and I_3S system a maximum at 35°. (This last condition is shifted slightly from $\beta = 30°$ by the presence of some single-quantum coherence which goes through a maximum at 90°). In fact, the curves for the dependence of observed signal on β are identical with those for the INEPT experiment if θ is replaced by β in eqn [1]. This analysis has the advantage that it explains why the 'reconversion' pulse must be $\beta(Y)$ rather than $\beta(X)$, and why 'cross-talk' normally occurs in the 'downward' direction, that is from I_3S systems to IS systems but not *vice versa*.

Editing thus consists of exploiting the differences between the curves for methine, methylene and methyl groups so as to be able to achieve a complete separation. With $\beta = 90°$ the methine subspectrum is obtained. With $\beta = 45°$ there are positive-going signals from all three sites, whereas with $\beta = 135°$ the spectrum is identical except that the CH_2 signals are inverted. Consequently, the methylene subspectrum may be obtained by subtraction of these two spectra. A methyl subspectrum is derived by adding the spectra obtained with $\beta = 45°$ and $\beta = 135°$ and then subtracting a suitable proportion of the spectrum obtained with $\beta = 90°$. In practice, since the $^1J_{CH}$ values are not all the same, the best separation of subspectra is achieved with slightly different linear combinations of the results of these three experiments. Routine carbon-13 assignments do not normally demand this degree of sophistication.

REFERENCES

1. H. J. Reich, M. Jautelat, M. T. Messe, F. J. Weigert and J. D. Roberts, *J. Am. Chem. Soc.* **91**, 7445 (1969).
2. R. Freeman, J. B. Grutzner, G. A. Morris and D. L. Turner, *J. Am. Chem. Soc.* **100**, 5637 (1978).
3. R. Freeman and G. A. Morris, *J. Magn. Reson.* **29**, 173 (1978).

4. A. J. Shaka, J. Keeler and R. Freeman, *J. Magn. Reson.* **53**, 313 (1983).
5. G. A. Morris, G. L. Nayler, A. J. Shaka, J. Keeler and R. Freeman, *J. Magn. Reson.* **58**, 155 (1984).
6. J. Keeler, *J. Magn. Reson.* **56**, 463 (1984).
7. G. A. Morris and R. Freeman, *J. Am. Chem. Soc.* **101**, 760 (1979).
8. F-K. Pei and R. Freeman, *J. Magn. Reson.* **48**, 318 (1982).
9. D. M. Doddrell, D. T. Pegg and M. R. Bendall, *J. Magn. Reson.* **48**, 323 (1982).
10. L. Müller, *J. Am. Chem. Soc.* **101**, 4481 (1979).
11. M. H. Levitt, *Progress in NMR Spectroscopy* **18**, 61 (1986).

Cross-references

Broadband decoupling
Chemical shift correlation
Composite pulses
Multiple-quantum coherence
Off-resonance decoupling
Polarization transfer
Spin echoes
Two-dimensional spectroscopy
Vector model

Nuclear Overhauser Effect

A. W. Overhauser (1) suggested an experiment whereby the intensity of a nuclear magnetic resonance signal is greatly enhanced by saturating the electron spin resonance signal from unpaired electrons in the sample. A prerequisite for this effect is that the spin–lattice relaxation* of the nuclei is dominated by dipolar coupling with the unpaired electrons. It is thus a *cross-relaxation* experiment. Although enhancements of up to two orders of magnitude are feasible in favourable cases, no widespread use of this has been made in high resolution NMR. One reason is that the Overhauser experiment requires, in addition to the usual radiofrequency circuitry, the hardware for microwave irradiation at the same site in the probe, and this tends to interfere with sensitivity and resolution. It would not be feasible to generate microwaves at a frequency high enough for the magnetic field strengths employed in present-day superconducting spectrometers. Even at moderate field strengths, the enhancements achieved are quite strongly dependent on the nature of the sample and the way in which the nuclei interact with the unpaired electrons in a free radical added to the solution.

An analogous experiment where saturation of one kind of nuclear spin I enhances the signal from a second kind of nuclear spin S, has had much more success in high resolution NMR spectroscopy. As before, the spin–lattice relaxation of S must be governed by dipole–dipole interaction with I; there must be appreciable cross-relaxation. This *nuclear Overhauser effect* is widely used for two main purposes: the enhancement of signals from low-sensitivity nuclei such as carbon-13, and the study of the dipolar mechanism for spin–lattice relaxation in order to obtain information about internuclear distances (2).

Suppose (3) there are two alpine villages A and B, each with a ski lift to the tops of the respective ski slopes, A′ and B′. Both ski lifts have the same length but since village B is slightly higher up the mountain than A, the top of ski-lift B′ is similarly higher than A′. Imagine that we are interested in the relative popularities of the two ski villages and that we attempt to measure this in terms of the ratio of the number of skiers at each village. There might well be a small preponderance at A since it is the first village on the approach road. This 'natural' population ratio might also be measured between A′ and B′ since the two lifts are assumed to have identical passenger-carrying capacities and the ratio would not be affected by the normal flow of skiers down the slopes (relaxation). We might, however, suppose that towards the end of the day many skiers at B′ might choose to make their final run on the longer *piste* from B′ to A, having no particular preference for the *après ski* activities at the two villages. A head count of the skiers at A and B would then show a considerably enhanced ratio, a consequence of the asymmetry introduced into the problem by the cross-relaxation path B′–A. In the limit the entire population of village B might end up at A. This is the essence of the nuclear Overhauser effect, in which pumping of two proton transitions (ski-lifts) combined with strong cross-relaxation of carbon-13 by protons has the effect of transferring a strong proton spin polarization to carbon-13.

The mechanism by which this intensity enhancement is accomplished (4) may be appreciated by considering an IS spin system (both spin-1/2 nuclei), assuming initially that the relaxation is entirely dipolar in origin. This four-level system is illustrated in Fig. 1. It is readily seen that the S spin intensities are not affected by saturation of the I spin resonances, provided that both I transitions are affected equally. The Overhauser enhancement is created only if cross-relaxation effects come into play in an effort to restore equilibrium. In fact the result is a steady-state distribution, reflecting the competition between the pumping effect of the B_2 irradiation and the several relaxation pathways.

Relaxation operates through the single-quantum transition probabilities W_I (where only the I spin flips) and W_S (S spin flip), the zero-quantum transition probability W_0 (where I and S flip in opposite senses) and the double-quantum probability W_2 (where I and S flip in the same sense). The cross-relaxation terms W_2 and W_0 are only important in the dipole–dipole mechanism. In the practical situation the spin populations are perturbed symmetrically – both I transitions are fully saturated and both S transitions are equally affected. The deviations of the spin populations from Boltzmann equilibrium can then be represented by two numbers, a and b, where

$$2a = S_Z - S_0 \qquad 2b = I_Z - I_0. \qquad [1]$$

Here I_0 and S_0 are the respective intensities at Boltzmann equilibrium and they are necessarily related by

$$I_0/S_0 = \gamma_I/\gamma_S. \qquad [2]$$

When the spin population of any level deviates from equilibrium, there is a resulting flux of spins in such a direction as to return the system toward equilibrium, and this flux is proportional to the appropriate population differences and the relaxation transition probabilities

$$dn_1/dt = 2aW_S + 2bW_I + 2(a + b)W_2 \qquad [3]$$

$$dn_2/dt = -2aW_S + 2bW_I - 2(a - b)W_0. \qquad [4]$$

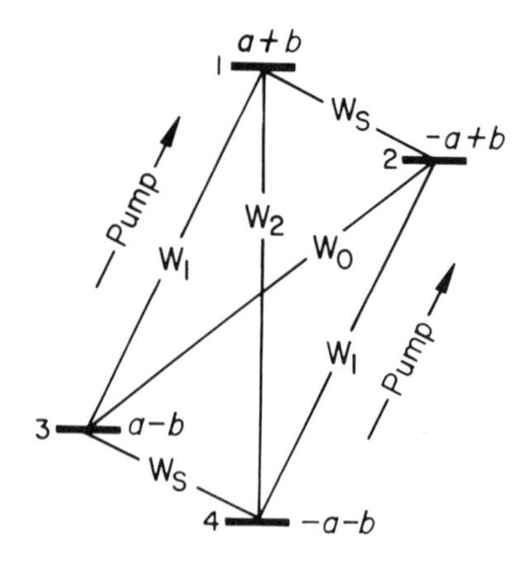

Fig. 1. Energy-level diagram showing transition probabilities for relaxation processes W_I, W_S, W_0 and W_2, and the deviations of the populations from the values at Boltzmann equilibrium. The second radiofrequency field B_2 saturates the I-spin transitions.

Consequently,

$$d(n_1 - n_2)/dt = 2a(2W_S + W_2 + W_0) + 2b(W_2 - W_0). \quad [5]$$

For a steady-state condition this derivative must be zero, giving

$$a/b = (W_2 - W_0)/(2W_S + W_2 + W_0). \quad [6]$$

This is a general expression; the condition that the I spins are saturated may now be inserted, $I_Z = 0$, giving $2b = -I_0$. Using eqn [2]

$$\eta = \frac{S_Z - S_0}{S_0} = \frac{W_2 - W_0}{(2W_S + W_2 + W_0)}\frac{\gamma_I}{\gamma_S}. \quad [7]$$

The parameter η is called the nuclear Overhauser enhancement factor. It is more useful to calculate the ratio by which the S signal intensity is increased

$$S_Z/S_0 = 1 + \eta. \quad [8]$$

Now for purely dipolar relaxation the relative importance of the various relaxation probabilities is well known to be

$$W_2 : W_S : W_0 :: 12 : 3 : 2. \quad [9]$$

This gives the simplified expression

$$1 + \eta = 1 + \tfrac{1}{2}\gamma_I/\gamma_S. \quad [10]$$

In the original Overhauser experiment I was an electron spin (which has a negative gyromagnetic ratio) and the maximum enhancement of a proton resonance was about -600, giving an inverted proton signal. A much more important case for the high resolution NMR spectroscopist is where I represents proton spins and the observed S resonance is from a carbon-13 spin. Equation [10] predicts an almost threefold improvement in signal-to-noise ratio if the relaxation is predominantly dipolar. As a result, almost all carbon-13 spectra are obtained with broadband proton irradiation during a preparation period just before the radiofrequency excitation pulse (this may be continued during acquisition of the free induction decay, but then it decouples the spectrum from protons, the nuclear Overhauser enhancement is not affected). The only real drawback of this technique is that the nuclear Overhauser enhancement factors may not all be the same for the various carbon-13 sites, leading to a distortion of the relative intensities. Where this quantitative information is important, gated decoupling* or a relaxation agent is employed and the nuclear Overhauser enhancement is sacrificed.

The corresponding enhancement for nitrogen-15 is even larger, but there is a problem, because nitrogen-15 has a negative gyromagnetic ratio ($\gamma_I/\gamma_S \simeq -10$), so the enhanced signal is in the opposite sense to the unenhanced signal, giving a maximum effect of $1 + \eta = -4$. This can be a serious drawback in cases where the relaxation mechanism is not entirely dipolar and the nuclear Overhauser effect is incomplete. For this reason, polarization transfer* experiments are sometimes preferred since they can enhance the nitrogen-15 intensities by a factor γ_I/γ_S (approximately tenfold) and they do not depend on the relaxation mechanism being dipolar.

COMPETITION WITH OTHER RELAXATION MECHANISMS

When the dipole–dipole interaction is one of several different competing mechanisms of relaxation, the full nuclear Overhauser enhancement is not observed. Equation [7] has to be modified to incorporate another single-quantum transition probability W_S^* which acts as a 'leakage' path and attenuates the nuclear Overhauser enhancement.

$$\eta = \frac{S_Z - S_0}{S_0} = \frac{W_2 - W_0}{(2W_S^* + 2W_S + W_2 + W_0)} \frac{\gamma_I}{\gamma_S}. \qquad [11]$$

In the limit where $2W_S^* \gg W_2 - W_0$ the enhancement factor goes to zero. Relaxation reagents have sometimes been used in this way to quench the nuclear Overhauser effect on carbon-13 in order to restore uniform intensities across the spectrum (5). In most applications, however, the aim is to keep such leakage terms to a minimum.

An entirely different application of the nuclear Overhauser effect is in the determination of internuclear distances, usually interproton distances. The cross-relaxation transition probabilities are related to internuclear distance r_{IS} through the expression

$$W_2 - W_0 = \tfrac{1}{2}\gamma_I^2\gamma_S^2/\hbar^2\tau_c(r_{IS})^{-6}, \qquad [12]$$

where τ_c is the correlation time for molecular reorientation. Because of the inverse sixth power dependence on the distance this should give an accurate determination of r_{IS}. The problem is in the leakage term of eqn [11], which is not known. Relative internuclear distances can, however, be measured if two nuclear Overhauser enhancements are determined for the same observed spin S, while saturating first one neighbour I, then another neighbour R. The key is that the leakage factor remains constant, and if the I–S–R spin system is in a rigid part of the molecule, the correlation times are the same. Thus, provided there is no nuclear Overhauser effect between I and R, the ratio of the Overhauser effects is given by

$$\eta_S(I)/\eta_S(R) = (r_{SR}/r_{IS})^6. \qquad [13]$$

In many practical cases the situation is complicated by the presence of other interacting spins and more general expressions (6) are appropriate, but the principle is the same.

The treatment given above presupposes the condition that molecular reorientation is fast compared with the Larmor precession frequency. This is not the case for some macromolecules in solution. Equation [9] is then no longer valid and negative nuclear Overhauser effects may be observed on protons when other nearby protons are irradiated.

Nuclear Overhauser measurements are at their most effective when used to distinguish between two alternative structures. In such applications many of the possible complicating factors are the same in the two cases and do not affect the decision. These tend to be experiments on small molecules in the realm of organic chemistry.

Nevertheless, a great deal of work has been done in biochemistry, where proton–proton Overhauser measurements are used to determine the sequencing of aminoacids and to establish the tertiary structure of macromolecules. Here the conclusions are less secure. First of all, these molecules can hardly be regarded as rigid, so the nuclear Overhauser enhancement represents a complicated average over all possible conformations (remember the r^{-6} dependence). Second, we are unlikely to be dealing with an isolated two-spin system, so the relaxation by neighbouring protons cannot be neglected. This effect is often loosely referred to as *spin diffusion*, a kind of leakage that is propagated through the surrounding cloud of protons. This problem can be mitigated by measuring the transient nuclear Overhauser effect – the initial time evolution of the enhancement – before there is any appreciable spin diffusion.

In spite of these daunting complications, the method has been surprisingly successful in some applications. After all, if an enhancement is observed at all, the two protons involved must lie within a distance of a few Ångstroms and not at opposite ends of the molecule. Thus by establishing *distance constraints*, the method provides valuable complementary evidence about the three-dimensional structure, although it cannot in general determine the structure from scratch.

REFERENCES

1. A. W. Overhauser, *Phys. Rev.* **92**, 411 (1955).
2. J. H. Noggle and R. E. Schirmer, *The Nuclear Overhauser Effect.* Academic Press: New York, 1971.
3. Adapted from a story in A. Abragam, *Réflexions d'un Physicien*, Hermann: Paris, 1983.
4. I. Solomon, *Phys. Rev.* **99**, 559 (1955).
5. R. Freeman, K. G. R. Pachler and G. N. LaMar, *J. Chem. Phys.* **55**, 4586 (1971).
6. R. Freeman, H. D. W. Hill, B. L. Tomlinson and L. D. Hall, *J. Chem. Phys.* **61**, 4466 (1974).

Cross-references

Gated decoupling
Polarization transfer
Spin–lattice relaxation

Off-Resonance Decoupling

Irradiation of one set of spins X with an intense B_2 field far from resonance has the effect of scaling down heteronuclear splittings on a coupled set of spins A, while retaining the same form for the A-spin multiplets and the same binomial intensity distribution. This has found considerable use in carbon-13 spectroscopy for assignments based on the number of directly attached protons (1). While the same information is contained in the ordinary proton-coupled spectrum, there the overlap of adjacent spin multiplets often makes interpretation very difficult. For off-resonance decoupling, the B_2 field is coherent and is usually used at full intensity, the proton resonance offset being adjusted to give the desired order-of-magnitude for the residual splittings. This may be as small as a few hertz, provided that the multiplet structure can be resolved.

There are certain circumstances where this method of assignment could be ambiguous. The most serious problem arises with certain CH_2 groups where the protons are strongly coupled to other protons. In the presence of the strong B_2 field, effective proton chemical shifts may be reduced, making the proton spin system more strongly coupled than in the conventional spectrum. The resulting carbon-13 triplets are then poorly defined, with the outer lines split or broadened. Some of these problems are evident in the set of off-resonance decoupled spectra of carbon-13 illustrated in Fig. 1. A complicating factor is introduced by spatial inhomogeneity of the B_2 field which broadens the outer lines of the triplets but leaves a sharp central line; a similar phenomenon is observed for quartets, where the outer lines are three times more sensitive to this kind of broadening than the inner lines (2).

The practical advantage of off-resonance decoupling is that it is readily implemented on existing equipment; however, its application to the determination of carbon-13 multiplicity has been largely superseded by more sophisticated techniques. For example, J-scaling (3) also reduces the CH splittings in a carbon-13 spectrum but has the important advantage that the scaling factor is virtually constant across the entire spectrum. Two-dimensional J-spectroscopy* displays all the carbon-13 spin multiplets in a second frequency dimension, thus neatly avoiding the problem of overlap. Then there is a family of methods based on polarization transfer* from protons to carbon-13 employing the 'INEPT' or 'DEPT' sequences. These alternative assignment techniques are examined under Multiplicity determination*.

An off-resonance decoupling experiment can also be used to give an indication of the chemical shift of the protons coupled to a given carbon-13 nucleus, a crude form of chemical shift correlation*. The method depends on being able to calculate the proton resonance offset from the observed residual splitting and the known values of $\gamma_H B_2/2\pi$ and J_{CH}. The required expression is derived (4) from effective field calculations using Δ to represent the offset of the decoupler frequency from the proton chemical shift. We consider the two proton satellite signals at frequencies $\Delta + \frac{1}{2}J$ and $\Delta - \frac{1}{2}J$. The effective fields in the rotating reference frame*, expressed in frequency units, are given by

$$E_1 = [(\Delta + \tfrac{1}{2}J)^2 + (\gamma_H B_2/2\pi)^2]^{1/2} \quad (\text{Hz})$$
$$E_2 = [(\Delta - \tfrac{1}{2}J)^2 + (\gamma_H B_2/2\pi)^2]^{1/2} \quad (\text{Hz}). \qquad [1]$$

Provided that the offsets are large compared with $\frac{1}{2}J$, it is easy to show that the residual splitting $E_1 - E_2$ is given by

$$E_1 - E_2 \approx \pm J\Delta/[\Delta^2 + (\gamma_H B_2/2\pi)^2]^{1/2} \quad (\text{Hz}). \qquad [2]$$

In most practical situations the further approximation can be made that $\gamma_H B_2/2\pi \gg |\Delta|$ giving the simpler expression

$$E_1 - E_2 \approx \pm J\Delta/(\gamma_H B_2/2\pi) \quad (\text{Hz}). \qquad [3]$$

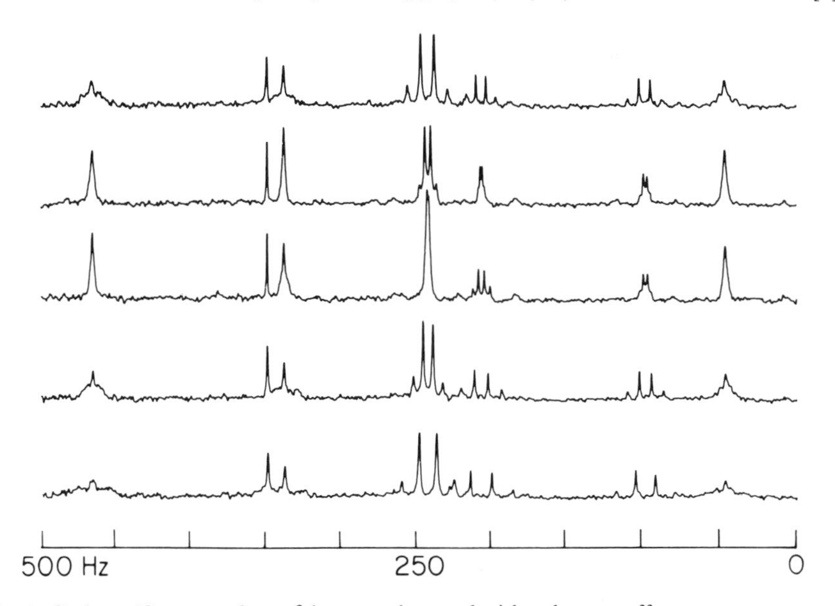

Fig. 1. Carbon-13 spectra from β-ionone observed with coherent off-resonance decoupling in order to determine the multiplicity of each resonance. Since the multiplets are not always well-defined (particularly triplets) it may be necessary to record the spectrum at several different proton resonance offsets.

This form of the residual splitting expression is widely used, bearing in mind that the estimation of the proton offset is only rather approximate. For assignment purposes it may only be necessary to note that one residual splitting is significantly smaller than another, while the J values are known to be similar. The ambiguity in sign of the offset Δ is not usually a problem provided that the decoupler frequency is set well above or below the frequencies of all the relevant proton resonances so that all offsets have the same sign. Alternatively, more than one off-resonance decoupling experiment is performed, for different proton offsets. For accurate determinations of proton shifts, however, a series of selective decoupling* experiments is to be preferred.

REFERENCES

1. H. J. Reich, M. Jautelat, M. T. Messe, F. J. Weigert and J. D. Roberts, *J. Am. Chem. Soc.* **91**, 7445 (1969).
2. R. Freeman, J. B. Grutzner, G. A. Morris and D. L. Turner, *J. Am. Chem. Soc.* **100**, 5637 (1978).
3. G. A. Morris, G. L. Nayler, A. J. Shaka, J. Keeler and R. Freeman, *J. Magn. Reson.* **58**, 155 (1984).
4. W. A. Anderson and R. Freeman, *J. Chem. Phys.* **37**, 85 (1962).

Cross-references

Coherent decoupling
J-spectroscopy
Multiplicity determination
Polarization transfer
Rotating frame
Selective decoupling
Shift correlation

Phase Cycling

A key feature of many modern pulse sequences is a *phase-cycling* routine. This involves repeating the pulse sequence and signal acquisition with all parameters unchanged except radiofrequency phase. Commonly, four stages might be used, the appropriate pulse phases being shifted in 90° steps, together with changes in the receiver reference phase. The aim is to suppress certain undesirable signal components at the completion of the cycle, while the signal of interest is improved by time averaging*. Some phase cycles have become enormously complicated through the nesting of cycles one within the other, so it is important to keep the guiding principles very clear. We consider here two simple illustrative examples, 'CYCLOPS' and 'EXORCYCLE'.

CYCLOPS was introduced by Hoult and Richards (1) as a method of improving the performance of quadrature detection* schemes, now employed in almost all high-resolution spectrometers to improve sensitivity and to allow the transmitter frequency to be set in the centre of the spectrum. Quadrature detection schemes employ two phase-sensitive detectors instead of one, and this introduces some new practical problems. One is that the two 'channels' may not be properly matched in gain, and this leads to weak spurious responses at mirror-image frequencies with respect to the centre of the spectrum (zero frequency). The second is that the baseline level of the free induction decay acquired in one channel may not be exactly balanced with respect to that in the other channel. This gives rise to an anomalous response in the centre of the spectrum, usually known as the *quadrature glitch*.

Now suppose we set out to eliminate these problems one at a time using the principle of difference spectroscopy*. We perform two experiments: in the first we use a 90°(+X) pulse, while in the second we use a 90°(−Y) pulse and interchange the roles of the two phase-sensitive detector channels, equivalent to a 90° receiver phase shift. (For the details, see Quadrature detection). It is then easy to see that any difference in gain between the two channels is compensated; interestingly, the errors due to an inaccurate 90° phase shift also cancel.

Alternatively, we can attack the second problem, performing one experiment with a 90°(+X) pulse and the second with a 90°(−X) pulse and reversed receiver reference phase. Since the baseline error should be independent of the phase of the excitation pulse, it can be cancelled out, whereas the NMR signal is reinforced since it follows the pulse phase. Fourier transformation then gives a spectrum with no quadrature glitch.

When both types of imperfection need to be compensated, four experiments are performed and it is convenient to write them in the form of a *cycle*. If the

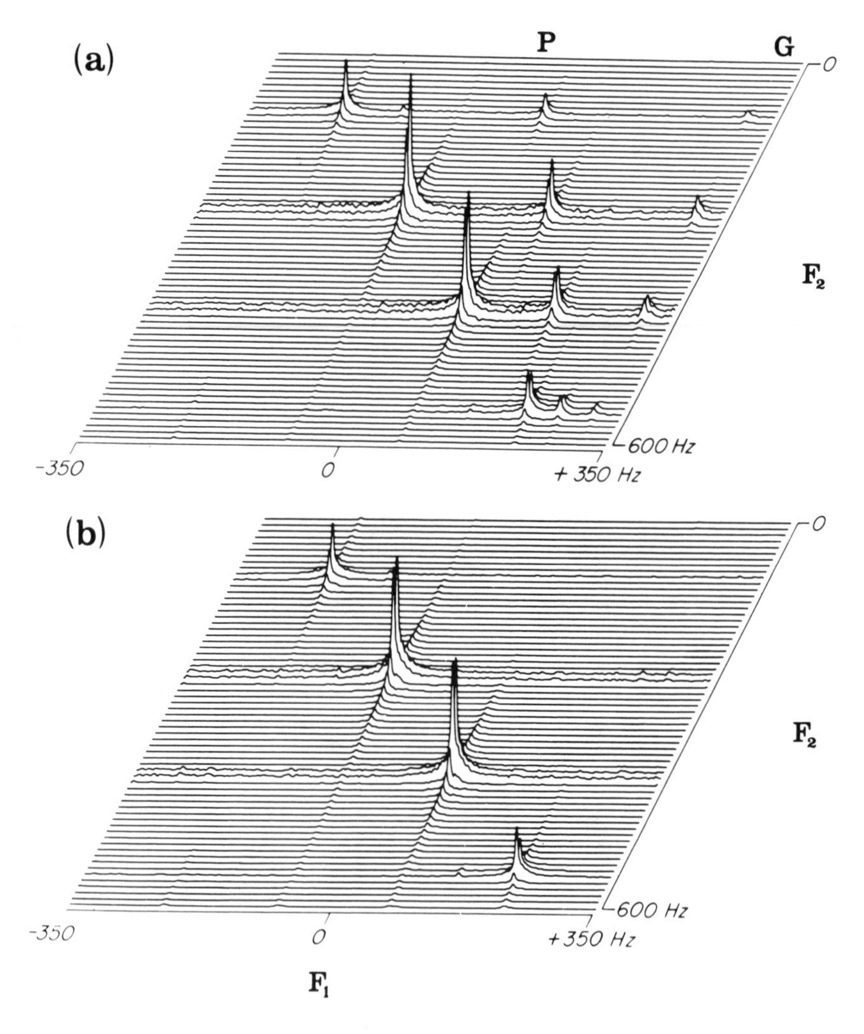

Fig. 1. (a) J-spectroscopy of carbon-13 in methyl iodide carried out with the radiofrequency pulse lengths deliberately mis-set in order to emphasize the spurious responses called ghosts (G) and phantoms (P). (b) The same experiment with the EXORCYCLE scheme which is designed to suppress ghost and phantom responses.

two channels are labelled (a) and (b) then the corresponding signals are combined as in the following table, giving two components, conveniently called 'absorption mode' and 'dispersion mode'.

Pulse	'Absorption mode'	'Dispersion mode'
90°(+X)	+(a)	+(b)
90°(−Y)	+(b)	−(a)
90°(−X)	−(a)	−(b)
90°(+Y)	−(b)	+(a)

In some spectrometers the quadrature glitch seems to be the dominant imperfection; then it is usually sufficient to operate only a two-stage *phase-alternation* sequence.

The EXORCYCLE scheme (2) was introduced to solve the problem of some spurious responses observed in two-dimensional spectroscopy*. These experiments involved J-modulated carbon-13 spin echoes* and it was observed that the expected two-dimensional spectrum was accompanied by some weak responses at coordinates where there should have been no responses at all (Fig. 1). These 'ghost' and 'phantom' signals could be traced to the influence of imperfections in the radiofrequency pulses. Ghost signals arise from imperfections in the 180° pulse, where some transverse magnetization escapes the effect of the pulse and it is not refocused. Consequently, it is possible to shift the phase of this pulse without affecting the ghost signals, and since it is well known that a 90° phase shift inverts the sense of a spin echo (3) the ghost responses can be eliminated by difference spectroscopy.

Phantom signals arise from the combined effects of imperfections in the 90° and 180° pulses, some longitudinal magnetization after the 90° pulse being converted by the imperfect 180° pulse into transverse magnetization. Phase inversion of the 180° pulse reverses the sense of this signal but not that of the spin echo, hence addition of the results of the two experiments eliminates the phantom signals. Both ghost and phantoms are 'exorcized' by the phase cycle.

90°(+X)	180°(+X)	Acq (+)
90°(+X)	180°(−Y)	Acq (−)
90°(+X)	180°(−X)	Acq (+)
90°(+X)	180°(+Y)	Acq (−)

Figure 1 demonstrates the operation of EXORCYCLE in a two-dimensional experiment where the pulse flip angles were deliberately mis-set in order to emphasize the ghost and phantom responses.

It could be argued that CYCLOPS and EXORCYCLE are not true cycles in the strictest sense but merely the result of combining two kinds of difference spectroscopy. There are, however, other experiments in which the procedure is undoubtedly a cycle. For example, NMR signals may be filtered so as to retain only those that have passed through a particular order n of multiple-quantum coherence by exploiting the fact that n-quantum coherence is n times as sensitive to radiofrequency phase shifts as single-quantum coherence. Rotation of the phases of the excitation pulses in small steps thus rotates the observed signal phase n-times as fast, and the receiver reference phase is programmed to follow at just the required rate. In these experiments it is no longer a question of eliminating small anomalous signals; in the double-quantum 'INADEQUATE' experiment (4) the signals of interest have to be extracted from signals 200 times more intense. Suppression ratios as high as 1000:1 can be achieved by filtration through double-quantum coherence, but the necessary phase cycle becomes quite elaborate.

Phase-cycling experiments are perhaps most clearly visualized in terms of *coherence transfer pathways* (5). This is a method for sketching out the changes in order of coherence caused by the various radiofrequency pulses in a sequence. The resulting coherence transfer map provides an invaluable guide to the design of a phase cycle to accomplish a specific purpose, for example to promote coherence transfer through four-quantum coherence while blocking other paths (Fig. 2). It turns out to be useful to distinguish the *sign* of a particular order of coherence. In homonuclear chemical shift correlation spectroscopy (COSY) there are two coherence transfer pathways, one involving +1 order in the evolution period, the other −1 order. They can be separated by setting up a phase cycle which relates them to the order of coherence (+1) observed during the detection period. (This is the same as separation of *echo* and *antiecho* responses.) One very practical advantage of the coherence transfer pathway viewpoint is that it demonstrates that phase cycles can sometimes be simplified. For example, the common four-step cycle which samples all four orthogonal phases may sometimes be replaced with a three-step cycle which increments the phase in 120° steps. This is the case when the phase cycling is used to eliminate all transverse components of magnetization at some point in the sequence.

A rather different application of phase cycling is the technique of *time-proportional phase increments* (TPPI) used in multiple-quantum spectroscopy and two-dimensional spectroscopy (6, 7). In this scheme, if n is the order of coherence evolving during t_1, the radiofrequency phase is incremented by 90°/n for each increment in t_1. This may be thought of as a shift in the frequency Ω_0 of the rotating reference frame to a frequency $\Omega_0 - \Omega_c$ where Ω_c is half the spectral width in the F_1 dimension. For a negative order of coherence −n, the

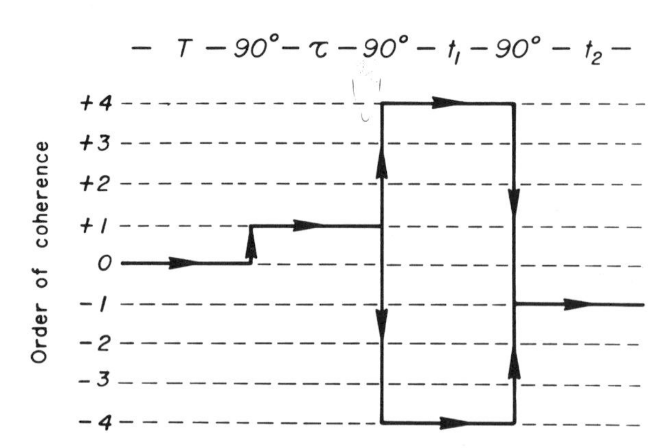

Fig. 2. The coherence transfer pathways appropriate to a homonuclear shift correlation (COSY) experiment filtered through 4-quantum coherence. The experiment starts with equilibrium Z-magnetization and ends with single-quantum coherence.

frequency shift is in the opposite sense. This is therefore a method of separating F_1 spectra with opposite signs of n, which would normally overlap and lead to confusion – the equivalent of quadrature detection in the F_1 dimension. It has been shown to be important for discriminating the sense of precession during the evolution period of two-dimensional experiments and for obtaining pure absorption-mode lineshapes (8). (See Lineshapes in two-dimensional spectra*.)

Optimum spectrometer sensitivity is usually preserved during phase-cycling, provided that signals are acquired at each stage of the cycle, and after allowance is made for signals deliberately rejected. However, it becomes very inefficient in the use of spectrometer time for nuclei and samples where adequate signal-to-noise can be achieved from a single free induction decay, where time averaging would not normally be used. Then it is important to bear in mind that there are alternative schemes for suppressing single-quantum transverse magnetization or for discriminating between different orders of multiple-quantum coherence*. One such alternative is to apply pulsed field gradients. If these are sufficiently intense they disperse single-quantum coherence vectors essentially uniformly in the transverse plane so that no net signal is detected. If they are used in suitably matched pairs (pulse widths in the ratio 1 : n) they can be used to select out n-quantum coherence (9, 10). Discrimination between echo and antiecho responses in a homonuclear shift correlation (COSY) experiment can be achieved by inserting matched field gradient pulses at the end of the evolution period and the beginning of the detection period (10).

REFERENCES

1. D. I. Hoult and R. E. Richards, *Proc. Roy. Soc. (Lond.) A* **344**, 311 (1975).
2. G. Bodenhausen, R. Freeman and D. L. Turner, *J. Magn. Reson.* **27**, 511 (1977).
3. S. Meiboom and D. Gill, *Rev. Sci. Instr.* **29**, 688 (1958).
4. A. Bax, R. Freeman and T. Frenkiel, *J. Am. Chem. Soc.* **103**, 2102 (1981).
5. G. Bodenhausen, H. Kogler and R. R. Ernst, *J. Magn. Reson.* **58**, 370 (1984).
6. G. Drobny, A. Pines, S. Sinton, D. Weitekamp and D. Wemmer, *Faraday Division Chem. Soc. Symp.* **13**, 49 (1979).
7. G. Bodenhausen, R. L. Vold and R. R. Vold, *J. Magn. Reson.* **37**, 93 (1980).
8. D. Marion and K. Wüthrich, *Biochem. Biophys. Res. Commun.* **113**, 967 (1983).
9. A. Bax, P. G. de Jong, A. F. Mehlkopf and J. Smidt, *Chem. Phys. Lett.* **69**, 567 (1980).
10. P. B. Barker and R. Freeman, *J. Magn. Reson.* **64**, 334 (1985).

Cross-references

Difference spectroscopy
Lineshapes in two-dimensional spectra
Multiple-quantum coherence
Quadrature detection
Spin-echoes
Time averaging
Two-dimensional spectroscopy

Polarization Transfer

Many interesting nuclei, for example carbon-13, nitrogen-15 and silicon-29, suffer from poor inherent sensitivity* compared with proton NMR. This is partly a result of low natural isotopic abundance and partly because of their low magnetogyric ratios. Signal intensity increases with γ^3, since the precession frequency (and hence the induced voltage), the magnetic moment and the Boltzmann populations are all proportional to γ. From the earliest days of NMR it was realized that considerable improvements in sensitivity could be achieved by artificially increasing the Boltzmann factors. These methods include prepolarization in an intense magnetic field, using the Overhauser effect to take advantage of the much higher Boltzmann factor for electron spins, or polarization transfer through Hartmann–Hahn contact in the rotating frame. (See Hartmann–Hahn experiment*.)

The instrumental requirements for the electron–nucleus Overhauser effect are too complicated to be generally applicable to high resolution NMR, since they require simultaneous irradiation with a radiofrequency and with microwaves. The corresponding nuclear Overhauser effect* is widely used for carbon-13 spectroscopy, but suffers some practical drawbacks, particularly for nitrogen-15 and silicon-29. Since it requires that the weak nucleus be relaxed predominantly by its dipolar coupling to protons, the actual enhancement often falls short of the theoretical value. Now, nitrogen-15 and silicon-29 have negative magnetogyric ratios so that enhancement is always accompanied by signal inversion; thus incomplete enhancement can yield a near-zero signal.

The expression for the maximum nuclear Overhauser enhancement factor

$$E = 1 + \tfrac{1}{2}(\gamma_I/\gamma_S)$$

turns out to be less advantageous than the enhancement expected (γ_I/γ_S) for the polarization transfer methods described below. For the example of nitrogen-15, the maximum nuclear Overhauser effect would be -4, whereas polarization transfer from protons would give an enhancement of 10.

We consider here the polarization transfer experiment known as 'INEPT' (Insensitive Nuclei Enhanced by Polarization Transfer); a closely related technique, DEPT (Distortionless Enhancement by Polarization Transfer) involves the creation of multiple-quantum coherence* and is rather more difficult to analyse. This experiment is analysed in the section on multiplicity determination*. But first, it is helpful to consider a simpler experiment, usually called selective population transfer (SPT) where a selective 180° spin inversion pulse is applied to a single NMR transition (1). Consider a coupled heteronuclear two-spin system consisting of four energy levels (Fig. 1). At Boltzmann equilibrium the populations on these levels can be written (to a very good approximation) as

$$\begin{aligned}
&(1)\ -\Delta - \delta\\
&(2)\ -\Delta + \delta\\
&(3)\ +\Delta - \delta\\
&(4)\ +\Delta + \delta,
\end{aligned}$$

where 2Δ represents the population difference across proton (I) transitions and 2δ the population difference across the transitions of the weak nucleus (S). Clearly, $\Delta/\delta = \gamma_I/\gamma_S$. Suppose we invert the populations across one proton transition, say 2–4, leaving transition 1–3 unaffected. The new, non-equilibrium populations will be

$$\begin{aligned}
&(1)\ -\Delta - \delta\\
&(2)\ +\Delta + \delta\\
&(3)\ +\Delta - \delta\\
&(4)\ -\Delta + \delta.
\end{aligned}$$

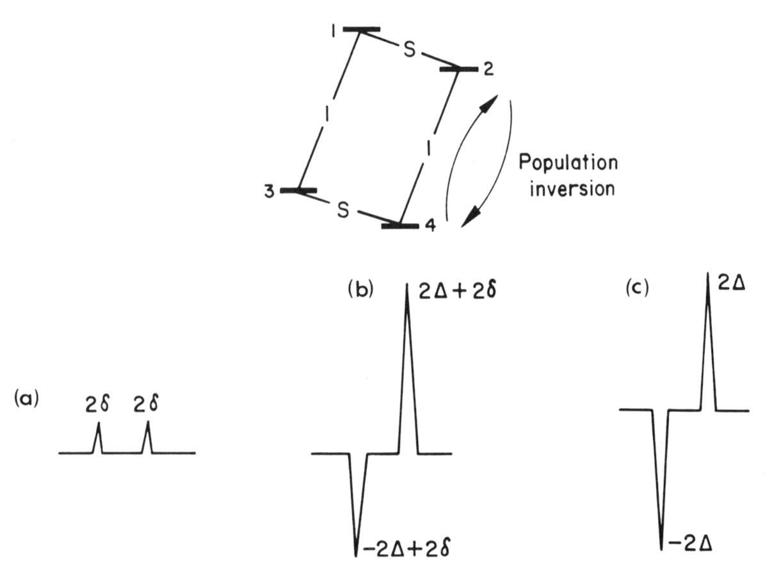

Fig. 1. Energy level diagram for an IS spin system. A selective 180° pulse causes a population inversion across one of the I-spin transitions perturbing the intensities of the S-spin transitions. From an initial intensity pattern (a) this causes the perturbed intensity pattern (b). The difference is shown in (c). The sensitivity advantage is $\Delta/\delta = \gamma_I/\gamma_S$.

Since they share energy levels in common with the proton transitions, the S transitions are considerably enhanced in intensity, the 1–2 transition showing a population difference $+2\Delta + 2\delta$ and the 3–4 transition a population difference $-2\Delta + 2\delta$. The *changes* in intensity due to the SPT effect are therefore $+\Delta/\delta$ and $-\Delta/\delta$ relative to the unperturbed intensities. There has been an enhancement of γ_I/γ_S in one case, and an equal enhancement plus inversion in the other. Many of the polarization transfer methods and related experiments in two-dimensional spectroscopy* show this 'up–down' intensity pattern with enhancements given by the ratio of the gyromagnetic ratios.

Selective population transfer has been used as a method of assigning the protons directly attached to a given carbon-13 spin, a type of chemical shift correlation* experiment (2), normally carried out nowadays by two-dimensional spectroscopy. It can also be used to determine the relative signs of spin coupling constants by a method analogous to that described under selective decoupling*.

Selective population transfer is hardly a general method of signal enhancement since the appropriate proton transition has to be found by a tedious trial-and-error search (it is one of the very weak satellites of the main proton

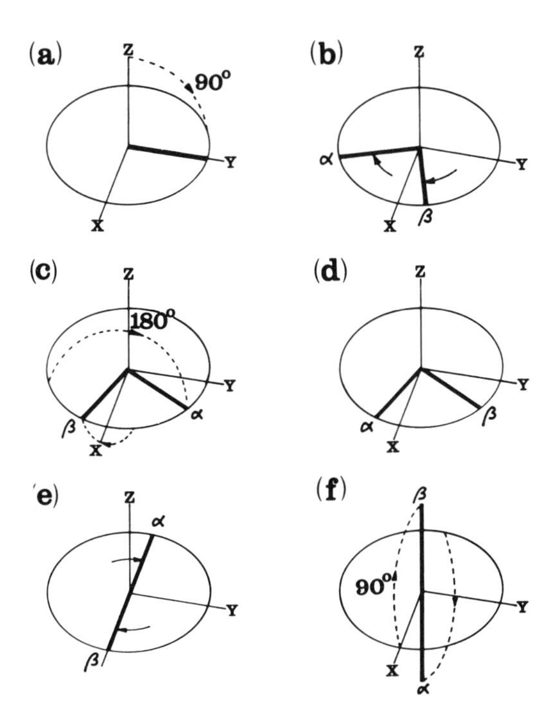

Fig. 2. Vector model representing the I-spin preparation in the 'INEPT' experiment. The two I-spin vectors α and β are allowed to precess for a time $\tau = 1/(4J_{IS})$ until they are 90° out of phase (b). Then a 180° refocusing pulse (c) and a 180° S-spin inversion pulse (d) are applied simultaneously. After a further period τ the two I-spin vectors are antiparallel (e) and a 90°(Y) pulse realigns them along the $\pm Z$ axes. This corresponds to a selective population inversion as in Fig. 1.

resonance). The INEPT experiment enables the same result to be achieved without any frequency adjustment of the proton excitation – hard pulses are used throughout. The vector model* is the simplest way to analyse the experiment (Fig. 2). In a frame rotating at the proton transmitter frequency the two proton satellite resonances may be represented by two vectors, α and β. An initial 90°(X) pulse aligns them along the Y axis and they then precess in the X–Y plane at frequencies $(\delta + \frac{1}{2}J)$ and $(\delta - \frac{1}{2}J)$. After an interval $\tau = 1/(4J)$ they will have precessed to positions which depend on both δ and J, but the angle between the two vectors will be 90°. A 180° pulse applied to protons flips the vectors into mirror-image positions and at the same instant a 180° pulse applied to the S spins interchanges the α and β labels on the proton vectors, so that the one which initially precessed at $(\delta + \frac{1}{2}J)$ now precesses at $(\delta - \frac{1}{2}J)$ and vice versa. An equal time τ later, the two vectors are aligned along the ±X axes of the rotating frame; the chemical shift effect is refocused but the divergence due to J-coupling persists. This is why the enhancement is found to be independent of chemical shift. Now a 90° pulse about the Y axis converts the situation into a mere population disturbance – one vector is along +Z, corresponding to Boltzmann equilibrium, the other is along −Z, corresponding to a population inversion. This is now the same situation as in the selective population transfer experiment, and a 90° read pulse applied to the S spins reveals an 'up–down' intensity pattern with enhanced intensities. Usually the natural (unenhanced) signal of the S spins is cancelled by subtracting free induction decays obtained with 90°(+Y) and 90°(−Y) for the final proton pulse. The signal intensities are then enhanced by $\pm(\gamma_I/\gamma_S)$. The pulse sequence is set out in Fig. 3. Carbon-13 spectra of pyridine are illustrated in Fig. 4 in order to compare the enhancement achieved by the nuclear Overhauser effect with that obtained by the INEPT experiment (3).

If the coupled S spin spectrum is to be examined, the 'up–down' patterns seldom cause any difficulty; occasionally two adjacent multiplets may overlap and cause partial cancellation. However, if the proton-decoupled S-spin spectrum is required, switching on the decoupler causes signal cancellation unless a further time interval $2\tau_2$ is introduced to allow the antiphase S-spin vectors to

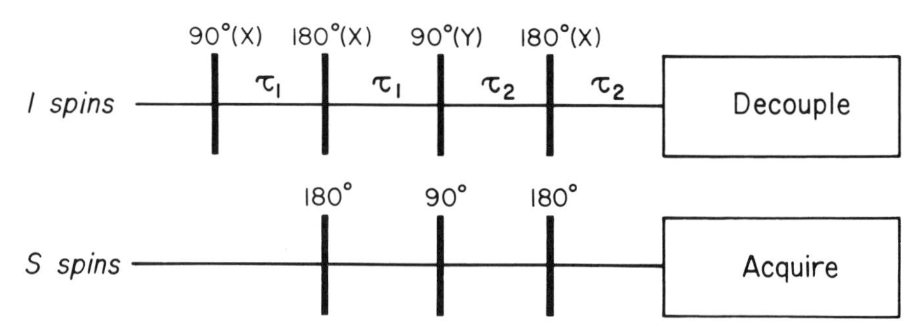

Fig. 3. Pulse sequence for the 'refocused INEPT' experiment. The vector model of Fig. 2 accounts for the disturbance of the energy level populations; the 90° pulse on the S-spins converts this into antiphase S-spin magnetization. The second period $2\tau_2$ brings these S-spin vectors into phase so that a decoupling field can be applied without signal cancellation.

precess into a parallel orientation. In order to avoid phase shifts which depend on the S-spin chemical shift, it is preferable to introduce refocusing and spin inversion 180° pulses at the centre of this interval (Fig. 3). If there is only a single proton attached to the S spin, then the condition for optimum signal strength is $\tau_2 = 1/(4J)$ but if two or three protons are involved at some sites, then a compromise setting $\tau_2 = 1/(6J)$ is to be preferred. Efficient polarization transfer depends on choosing a suitable τ_1 delay, based on an estimate of the J_{IS} coupling constant. A small mis-set can be tolerated, since the intensity, which depends on $\sin(2\pi J_{IS}\tau_1)$, is only changing slowly at this point, but if there is a wide range of J_{IS} values in the molecule, the relative intensities are uneven and some lines may vanish. One expedient is to set τ_1 shorter than $1/(4J)$ for the largest J value so that all sites yield intensities appropriate to the early linear region of the sinewave, thus avoiding any null conditions. Another remedy is to repeat the experiment at two or more different τ_1 values.

It is more difficult to obtain polarization transfer through weak long-range couplings since the corresponding τ_1 value must be proportionately longer and appreciable magnetization may be lost by spin–spin relaxation. Clearly, polarization transfer by this technique presupposes a resolved J-coupling in the

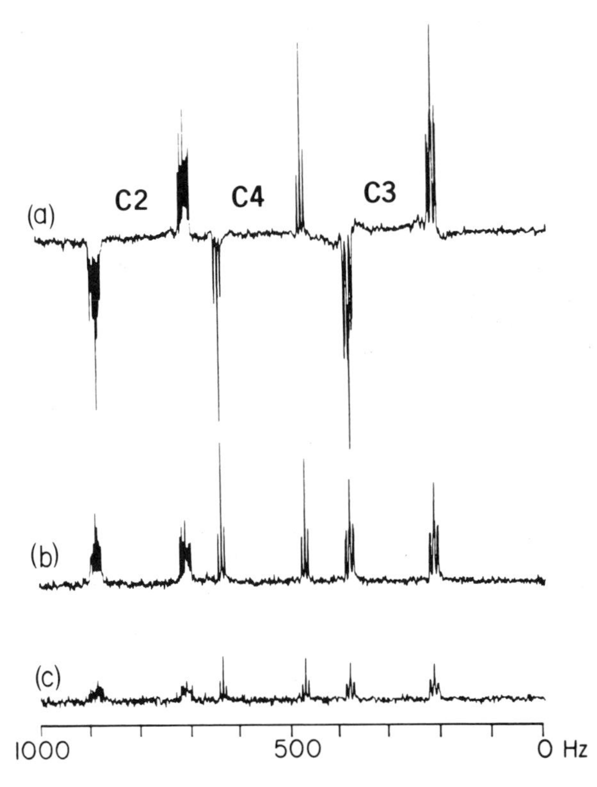

Fig. 4. (a) INEPT polarization transfer from protons to carbon-13 in pyridine (without proton decoupling and the refocusing stage $2\tau_2$). For comparison, (b) shows the same spectrum with the enhancement by the nuclear Overhauser effect and (c) shows the conventional unenhanced spectrum.

conventional I-spin spectrum. Furthermore, the proton–proton couplings can interfere with the experiment since they are comparable in magnitude to the long-range J_{IS} coupling.

Sensitivity enhancement is not the only application of the INEPT experiment. It can be used to measure the spin–lattice relaxation times of protons indirectly by observing the enhancement of the attached carbon-13 resonance (4). This might be useful where the proton spectrum is overcrowded but the carbon-13 spectrum well-resolved.

When the analysis of the polarization transfer experiment is extended to cases where there are two or three attached protons, it is easy to see that the corresponding S-spin vectors diverge two or three times as rapidly as those appropriate to the IS case. This has been made the basis of a whole family of techniques for multiplicity determination in carbon-13 spectroscopy.

Polarization transfer also provides a very efficient means of discrimination between coupled spin systems and non-coupled spins. For example a reversed version of INEPT can be used to detect carbon-13 or nitrogen-15 satellites in proton spectra while suppressing the much stronger signals from protons attached to carbon-12 or nitrogen-14 nuclei (5). In a similar manner, suppression of the strong water peak can be achieved by polarization transfer from enriched carbon-13 molecules, albeit with a severe sensitivity loss.

RELAXATION EFFECTS

In practice, the sensitivity enhancement factor can be significantly higher than the predicted ratio γ_I/γ_S. This is because time averaging* is used and the rate at which the experiment can be repeated is a critical factor. Nitrogen-15 and silicon-29 have quite long spin–lattice relaxation* times compared with protons in the same material, so the conventional experiment can only be repeated at a rate which allows adequate S-spin relaxation. In the polarization transfer experiment, this natural signal is cancelled anyway, all the intensity being obtained by transfer from protons which relax relatively fast, permitting high repetition rates. The overall sensitivity enhancement* is then $(\gamma_I/\gamma_S)(T_1^S/T_1^I)^{1/2}$. This advantage has been demonstrated by Rinaldi and Baldwin (6) who have shown that polarization transfer from deuterium spins (which have an unfavourably low magnetogyric ratio) is nevertheless advantageous because of the rapid spin–lattice relaxation of deuterium (attributable to the quadrupole moment).

REFERENCES

1. K. G. R. Pachler and P. L. Wessels, *J. Magn. Reson.* **12**, 337 (1973).
2. A. A. Maudsley and R. R. Ernst, *Chem. Phys. Lett.* **50**, 368 (1977).

3. G. A. Morris and R. Freeman, *J. Amer. Chem. Soc.* **101**, 760 (1979).
4. J. Kowalewski and G. A. Morris, *J. Magn. Reson.* **47**, 331 (1982).
5. T. H. Mareci and R. Freeman, *J. Magn. Reson.* **44**, 572 (1981).
6. P. L. Rinaldi and N. J. Baldwin, *J. Am. Chem. Soc.* **104**, 5791 (1982).

Cross-references

Hartmann–Hahn experiment
Multiple-quantum coherence
Multiplicity determination
Nuclear Overhauser effect
Shift correlation
Selective decoupling
Sensitivity
Sensitivity enhancement
Spin–lattice relaxation
Time averaging
Two-dimensional spectroscopy
Vector model

Product Operator Formalism

Wherever possible in this book, the simplest, non-mathematical treatment has been adopted. The majority of pulsed NMR experiments have been described in terms of extensions of the vector model* first introduced by Bloch. In a few applications, notably those involving multiple-quantum coherence*, this model breaks down, or at least has to be extended in an *ad hoc* manner. The general theory to describe the response to an arbitrary pulse sequence is the *density matrix* or *density operator* treatment (1, 2). Unfortunately, this becomes very unwieldy for systems of several coupled spins, and very quickly gets out of touch with physical intuition which has been our principal guide in this book. It may be that the forbidding aspect of density matrix theory has been responsible for the initial reluctance of many chemists and biochemists to take up the new techniques of two-dimensional spectroscopy.

Fortunately, there is a more pictorial approach, recently championed by Sørensen *et al.* (3), which allows the new spin gymnastics to be treated formally without losing sight of the physical interpretation so important for our sanity. It is based on the decomposition of the density operator into a linear combination of products of spin angular momentum operators (4). It is applicable to weakly-coupled spin systems. With this shorthand algebra, the fate of the various operators can be followed throughout a complex sequence of pulses and free precessions, throwing light on the details of the time evolution of the particular experiment. Lallemand (5) has suggested a tree-like pictorial representation to aid this kind of visualization.

For simplicity, we restrict ourselves here to the weakly coupled two-spin system IS, writing down the 16 product operators,

E/2 (where E is the unity operator)
I_X X-component of I-spin magnetization
I_Y Y-component of I-spin magnetization
I_Z Z-component of I-spin magnetization (populations)
S_X X-component of S-spin magnetization

S_Y Y-component of S-spin magnetization
S_Z Z-component of S-spin magnetization (populations)
$2I_XS_Z$ Antiphase I-spin magnetization
$2I_YS_Z$ Antiphase I-spin magnetization
$2I_ZS_X$ Antiphase S-spin magnetization
$2I_ZS_Y$ Antiphase S-spin magnetization
$2I_ZS_Z$ Longitudinal two-spin order
$2I_XS_X$ Two-spin coherence
$2I_YS_Y$ Two-spin coherence
$2I_XS_Y$ Two-spin coherence
$2I_YS_X$ Two-spin coherence.

The term $2I_XS_Z$ represents the X component of the I-spin magnetization split into two antiphase components corresponding to the two possible spin states of S. Such operators can be represented by the vector model but the remaining five product operators cannot be easily represented by vectors.

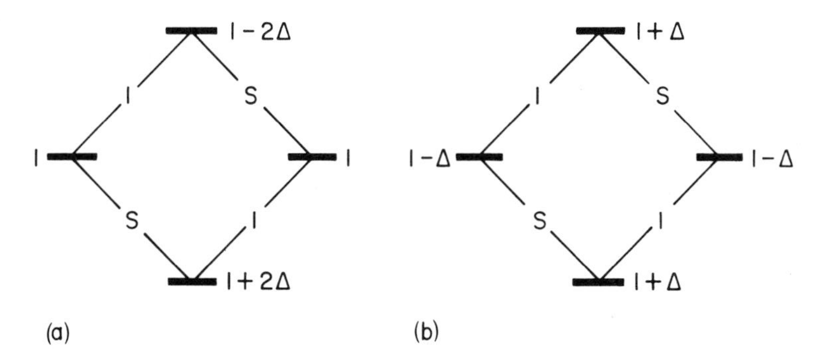

Fig. 1. (a) Energy level populations appropriate to a homonuclear IS spin system at Boltzmann equilibrium ($\Delta \ll 1$). (b) Populations corresponding to longitudinal two-spin order represented by the product operator term $2I_ZS_Z$.

Longitudinal two-spin order $2I_ZS_Z$ is a specific disturbance of the populations of the four energy levels, having no net polarization. If the normal Boltzmann populations are represented as in Fig. 1(a), with population differences of 2Δ across each transition, then this *J-ordered state* has the populations indicated in Fig. 1(b). Both the I-spin doublet and the S-spin doublet have population disturbances such that a small flip angle read pulse would indicate an 'up–down' pattern of intensities. This is a common occurrence in certain polarization transfer* experiments.

Two-spin coherence $2I_XS_X$ is a concerted motion of the I and S spins that induces no signal in the NMR receiver coil, but can only be detected indirectly by two-dimensional spectroscopy*. It is a superposition of zero-quantum coherence (simultaneous I and S spin flips in opposite senses) and double-quantum coherence (flips in the same sense). Pure zero-quantum coherence corresponds to linear combinations of these product operators

$$2I_XS_X + 2I_YS_Y \quad \text{or} \quad 2I_YS_X - 2I_XS_Y.$$

Pure double-quantum coherence corresponds to the alternative linear combinations

$$2I_XS_X - 2I_YS_Y \quad \text{or} \quad 2I_XS_Y + 2I_YS_X.$$

We shall see below that one of the great strengths of the product operator formalism is its ability to account for experiments which involve multiple-quantum coherence.

SIGN CONVENTIONS FOR ROTATIONS

For the vast majority of NMR experiments, the outcome is independent of the choice of the direction of precession of spins about magnetic fields. When using the vector model we adopted the widely used convention that (for a positive gyromagnetic ratio) a vector M rotates about a field in the rotating frame* as in Fig. 2. Thus for a radiofrequency field B_1 applied along the +X axis, a 90° pulse rotates $+M_Z$ to $+M_Y$

$$+M_Z \xrightarrow{90°(+X)} +M_Y. \qquad [1]$$

Similarly, we chose to take the sense of free precession to be clockwise looking down on the X–Y plane

$$+M_Y \rightarrow +M_X \rightarrow -M_Y \rightarrow -M_X \qquad [2]$$

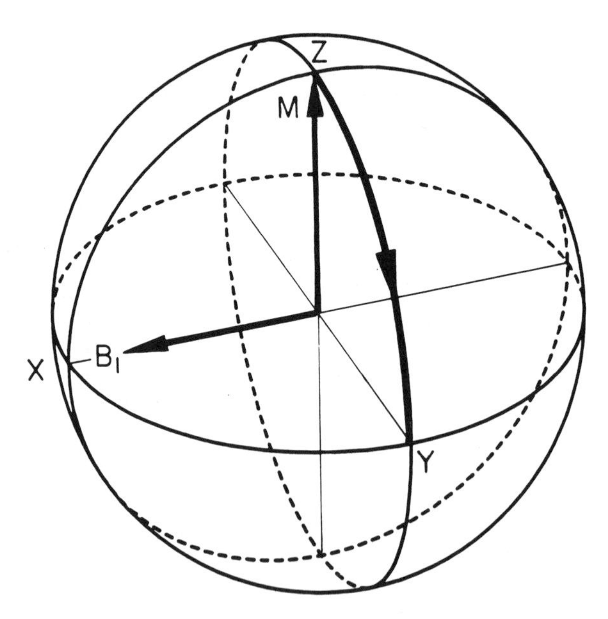

Fig. 2. Convention adopted for the sense of rotation of a magnetization vector M about a radiofrequency field B_1.

for a Larmor frequency higher than the frequency of the rotating frame (ΔB positive). This convention simplifies diagrams of magnetization trajectories by concentrating on the front quadrant of the unit sphere.

When it comes to mathematical treatments using density operators or product operators, the sense of rotation is rather less of an academic point, and two opposite schools of thought persist. Since the treatise of Sørensen *et al.* has become the standard article on the use of product operators in NMR pulse experiments, we adopt their sign convention, which is opposite to that of several other authors (2, 5, 6). In the product operator nomenclature, an operator (I_Z) is acted on by another operator I_X and the sign convention is *opposite* to that used above for magnetization vectors and fields (Fig. 3)

$$+I_Z \xrightarrow{+I_X} -I_Y. \qquad [3]$$

Similarly, a resonance offset effect causes a counterclockwise rotation looking down on the X–Y plane

$$+I_Y \xrightarrow{+I_Z} -I_X \xrightarrow{+I_Z} -I_Y \xrightarrow{+I_Z} +I_X.$$

Finally, an operator $2I_ZS_Z$ has a specific sense of rotation

$$+I_X \xrightarrow{+2I_ZS_Z} +2I_YS_Z$$

$$+I_Y \xrightarrow{+2I_ZS_Z} -2I_XS_Z. \qquad [4]$$

These conventions are illustrated pictorially in Fig. 3 and embodied in Table 1.

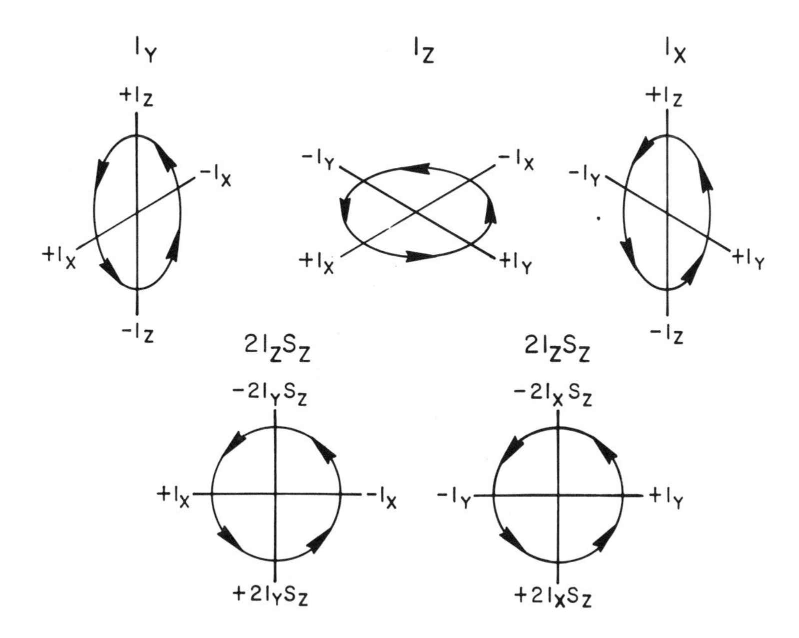

Fig. 3. Sign conventions for the evolution operators I_Y, I_Z, I_X and $2I_ZS_Z$ acting on the product operators I_X, I_Y, I_Z, $2I_XS_Z$ and $2I_YS_Z$. This schematic diagram is equivalent to Table 1.

MANIPULATION OF PRODUCT OPERATORS

For the majority of pulsed NMR experiments in liquids, we are concerned with three main types of evolution – rotation by a radiofrequency pulse, rotation due to chemical shift, and rotation due to spin–spin coupling. Although the operation of a given pulse sequence clearly depends on the time-ordering of the pulses and the intervening periods of free precession, during these latter periods we are at liberty to change the ordering of chemical shift and spin coupling evolutions, provided that the spin system is weakly coupled. The corresponding terms in the Hamiltonian are said to *commute*. We may speak of a *cascade* (7) of chemical shift or spin coupling terms where the time-ordering is immaterial. Furthermore, a non-selective radiofrequency pulse acting on both the I and S spins may be broken down into a cascade of two pulses acting selectively on the I spins and the S spins, and the relative ordering does not matter.

Table 1. The effect of one of the evolution operators (top row) acting on one of the operators describing the state of the spin system (left-hand column).

	I_X	I_Y	I_Z	S_X	S_Y	S_Z	$2I_ZS_Z$
I_X	E/2	$-I_Z$	I_Y	E/2	E/2	E/2	$2I_YS_Z$
I_Y	I_Z	E/2	$-I_X$	E/2	E/2	E/2	$-2I_XS_Z$
I_Z	$-I_Y$	I_X	E/2	E/2	E/2	E/2	E/2
S_X	E/2	E/2	E/2	E/2	$-S_Z$	S_Y	$2I_ZS_Y$
S_Y	E/2	E/2	E/2	S_Z	E/2	$-S_X$	$-2I_ZS_X$
S_Z	E/2	E/2	E/2	$-S_Y$	S_X	E/2	E/2
$2I_ZS_Z$	$-2I_YS_Z$	$2I_XS_Z$	E/2	$-2I_ZS_Y$	$2I_ZS_X$	E/2	E/2
$2I_XS_Z$	E/2	$-2I_ZS_Z$	$2I_YS_Z$	$-2I_XS_Y$	$2I_XS_X$	E/2	I_Y
$2I_YS_Z$	$2I_ZS_Z$	E/2	$-2I_XS_Z$	$-2I_YS_Y$	$2I_YS_X$	E/2	$-I_X$
$2I_ZS_X$	$-2I_YS_X$	$2I_XS_X$	E/2	E/2	$-2I_ZS_Z$	$2I_ZS_Y$	S_Y
$2I_ZS_Y$	$-2I_YS_Y$	$2I_XS_Y$	E/2	$2I_ZS_Z$	E/2	$-2I_ZS_X$	$-S_X$
$2I_XS_X$	E/2	$-2I_ZS_X$	$2I_YS_X$	E/2	$-2I_XS_Z$	$2I_XS_Y$	E/2
$2I_XS_Y$	E/2	$-2I_ZS_Y$	$2I_YS_Y$	$2I_XS_Z$	E/2	$-2I_XS_X$	E/2
$2I_YS_X$	$2I_ZS_X$	E/2	$-2I_XS_X$	E/2	$-2I_YS_Z$	$2I_YS_Y$	E/2
$2I_YS_Y$	$2I_ZS_Y$	E/2	$-2I_XS_Y$	$2I_YS_Z$	E/2	$-2I_YS_X$	E/2

RADIOFREQUENCY PULSES

During a radiofrequency pulse, the chemical shifts and spin–spin coupling constants can be imagined to be 'switched off' and the rotation is about an axis in the X–Y plane, normally the X axis. If necessary, we can consider rotation about a tilted radiofrequency field B_{eff}. Consider, first of all, an excitation pulse $\beta(X)$ acting on the Z-magnetization of the I spins, represented by I_Z. Thus

$$I_Z \xrightarrow{\beta I_X} I_Z \cos\beta - I_Y \sin\beta. \qquad [5]$$

In the common example of a 90° pulse, this generates pure −Y magnetization; if it is a 180° pulse, there is a population inversion ($-I_Z$). Analogous expressions apply to pulses applied to the S spins, and for a non-selective pulse we would cascade the two rotations

$$+I_Z \xrightarrow{(\pi/2)I_X} -I_Y \qquad [6]$$

$$+S_Z \xrightarrow{(\pi/2)S_X} -S_Y. \qquad [7]$$

A more complicated example occurs in the INEPT (8) experiment for polarization transfer in a heteronuclear IS system, commonly used to enhance the sensitivity of carbon-13 or nitrogen-15 spectra. In the key step of this sequence, I-spin magnetization vectors are prepared in an antiphase alignment along the ±X axes of the rotating frame and a $\pi/2$ pulse is applied to the I spins about the +Y axis. This rotation can be written

$$2I_XS_Z \xrightarrow{(\pi/2)I_Y} -2I_ZS_Z. \qquad [8]$$

This creates longitudinal two-spin order, usually represented by I-spin vectors aligned along the ±Z axes. These population disturbances affect the S spins through the common energy levels, and these perturbations can be 'read' by a $\pi/2$ pulse applied to the S spins

$$-2I_ZS_Z \xrightarrow{(\pi/2)S_X} 2I_ZS_Y. \qquad [9]$$

We observe that the S spin doublet has one line inverted and one line in the usual sense; in the case where the I-spins are protons and the S-spins are carbon-13, the 4 : 1 population advantage is transferred from protons to carbon-13, improving the sensitivity.

CHEMICAL SHIFTS

The evolution due to chemical shift effects may be represented by the operator equation

$$I_Y \xrightarrow{(2\pi\delta_I t)I_Z} I_Y \cos(2\pi\delta_I t) - I_X \sin(2\pi\delta_I t), \qquad [10]$$

where δ_I is the shift of the I-spin resonance measured from the transmitter frequency. Note the sense of rotation is opposite to that used in the vector model.

Chemical shifts of the S spins are handled in analogous fashion. For heteronuclear systems a separate rotating reference frame is assumed for each spin, the chemical shifts being measured with respect to the appropriate transmitter frequencies in their respective frames.

SPIN–SPIN COUPLING

According to the vector model, spin–spin coupling causes a divergence of I spin vectors at rates $\pm\frac{1}{2}J_{IS}$ with respect to a hypothetical vector precessing at the chemical shift frequency. In the product operator formalism coupling is represented by

$$I_Y \xrightarrow{(\pi J_{IS}\tau)2I_ZS_Z} I_Y \cos(\pi J_{IS}\tau) - 2I_XS_Z \sin(\pi J_{IS}\tau). \qquad [11]$$

If the interval τ is chosen such that $\tau = 1/(2J_{IS})$ then the cosine term is zero and we are left with

$$I_Y \xrightarrow{(\pi/2)2I_ZS_Z} -2I_XS_Z, \qquad [12]$$

that is to say, two I-spin magnetization vectors aligned in opposition along the $\pm X$ axes. We may then consider another period of free precession

$$-2I_XS_Z \xrightarrow{(\pi J_{IS}\tau)2I_ZS_Z} -2I_XS_Z \cos(\pi J_{IS}\tau) - I_Y \sin(\pi J_{IS}\tau). \qquad [13]$$

If we make this second interval $\tau = 1/(2J_{IS})$ then we find that the two vectors are realigned along the $-Y$ axis

$$-2I_XS_Z \xrightarrow{(\pi/2)2I_ZS_Z} -I_Y. \qquad [14]$$

With these simple rules the evolution of spin systems under the influence of a pulse sequence can be followed by evaluating the effect of the seven evolution operators I_X, I_Y, I_Z, S_X, S_Y, S_Z and $2I_ZS_Z$ on the operators describing the state of the spin system (15 in all). Table 1 shows the results. Then an 'evolution tree' can be constructed (5) where by convention each left-hand branch represents the cosine term of the evolution equations [5], [10], [11] or [13] while the right hand side represents the sine term (evaluated from Table 1). When the two operators commute (E/2 in Table 1) then there is no change in that term. Note that the unaffected term is always associated with cosine, the affected term is associated with sine.

CORRELATION SPECTROSCOPY (COSY)

For the worked example we take the homonuclear chemical shift correlation experiment (COSY) for a system of two coupled spins I and S. This simple system illustrates the essential points, additional spins merely make the spectrum more complicated by increasing the number of resonances and by splitting the IS peaks through 'passive' couplings J_{IQ} and J_{SQ}, etc. A second important simplification is to drop S_Z from the initial density matrix, concentrating our attention on what happens to I_Z, since the problem is symmetrical with respect to the two spin systems.

The pulse sequence is deceptively simple

$$90°(+X) - t_1 - 90°(+X) - \text{Acquisition } (t_2). \qquad [15]$$

For the present purposes we may ignore the phase cycling* that is normally employed.

Chemical shift (I_Z) and spin coupling operators ($2I_ZS_Z$) may be applied in any order; in the acquisition period t_2 precession of the S spins is also considered, since by then there has been some transfer of coherence from the I spins. The evolution tree is set out in Fig. 4 showing the eight stages of branching, leading to thirteen terms in the final density operator. Of these, nine represent unobservable quantities – longitudinal magnetization (Z), multiple-quantum coherence (M) and antiphase magnetizations (A). It is the remaining four terms that are important; they can be grouped in pairs

$$D = \sin(2\pi\delta_I t_1)\cos(\pi J_{IS}t_1)\cos(\pi J_{IS}t_2)[I_X\cos(2\pi\delta_I t_2) + I_Y\sin(2\pi\delta_I t_2)] \quad [16]$$

$$C = \sin(2\pi\delta_I t_1)\sin(\pi J_{IS}t_1)\sin(\pi J_{IS}t_2)[S_X\cos(2\pi\delta_S t_2) + S_Y\sin(2\pi\delta_S t_2)]. \quad [17]$$

It is clear that D represents coherence that has processed at frequencies close to the chemical shift δ_I in both t_1 and t_2. These are the *diagonal* peaks. The term in square brackets indicates that there is phase modulation in the t_2 interval. The significance of the two J-modulation terms, $\cos(\pi J_{IS}t_1)$ and $\cos(\pi J_{IS}t_2)$ may not be immediately apparent. They may be converted by means of trigonometrical identities

$$\sin(2\pi\delta_I t_1)\cos(\pi J_{IS}t_1) = 0.5[\sin(2\pi\delta_I + \pi J_{IS})t_1 + \sin(2\pi\delta_I - \pi J_{IS})t_1]. \quad [18]$$

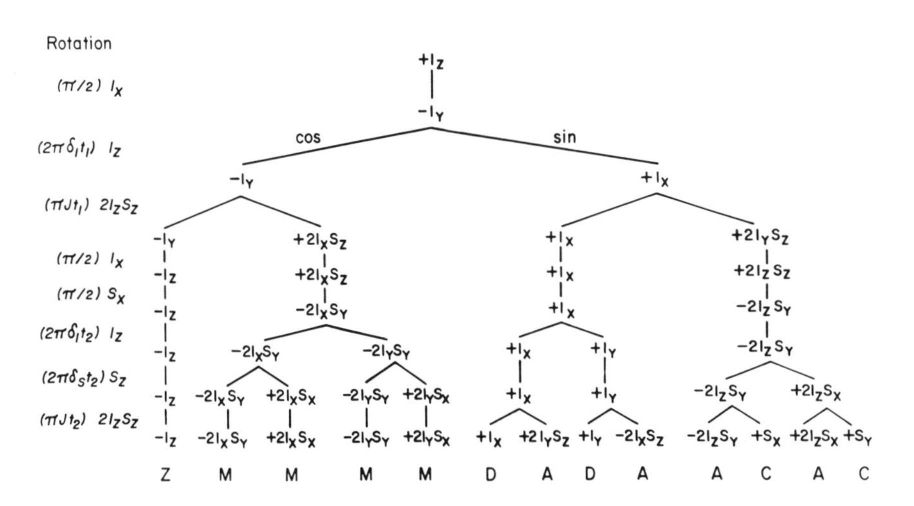

Fig. 4. Evolution of product operators appropriate to the homonuclear shift correlation experiment 'COSY'. For simplicity, the evolution of S_Z is omitted; it may be deduced from considerations of symmetry. Each left-hand branch implies multiplication by the cosine of the argument shown in the left-hand column, for example $-I_Y\cos(2\pi\delta_I t_1)$, while each right-hand branch implies multiplication by the corresponding sine term. The final 13 product operators are identified as Z-magnetization (Z), multiple-quantum coherence (M), antiphase magnetization (A), diagonal peaks (D) or cross-peaks (C).

This represents a response in the F_1 dimension, centred at δ_I and split into a doublet (J_{IS}), both lines having the same phase. Similarly, the terms in t_2 may be combined to show that there is an in-phase doublet in the F_2 dimension. This is the familiar square pattern of lines straddling the principal diagonal.

By contrast, eqn [17] represents coherence that originated at frequencies near the chemical shift δ_I but which was detected at frequencies near δ_S, and thus describes one of the *cross-peaks*. (The other cross-peak would have been predicted by following the fate of S_Z, neglected in our calculation.) In this case the trigonometrical identity is

$$\sin(2\pi\delta_I t_1)\sin(\pi J_{IS}t_1) = 0.5\,[\cos(2\pi\delta_I - \pi J_{IS})t_1 - \cos(2\pi\delta_I + \pi J_{IS})t_1]. \qquad [19]$$

This represents a response, centered at δ_I in the F_1 dimension, which is an antiphase doublet (J_{IS}). A similar identity shows that it is also an antiphase doublet in the F_2 dimension, centred at δ_S. The cross-peak is therefore a square pattern with the usual intensity alternation. Normally we adjust the spectrometer phase so that the cross-peaks are in the absorption mode; then the diagonal peaks are in dispersion (sine modulation).

The presence of the terms $\sin(\pi J_{IS}t_1)\sin(\pi J_{IS}t_2)$ has another interesting consequence. It predicts that cross-peaks will have low relative intensities unless both t_1 and t_2 are permitted to evolve for times comparable with $(\pi J_{IS})^{-1}$, whereas the diagonal peaks will be relatively strong. This is important when searching for correlations based on very small coupling constants. Sometimes, a fixed delay is introduced into the evolution period in order to emphasize the effects of very small couplings (9).

Since this has been an illustrative exercise, all the evolutions have been worked out explicitly. Once familiarity with product operator algebra has been acquired, it is not normally necessary to carry through the calculation to the bitter end. For example, we could choose to stop the COSY calculation immediately after the second pulse ($t_2 = 0$), recognizing that the I_X term will give an in-phase doublet in the F_2 dimension and that the $-2I_ZS_Y$ term will evolve to given an antiphase doublet in F_2.

REFERENCES

1. A. Abragam, *The Principles of Nuclear Magnetism.* Oxford University Press, 1961.

2. C. P. Slichter, *The Principles of Magnetic Resonance.* Springer: Berlin, 1978.

3. Ø. W. Sørensen, G. W. Eich, M. H. Levitt, G. Bodenhausen and R. R. Ernst, *Progr. NMR Spectroscopy* **16**, 163 (1983).

4. U. Fano, *Rev. Mod. Phys.* **29**, 74 (1957).

5. J. Y. Lallemand, *École d'Été sur la Spectroscopie en Deux Dimensions*, Orléans, France, 1984.

6. U. Haeberlen, *High Resolution NMR in Solids, Selective Averaging.* Academic Press: New York, 1976.
7. G. Bodenhausen and R. Freeman, *J. Magn. Reson.* **36**, 221 (1979).
8. G. A. Morris and R. Freeman, *J. Am. Chem. Soc.* **101**, 760 (1979).
9. A. Bax and R. Freeman, *J. Magn. Reson.* **44**, 542 (1981).

Cross-references

Multiple-quantum coherence
Phase cycling
Polarization transfer
Rotating frame
Two-dimensional spectroscopy
Vector model

Quadrature Detection

Suppose the nuclear magnetization is represented by a vector M which rotates clockwise in the rotating reference frame* in the X–Y plane. A phase-sensitive detector which measures the Y component would pick up a signal M cos ωt and would not be able to determine the sense of the rotation. Indeed there might have been two counter-rotating vectors of amplitude M/2 each. The introduction of a second phase-sensitive detector which measures the X component, described by M sin ωt, permits the sense of rotation to be determined. Since the two phase detectors are fed with a reference frequency which is phase-shifted by 90° in one of the detectors, the method is known as *quadrature detection*. Now, for reasons of simplicity, the early Fourier transform spectrometers used a single phase-sensitive detector, and were therefore unable to detect the sense of rotation of the nuclear magnetization vectors (the sign of the precession frequency). At the time this did not seem to be an important limitation, it simply meant that the transmitter frequency had to be set at one extreme end of the spectrum, otherwise positive frequencies were aliased and superimposed on negative frequencies. Unfortunately, it also meant that sensitivity was reduced by a factor $2^{1/2}$ because noise components were accepted over a bandwidth spanning $\pm\Delta f$ with respect to the transmitter frequency, whereas signals were gathered from a bandwidth Δf on only one side of the transmitter.

With quadrature detection, each sampling operation on the free induction decay* involves two measurements, and the signal is usually represented by a complex number $S_a + iS_b$. The transmitter frequency is set in the centre of the spectrum. The sampling rate is set equal to the spectral width SW, so that all NMR responses within the range $-\frac{1}{2}$SW to $+\frac{1}{2}$SW are properly digitized according to the Nyquist criterion (see Digitization*). Suitable audiofrequency filters remove noise components outside this range. Noise can only originate from regions of frequency space which encompass the NMR spectrum. A second advantage is that the maximum offset from resonance is now halved, so that the B_1 intensity is used more effectively; for the same maximum tilt of the effective field, the radiofrequency power may be reduced by a factor of four. Some schemes for solvent suppression* require that the transmitter frequency be set very close to the strong solvent peak; this is easy to achieve with quadrature detection. Indeed, modern Fourier spectrometers invariably incorporate quadrature detection.

CHANNEL IMBALANCE

It is no simple matter to construct two phase-sensitive detectors that are properly balanced in amplitude and exactly 90° out of phase. Imbalance leads to a slight admixture of the real component in the imaginary channel and vice versa. After a complex Fourier transformation this generates *quadrature images*, weak spurious signals symmetrically disposed on the other side of the transmitter frequency (zero frequency in the spectrum) with respect to the parent signal. If one phase sensitive detector is put out of action completely, then parent signals and images are of equal intensity; the system reverts to single-channel detection. There is also a spurious response which appears at zero frequency in the centre of the spectrum, usually known as the *quadrature glitch* and attributable to an incorrect d.c. balance of the free induction decays in the two channels.

It is important to reduce quadrature images and the quadrature glitch to a very low level compared with the parent signals. Fortunately, this can be achieved very effectively by CYCLOPS (Cyclically Ordered Phase Sequence) introduced by Hoult and Richards (1). This is described in detail in the section on phase cycling*. This scheme compensates very well for a small amplitude imbalance between the channels and for lack of exact orthogonality, so that the quadrature images are greatly attenuated. It also eliminates the zero-frequency quadrature glitch. There is no significant loss of sensitivity compared with four cycles of conventional time averaging*, but the method does inevitably increase the minimum time required to perform the experiment.

'PSEUDO-QUADRATURE' DETECTION

It is quite feasible to update a single-channel spectrometer so that it behaves as a quadrature detection system (2), at the same time avoiding the problems of channel imbalance. The transmitter frequency may be placed anywhere within the spectral width without introducing ambiguities in the sign of precession frequencies. This is achieved by stepping the reference phase of the receiver by 90° for each new sampling operation on the free induction decay; this is equivalent to a shift in the *apparent* precession frequencies of the nuclei by $\frac{1}{2}$SW Hz (2). This may be thought of as a shift in the frequency of the rotating reference frame by $\frac{1}{2}$SW Hz in one sense or the other. Then the transmitter frequency can be placed near the centre of the NMR spectrum while the effective receiver frequency is near one edge, so that all precession frequencies are perceived to have the same sign. Noise is still accepted from a bandwidth spanning $\pm$SW Hz, so no sensitivity advantage is enjoyed over a single-channel phase-sensitive detector.

This concept can be extended into the evolution period of a two-dimensional Fourier transform experiment, where it is usually called the *time-proportional phase increments* (TPPI) method. This is important in one scheme for obtaining pure absorption-mode lineshapes in two-dimensional spectra.

REFERENCES

1. D. I. Hoult and R. E. Richards, *Proc. Roy. Soc. (Lond.)* **A 344**, 311 (1975).
2. A. G. Redfield and S. D. Kunz, *J. Magn. Reson.* **19**, 250 (1975).

Cross-references

Digitization
Free induction decay
Lineshapes in two-dimensional spectra
Phase cycling
Rotating frame
Solvent suppression
Time averaging

Radiation Damping

The NMR phenomenon is an exceedingly small effect. Under normal circumstances the absorption of energy by the nuclear spins is very much smaller than the energy associated with the driving radiofrequency field. However, in a few extreme cases this inequality is not satisfied and then a new phenomenon may be observed known as *radiation damping*. When all the individual nuclear spins in a sample are excited by a 90° pulse and precess in phase, a current is induced in the receiver coil; this in turn generates a weak radiofrequency field which reacts on the nuclear spins. This field is in such a sense as to rotate a magnetization vector M_{XY} back towards the +Z axis, thus returning the spin system to equilibrium rather faster than expected from the relaxation processes alone. The result is that the free induction signal decays more rapidly than normal (see Free induction decay*), and the lines in the frequency-domain spectrum are slightly broadened. In many practical situations this effect is very small and may be neglected, but for high-sensitivity nuclei such as protons or fluorine, for high B_0 fields and concentrated samples, and for spectra where the intensity is concentrated into a single line, the radiation damping effect can be significant (1).

For a correctly tuned receiver coil the decay time constant due to radiation damping can be shown to be given by

$$T_{RD} = (2\pi\eta Q\gamma M_0)^{-1}, \qquad [1]$$

where η is the receiver coil filling factor, Q is the quality factor, and M_0 is the equilibrium nuclear magnetization. If this magnetization is distributed over many different lines in the high resolution spectrum, only the contribution from the line in question enters the formula for T_{RD}. A simple test for radiation damping is to detune the receiver coil, thus reducing the induced voltage by a factor approaching Q, and see if the line becomes narrower.

A quite common manifestation of radiation damping effects is the perturbation of the relative intensities of spin multiplet components from the expected binomial coefficients. Actually, it is the peak heights that are affected, the stronger multiplet components being preferentially broadened by radiation damping. Thus a 1:3:3:1 quartet shows relatively weaker central components as illustrated in Fig. 1 for the protons in a strong sample of ethanol observed at 500 MHz (2). The correct ratio of peak heights is observed for the carbon-13 satellites where radiation damping is negligible. The most likely case for line broadening by radiation damping is the H_2O signal from aqueous solutions. The tails of this very strong signal often extend over a considerable range of frequencies, probably because of broadening by radiation damping.

Radiation damping may also interfere with relaxation measurements when the detected signals are strong. Just after a slightly imperfect 180° inversion pulse, the current in the receiver coil due to radiation damping creates a continuous radiofrequency field which exerts a torque on the Z-magnetization (3). An increasing transverse component of magnetization is excited, increasing the radiation damping signal in an escalating process which eventually rotates the magnetization vector through the X–Y plane towards the +Z axis. Since a proper study of spin–lattice relaxation* requires the magnetization to be strictly longitudinal, precautions must be taken to avoid radiation damping.

Many high resolution NMR experiments now involve complicated pulse sequences which include periods of free nuclear precession. Radiation damping can then have quite insidious effects, because the nuclear precession is no longer strictly free if there is a continuous radiofrequency field created by the radiation damping current induced in the receiver coil.

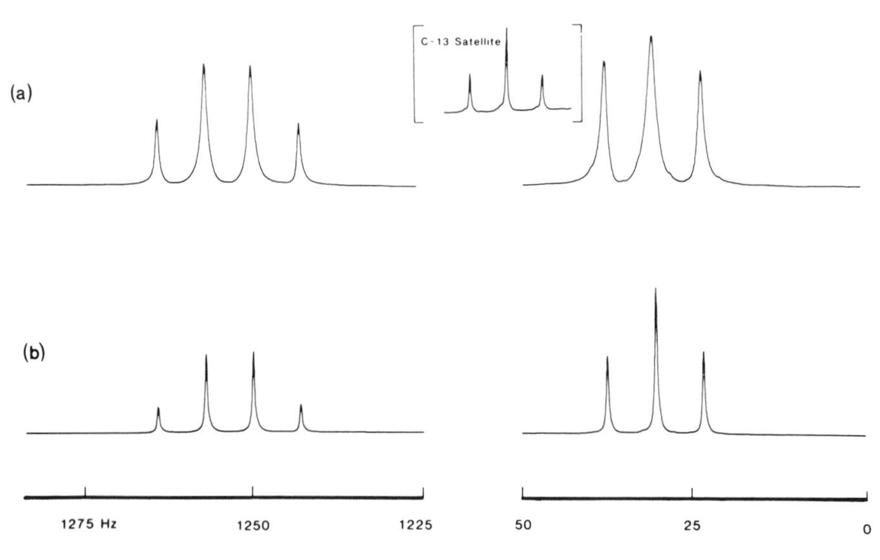

Fig. 1. 500 MHz proton spectrum of ethanol showing severe line broadening by radiation damping, most marked on the strongest lines. The expected binomial distribution of peak heights is grossly distorted (a) but returns to normal when the receiver coil is detuned (b) and for the weak carbon-13 satellites (inset). Dr T. Frenkiel, unpublished work.

The weak satellites in proton spectra (due to the presence of low-abundance carbon-13 or nitrogen-15 spins) are sometimes investigated by double-resonance difference spectroscopy* in order to remove the strong 'parent' peak from molecules containing carbon-12 or nitrogen-14. Radiation damping currents induced by this strong proton signal can have a secondary influence on the observed weak satellites if they lie close to the parent peak. This appears as a dephasing of the satellite signals giving them some dispersion-mode character (4).

Positive feedback onto the receiver coil amplifies the radiation damping current and greatly increases the consequent line broadening. This is the reason why regenerative oscillators (marginal oscillators) are not suitable for high resolution measurements. Very high levels of prepolarization of nuclei also induce strong radiation damping effects. Earth's field proton magnetometers which employ prepolarization in a high applied field exhibit free induction decays largely dominated by radiation damping.

REFERENCES

1. N. V. Bloembergen and R. V. Pound, *Phys. Rev.* **95**, 8 (1954).
2. T. Frenkiel, private communication.
3. A. Szöke and S. Meiboom, *Phys. Rev.* **113**, 585 (1959).
4. R. Freeman and W. A. Anderson, *J. Chem. Phys.* **42**, 1199 (1965).

Cross-references

Difference spectroscopy
Free induction decay
Spin–lattice relaxation

Radiofrequency Pulses

According to the vector model* of the magnetic resonance phenomenon, pulse excitation is regarded as a rotation of a macroscopic magnetization vector M about the X axis of the rotating frame through a 'flip angle' α rad. If resonance offset effects can be neglected, $\alpha = \gamma B_1 t_p$ where B_1 is the radiofrequency intensity and t_p is the pulse duration. This requires that $B_1 \gg \Delta B$ for all lines in the spectrum of interest. For proton NMR, a radiofrequency field intensity such that $\gamma B_1/2\pi \sim 10$ kHz might be typical; this necessarily entails a short duration pulse of the order of 10 μs. For carbon-13 spectroscopy the condition $B_1 \gg \Delta B$ is not always satisfied and lines distant from the transmitter frequency may not be properly excited. Quadrature phase detection mitigates this problem since the transmitter frequency may then be set near the centre of the chemical shift range rather than at one extreme.

For routine high-resolution spectroscopy the choice of flip angle α is not critical. For low-sensitivity nuclei such as carbon-13 it is advisable to pulse and acquire free induction decays as fast as permitted by the requirements for spin–lattice relaxation. There are two possible modes of operation. The first sets the flip angle to 90° and allows a suitable waiting period t_d between the end of the acquisition and the next pulse, adjusting t_d according to the spin–lattice relaxation times T_1. A compromise must be reached when there is a distribution of values of T_1 in the spectrum. The alternative mode of operation sets $t_d = 0$ and adjusts α to some value less than 90° so that a favourable steady-state M_Z magnetization is built up between pulses. The more efficient the relaxation the nearer the flip angle may be set to 90°. In such routine measurements spin–lattice relaxation information is seldom available to the operator, and this choice of t_d or flip angle is made on a trial-and-error basis. Fortunately, the sensitivity is not critically dependent on this choice. A quantitative treatment of these effects is given under Steady-state effects*.

Correct adjustment of the flip angle is much more important for relaxation studies. For example, it must be set to the 90° condition for progressive saturation or saturation recovery experiments, and at the 180° condition for inversion-recovery or spin echo experiments. Many two-dimensional spectroscopy* techniques also require precise adjustment of pulse flip angles. Particular care is required in repetitive experiments in which pulse length errors could be cumulative.

PULSE IMPERFECTIONS

Several instrumental shortcomings combine to make radiofrequency pulses behave imperfectly. Apart from simple misadjustment of the flip angle, this may also come about through the effects of spatial inhomogeneity of the radiofrequency field B_1 so that different regions of the sample experience different pulse flip angles. This inhomogeneity must be considered over the effective volume of the sample, restricted radially because the sample is confined within a cylindrical tube, and longitudinally by the decreasing coupling of the spins to the receiver coil as a function of distance from the coil. Thus in a single-coil probe (where transmitter and receiver coils are one and the same) the effects of B_1 inhomogeneity tend to be more important than in a crossed-coil probe (where the transmitter coil is separate from and larger than the receiver coil).

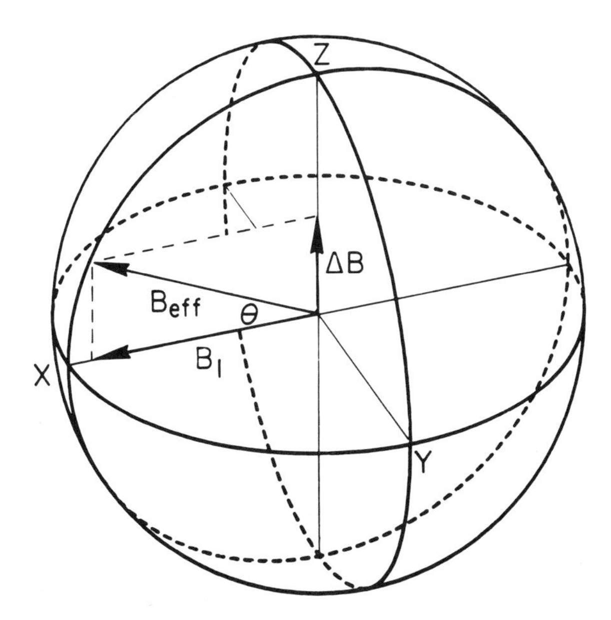

Fig. 1. In the rotating reference frame, a radiofrequency pulse of intensity B_1 at a resonance offset ΔB, acts as an effective field B_{eff} tilted through an angle θ with respect to the X axis.

A second source of pulse imperfection arises from resonance offset effects. In the general case the nuclei experience an effective field which is the resultant of B_1 and the offset from resonance ΔB. The flip angle is increased in the ratio B_{eff}/B_1 and the axis of rotation is tilted away from OX through θ radians towards the Z axis (Fig. 1). A significant tilt θ prevents proper inversion of M_Z whatever the flip angle. If the nominal flip angle is 90° there is a relatively uniform conversion of longitudinal magnetization M_Z into transverse magnetization M_{XY} since the tilt effect is compensated by the increase in flip angle. Figure 2 shows a family of magnetization trajectories for this condition for various values of $\Delta B/B_1$, illustrating that they all terminate quite close to the equatorial plane, but that large phase errors are introduced (deviation from the Y axis).

PULSE WIDTH CALIBRATION

The duration of the radiofrequency pulse t_p is almost invariably digitally controlled, so there is no difficulty in setting t_p or in assuring its reproducibility. Calibration is best carried out by searching for the null signal corresponding to the condition that there is a 360° rotation about the effective field B_{eff} for a resonance line where $\Delta B/B_1$ is not too large. The 180° condition is a less

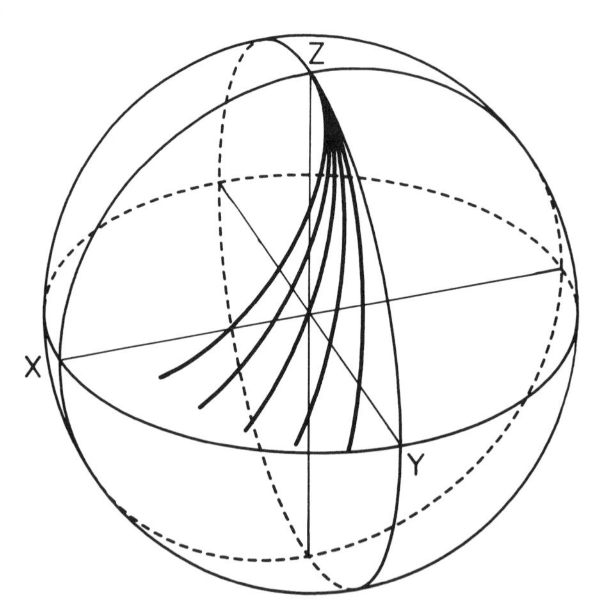

Fig. 2. Self-compensating property of a 90° pulse with respect to resonance offset. The tilt of the effective field is partially compensated by the increase in flip angle. Trajectories are shown for $\Delta B/B_1 = 0.1, 0.2, 0.3, 0.4$ and 0.5. The intensity of the excited signal is almost constant but it is appreciably shifted in phase.

satisfactory test because the tilt of B_{eff} introduces a significant M_X signal component (the dispersion mode) for this condition, making it difficult to recognize the null in M_Y. To search for the maximum M_Y component is not a sensitive test for a 90° pulse. A simple computer routine can be written to display the detected NMR signal as a function of t_p so that the 360° pulse can be recognized unambiguously (Fig. 3). Shorter pulses are then set by reducing t_p, assuming that the B_1 intensity does not fall off as t_p increases (*pulse droop*). Where pulse parameters are critical, recalibration may be necessary when samples are changed, owing to changes in the dielectric properties.

The calibration of radiofrequency pulses obtained with the B_2 field requires different techniques. One common method is to use the expression for residual splitting in an off-resonance decoupling* experiment

$$S_{resid} = J\Delta/(\gamma B_2/2\pi) \quad \text{(Hz)} \qquad [1]$$

after ensuring that the offset is sufficiently small that $\gamma B_2/2\pi \gg \Delta$. Another possibility is to use the expression for spin tickling* at the exact resonance condition (where the observed splitting is a minimum). Careful pulse calibration can be particularly important in two-dimensional spectroscopy* where quite small imperfections can give rise to spurious resonance peaks.

The adverse consequences of pulse imperfection can sometimes be minimized by compensation. A well-known example is the Meiboom–Gill (1) modification where the introduction of a 90° radiofrequency phase shift alters the symmetry of the problem in such a way that the effects of pulse length error or finite resonance offset are compensated on even-numbered spin echoes. The influence of pulse imperfections on spin–lattice relaxation measurements can be countered by alternating the phase of the population inversion pulse (2). Phase cycling* is used in many forms of two-dimensional Fourier transform spectroscopy as a means of cancelling the effects of imperfect pulses. A more satisfactory approach to the problem is provided by the concept of a composite pulse* (3). This is a cluster of radiofrequency pulses grouped closely together (usually without intervals between pulses) designed to compensate the imperfections inherent in the individual pulses. An example is the sequence 90°(X)

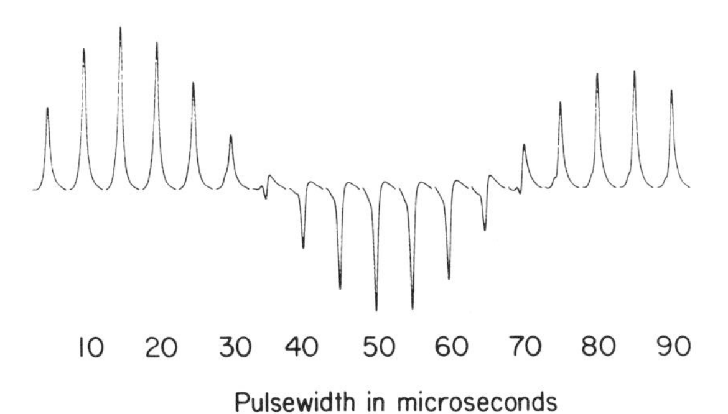

Fig. 3. Pulse width calibration by observing the intensity of the excited signal. The null-points where the flip angle is 180° or 360° are best located by interpolation since lineshapes near the null-points are distorted by the combined effects of B_1 and B_0 inhomogeneity. The decay in amplitude of the sinewave is a measure of the B_1 inhomogeneity.

180°(Y) 90°(X) which behaves like a simple 180°(Y) pulse but is compensated for pulse length error or resonance offset effects. Considerable progress has recently been made in the design of composite pulses; for example, spin inversion pulses can be devised with an operating bandwidth (2ΔB) several times larger than B_1 (4).

CLASSIFICATION OF RADIOFREQUENCY PULSES

Many recent innovations in high resolution NMR spectroscopy have involved the invention of new pulse sequences. In constructing a new pulse scheme it is often helpful to consider the various categories of radiofrequency pulses based on the manner in which the pulses act on nuclear spins.

Most sequences start with an *excitation pulse* which converts equilibrium Z-magnetization into transverse magnetization; often this is a 90°(X) pulse. A pulse which converts M_Z into $-M_Z$ is a 180° *inversion pulse*. It may be employed as the first step of an inversion-recovery experiment to measure spin–lattice relaxation* times, or it can be used to interchange the spin states of a nucleus X, thereby interchanging the appropriate multiplet components of a coupled A nucleus, an important trick for generating spin echo modulation. (See Spin echoes*.) A second type of 180° pulse is employed to carry precessing magnetization vectors from one position in the X–Y plane to a mirror-image position with respect to the pulse axis. This *refocusing pulse* is virtually indispensable for spin echo experiments.

Not all the effects of a radiofrequency pulse can be represented as the rotation of a vector in the rotating frame*. We encounter an example of this in the homonuclear shift correlation* experiment (COSY), where the second pulse induces a transfer of coherence from one group of spins to another, giving rise to cross-peaks in the two-dimensional spectrum. This is usually called a *mixing pulse*. The vector model also breaks down for a *conversion pulse* which converts ordinary precessing nuclear magnetization into unobservable multiple-quantum coherence*. Occasionally, conversion into multiple-quantum coherence is a device for suppressing undesirable magnetization components; the pulse in question would then be known as a *purging pulse*. Finally there are pulses whose only purpose is to convert unobservable quantities (population disturbances or multiple-quantum coherence) into observable precessing magnetization. These are known as *read pulses*.

REFERENCES

1. S. Meiboom and D. Gill, *Rev. Sci. Instr.* **29**, 688 (1958).
2. D. E. Demco, P. van Hecke and J. S. Waugh. *J. Magn. Reson.* **16**, 467 (1974).

3. M. H. Levitt and R. Freeman, *J. Magn. Reson.* **33**, 473 (1979).

4. A. J. Shaka, *Chem. Phys. Lett.* **120**, 201 (1985).

Cross-references

Composite pulses
Multiple-quantum coherence
Off-resonance decoupling
Phase cycling
Rotating frame
Shift correlation
Spin echoes
Spin–lattice relaxation
Spin tickling
Steady-state effects
Two-dimensional spectroscopy
Vector model

Resolution Enhancement

It is a considerable technological feat to achieve magnetic field uniformity approaching parts in 10^9 over volumes of the order of 1 cm^3, as in present day high resolution NMR spectrometers. This is accomplished only by very careful design of the magnet or solenoid, the use of a complex array of current shims, and by spinning the sample to average out transverse field gradients. For high-resolution work on small molecules, it would be attractive to go beyond these 'instrumental' linewidths and observe natural linewidths, for then much more fine structure would be discernible. The observed lineshape may be considered to be the natural lineshape convoluted* with the magnet field distribution function. Attempts to reverse this broadening effect are sometimes referred to as *deconvolution*. In fact, this process is not limited to the removal of field inhomogeneity broadening but is a general data manipulation technique known as resolution enhancement.

The convolution theorem states that the process of convoluting a curve A with another curve B in the frequency domain is equivalent to a multiplication of the Fourier transform of A by the Fourier transform of B in the time domain. It is therefore not surprising that most procedures for resolution enhancement operate on the transient free induction signal, multiplying it with a time-domain 'shaping function', a simpler operation than deconvolution in the frequency domain. When the linewidth is determined by field inhomogeneity, the free induction signal decays because of mutual interference between signals from different regions of the sample. In order to counteract this decay the free induction signal may be multiplied by a shaping function that increases with time, emphasizing the tail. This is therefore the converse of a sensitivity enhancement* function.

A shaping function that increases with time introduces two practical problems. It emphasizes any step-function discontinuity at the end of the free induction decay* (if zero filling* is used). Since the Fourier transform of a step-function is a sinc function, this process introduces some sinc function character into the frequency-domain lineshape, which is undesirable in many-line spectra. The shaping function should therefore rise with time but fall quite rapidly towards the end of the free induction decay in order to round off any discontinuity (see Apodization*). The second problem is that the shaping function emphasizes the noise in the tail of the free induction decay. Resolution can therefore only be improved at the expense of sensitivity. This is the practical limit on the extent of resolution enhancement which can be imposed in a given case; at some point the procedure enhances the high-frequency components of the noise, introducing spurious splittings or shoulders on the lines and false signal peaks elsewhere.

Many different shaping functions have been used for resolution enhancement. Possibly the simplest and most common is the rising exponential $\exp(at/T_2^*)$ followed by a suitable apodization function. An elegant alternative is the Lorentzian-to-Gaussian conversion which uses a time-domain shaping function of the form

$$\exp\,[at/T_2^* - b(t/T_2^*)^2].$$

Here T_2^* is the time constant of the unprocessed free induction decay, and a and b are variable parameters. If the original decay approximates to an exponential (which would give a Lorentzian lineshape) it can be counteracted by the first term of this expression through a suitable choice of the parameter a. The second term then imposes a Gaussian decay, and since the Fourier transform of this is another Gaussian, the lineshapes in the frequency domain are converted from Lorentzian into Gaussian. The smaller the choice of b, the narrower the resulting linewidth. An interesting extension of this procedure employs such a large value of a that the first term greatly overcompensates the raw decay, giving a free induction signal which rises very rapidly with time; the second term once again imposes a Gaussian shape, but now the maximum has moved from $t = 0$ to some point further along the time axis. There is particular interest in the case where the peak of the Gaussian envelope is set at the centre of the acquisition period. The free induction signal then resembles a spin echo in several respects, having an envelope which falls off essentially symmetrically on either side of the centre, and is therefore known as a *pseudo-echo* (1). Since the Fourier transform of an even function is also an even function, the resulting spectrum has no dispersion-mode components. However, unlike a true spin echo, a pseudo-echo does not have its constituent frequencies in phase at the centre, but at $t = 0$, and this introduces a strong frequency-dependent phase shift into the spectrum. It is nevertheless easy to sidestep this difficulty by computing the absolute-value mode spectrum which has lineshapes identical to pure absorption since the dispersion-mode components have been eliminated. Similar results can be achieved by any shaping function that forces the free induction signal to have approximate symmetry about its centre, for example the *sine-bell* (2). The price that has to be paid (in either case) is a rather severe loss in sensitivity.

A more pragmatic approach to resolution enhancement is offered by the technique of 'convolution-difference' (3). This simply imposes a sensitivity enhancement function on the free induction decay, for example $\exp(-at/T_2^*)$, and subtracts the result from the unprocessed free induction signal. This is particularly useful for the elimination of very broad lines from a spectrum which contains broad and narrow lines, since any broad component that is essentially unaffected by the sensitivity enhancement shaping function is simply cancelled by this process. The technique does, however, illustrate one drawback of certain resolution enhancement functions, in that the resulting spectra can no longer be properly integrated, and quite often the component lines have negative-going wings; this is a result of falsifying the initial ordinate of the free induction decay. In situations where linewidths vary within the

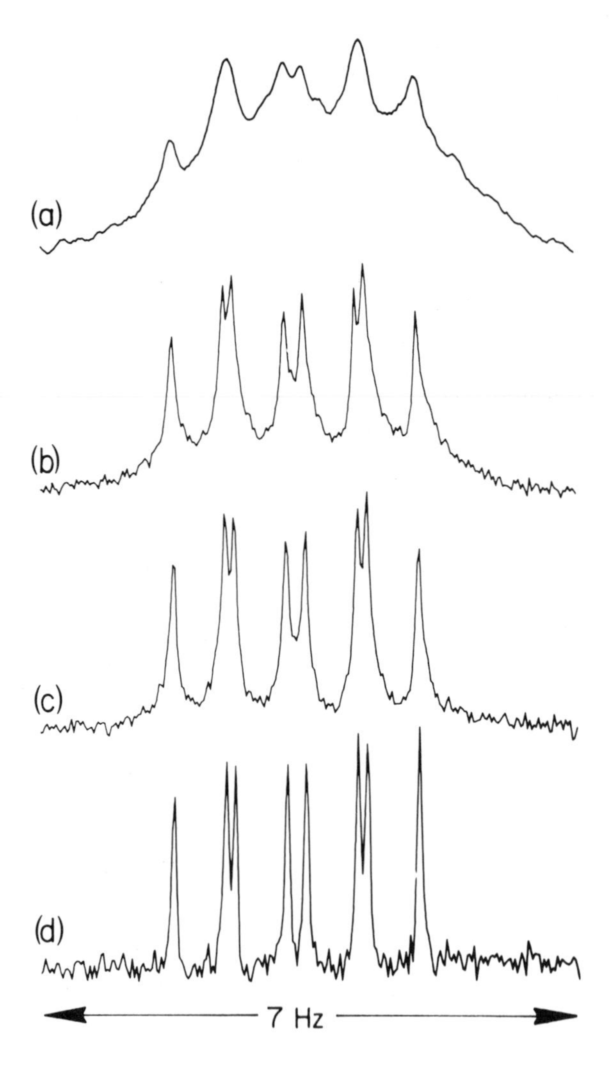

Fig. 1. Resolution enhancement by restricting the active sample volume. A selective pulse in an applied Z-gradient excites a flat disc-shaped volume of the sample. The free induction decay is acquired with this gradient switched off. The intensity of the applied gradient is increased in the series (a) to (d), reducing the active sample volume and thus increasing the resolution but decreasing the signal-to-noise ratio.

unprocessed spectrum, resolution enhancement may affect the relative peak heights, rendering quantitative analysis unreliable.

There are certain special techniques for improving resolution which do not rely on data manipulation. In principle, a very small sample would be insensitive to the inhomogeneity of the applied magnetic field; in practice there are discontinuities in magnetic susceptibility at the walls of the sample container which introduce field gradients which broaden the line. It is, however, possible to limit the *effective* volume of the sample by borrowing the technology of NMR imaging. For example, a signal may be excited by a frequency-selective pulse in a deliberately imposed field gradient, the free induction signal being acquired after this gradient is removed. Usually, the most critical gradient would be used; this is the Y gradient in the case of an iron magnet, the Z gradient for a superconducting solenoid. In this way only a thin disc-shaped volume of the sample is excited, and when the transient signal evolves in the absence of the imposed gradient, it is insensitive to the residual Y (or Z) gradients (4). Figure 1 shows an example of this method of resolution enhancement. Note that, as expected, the signal-to-noise ratio falls off as the resolution increases.

Certain phenomena based on spin–spin coupling can be used to circumvent the effect of magnetic field inhomogeneity since, by acting within a given molecule they are not sensitive to field gradients. This effect can be seen in the spin tickling* experiment where the doublets can be better resolved than the B_0 inhomogeneity would suggest. Similarly, a second radiofrequency field B_2 can be made to 'burn a hole' in a chosen resonance line, affecting only a limited region of the sample, and the resulting population disturbance may be observed on another line which shares a common energy level (the *regressive* case). A hole then appears in the second line and it may be much narrower than the instrumental linewidth. Spin echoes* provide another method of counteracting the effects of non-uniform magnetic fields, the signals from different volume elements of the sample getting out of phase with one another and then being refocused by a 180° radiofrequency pulse. Chemical shift effects are also refocused, but spin–spin coupling modulates the echo envelope, and if this modulation information is Fourier transformed, the lines of the resulting spectrum can approach their natural linewidths. Thus through J-spectroscopy* the spin multiplet structure can be resolved even though the conventional spectrum has broad lines.

In a similar way, the excitation of homonuclear zero-quantum coherence may be used to defeat the effects of magnetic field inhomogeneity. This requires a system of at least two coupled spins and a pulse sequence which excites a coherence between energy states which have the same magnetic quantum number. If the process is thought of as the absorption of one quantum of energy at frequency f_1 and simultaneous emission of another quantum at frequency f_2 then it is clear that the zero-quantum coherence is almost completely insensitive to field inhomogeneity, since this will only shift f_1 and f_2 together by the same amount.

For large molecules or very viscous solutions, linewidths are usually determined by spin–spin relaxation rather than field inhomogeneity, so these last-mentioned methods are not applicable.

REFERENCES

1. A. Bax, R. Freeman and G. A. Morris, *J. Magn. Reson.* **43**, 333 (1981).
2. A. de Marco and K. Wüthrich, *J. Magn. Reson.* **24**, 201 (1976).
3. I. D. Campbell, C. M. Dobson, R. J. P. Williams and A. V. Xavier, *J. Magn. Reson.* **11**, 172 (1973).
4. A. Bax and R. Freeman, *J. Magn. Reson.* **37**, 177 (1980).

Cross-references

Apodization
Convolution
Free induction decay
J-spectroscopy
Multiple-quantum coherence
Selective excitation
Sensitivity enhancement
Spin echoes
Spin tickling
Zero filling

Rotating Frame

In a magnetic resonance experiment we need to be able to predict the combined effect of immersing the sample in a static magnetic field B_0 and a radiofrequency field B_1. In practice, the radiofrequency is applied in the form of a linearly oscillating field $2B_1 \cos \omega t$, but this can be regarded as two counter-rotating fields of strength B_1 and only the component which rotates in the same sense as the nuclear precession has a significant influence on the spins.

The vector model* introduced by Bloch (1), translates the NMR phenomenon into a problem of classical mechanics by considering the net macroscopic magnetization arising from the resultant of all nuclear magnetization vectors present in the sample. At Boltzmann equilibrium this is simply a vector M_0 aligned along the magnetic field direction Z. Spins are either aligned with or against the field B_0 and for the great majority, their magnetizations cancel; only the very small excess population in the lower energy state contributes to M_0. Since the precession phases of the individual spins are random there is no transverse magnetization at equilibrium.

The problem is to calculate how B_1 interacts with the nuclear magnetization vector M_0. In particular, what happens if a short strong radiofrequency pulse is applied to a spin system at Boltzmann equilibrium? The trick is to transform the entire problem into a rotating coordinate frame in which B_1 is a static field, since then the resulting motion of M_0 is particularly simple, a rotation about the static field. If f_1 is the frequency of the B_1 field, the rotating frame is the usual Cartesian coordinate frame except that the X and Y axes are rotating about the Z axis at an angular frequency $\omega_1 = 2\pi f_1$ rad sec^{-1} in the sense of the nuclear precession.

Nuclei placed in a field B_0 precess at a frequency γB_0 rad sec^{-1}. Since the laws of physics must be independent of the reference frame, and since nuclei appear to precess more slowly in a rotating reference frame, the magnetic field must be correspondingly reduced to $B_0 - \omega_1/\gamma$. Thus if B_0 is adjusted for exact resonance, the nuclei cease to precess in the rotating frame and appear to experience zero applied field. Note that the normal choice for the frequency of the rotating frame is the frequency of the transmitter, not the Larmor precession frequency of the nuclei, although of course the two are close.

When a radiofrequency pulse is applied, for a few microseconds there is an intense B_1 field aligned along the +X axis of the rotating frame. For nuclei that are off-resonance by an amount ΔB, there are two fields at right angles, B_1 and ΔB which have a resultant B_{eff} acting in a direction which makes an angle θ with respect to the X axis in the X–Z plane with $\tan \theta = \Delta B/B_1$. The nuclear magnetization vector thus rotates about B_{eff} through an angle $\gamma B_{eff}t_p$, where t_p is the width of the pulse. (See Radiofrequency pulses*.)

This explains why a strong pulse can excite a wide range of chemical shifts. If $B_1 \gg \Delta B$ for the nuclei furthest from resonance, then θ is always very small and B_{eff} differs very little from B_1 so that all the different magnetization vectors rotate about the X axis through the same angle $\gamma B_1 t_p$ radians. In the fairly common case where $\gamma B_1 t_p = \pi/2$, all the individual magnetizations are left aligned along the Y axis where the induced signal is maximum. Once the radiofrequency pulse is extinguished these magnetization vectors are free to precess at their characteristic frequencies ω_k, and since the receiver coil is rotating at an angular frequency ω_1 rad sec^{-1} with respect to this frame, the induced signals are at $(\omega_k + \omega_1)/2\pi$ (hertz), in the radiofrequency region.

In a more mathematical sense, transformation into a rotating frame is a device for removing the time dependence from the problem. Essentially all of the vector pictures used in this book implicitly assume transformation into the rotating frame. This is by far the simplest way to visualize the effects of multiple-pulse sequences; it is applicable to the majority of high resolution NMR experiments with the exception of those that involve multiple-quantum coherence*. In this frame the motion of the nuclear magnetization breaks down into rotation about B_{eff} (during a pulse) and free precession about ΔB in the intervals between pulses. It is permissible to consider one nuclear species (say protons) in a frame rotating at the Larmor frequency of protons, while simultaneously considering another nuclear species (say carbon-13) in another frame rotating at the appropriate carbon-13 precession frequency. For the analysis of double resonance experiments it is convenient to transform to a coordinate system rotating in synchronism with the *second* radiofrequency field B_2. The concept may be extended to visualizing energy level diagrams in a rotating frame; such transformations have been used in discussing coherent decoupling* and spin tickling*.

Experiments which apply strong radiofrequency fields for relatively long periods cause continuous precession about the effective field B_{eff} in the rotating frame, sometimes described as *transient nutations** or 'Torrey oscillations' (2). In the simple case, B_{eff} is approximately the same as B_1 and nutation takes place about the X axis of the rotating frame. The macroscopic nuclear magnetization M_0 rotates in the Z–Y plane inducing a sinusoidally oscillating NMR signal of constant frequency $\gamma B_1/2\pi$ (Hz). This observed signal eventually decays due to spatial inhomogeneity of the B_1 field, as magnetization isochromats from different regions of the sample get out of phase with one another. If the B_1 field is suddenly reversed in phase, this loss of phase coherence is refocused and the result is a *rotary echo* (3). As with conventional spin echoes*, multiple refocusing is possible and then the envelope of the rotary echoes decays with a time constant T given by

$$1/T = 0.5[1/T_1 + 1/T_2]. \qquad [1]$$

Sometimes it may be necessary, while in the rotating frame, to make a further transformation into a second reference frame which rotates about B_{eff} at an angular frequency γB_{eff}. This provides a convenient picture for the 'rotating frame resonance' which occurs in the Hartmann–Hahn experiment* (4).

REFERENCES

1. F. Bloch, *Phys. Rev.* **102**, 104 (1956).
2. H. C. Torrey, *Phys. Rev.* **76**, 1059 (1949).
3. I. Solomon, *Phys. Rev. Lett.* **2**, 301 (1959).
4. S. R. Hartmann and E. L. Hahn, *Phys. Rev.* **128**, 2042 (1962).

Cross-references

Coherent decoupling
Hartmann–Hahn experiment
Multiple-quantum coherence
Radiofrequency pulses
Spin echoes
Spin tickling
Transient nutations
Vector model

Sample Spinning

When all precautions have been taken to obtain the most uniform applied field B_0, it is still possible to make a big improvement in resolution by spinning the sample about a suitable axis, and essentially all high-resolution spectrometers employ this technique (1). The principle is that if the nuclear spins pass through a range of magnetic field values at a rate that is fast compared with the linewidth of the stationary sample, then they behave as if they had experienced only the mean value of the field averaged over the entire trajectory. Thus the B_0 field must be reasonably homogeneous if resolution is to be improved by spinning at a practicable rate. It also follows that only the effects of transverse B_0 gradients are reduced, any gradients along the spinning axis retain their broadening effect. In an iron magnet it is customary to denote the spinning axis as the Y axis, and hence particular attention must be paid to residual Y-gradients. In a superconducting solenoid it is usual to spin about the Z axis (the direction of the applied static field) although occasionally a 'sideways spinning probe' is employed where the spinning axis is the radiofrequency coil axis (X) so that a solenoid coil may be used instead of a saddle coil.

Most spectrometers employ a simple air-driven turbine to spin the sample, usually supported on an air bearing. The use of precision bore sample tubes has alleviated some of the early difficulties with spinning samples, but it can still be a troublesome practical problem with variable temperature studies, since the mechanical tolerances can be critical. Monitoring the spinning rate with a tachometer is a useful guide; a light beam is reflected from a panel on the turbine and induces a modulated signal on a photodiode; this frequency is displayed on a meter. In spectrometers with internal field/frequency regulation*, a failure of the spinner may entail loss of the lock.

Heroic attempts have been made to extend this simple idea to motion about more than one axis, without any notable success. Several spectroscopists have, however, noted an interesting property of conventional spinning samples – while the spinner is accelerating from rest, the resolution appears to be significantly better than that observed for a constant spinning speed. During such an acceleration phase, the liquid layer near the wall is moving faster than layers nearer the tube axis. Presumably during this non-steady-state period, a form of cyclic motion is induced with components along the tube axis, rather like the motion in a washing machine (Fig. 1(a)). In this way some of the effects of axial gradients could be averaged, along with the expected averaging of transverse gradients. Recent experiments (2) with a stationary sample tube containing an inner coaxial tube that is rotating, seem to provide some support for these ideas. At a suitably high rotation speed, the liquid in the annular region breaks up into discrete cells which are square in cross-section and in which the liquid circulates as shown in Fig. 1(b). This is called *Taylor vortex flow* (3). Since similar shear forces might be expected in an ordinary sample tube during acceleration from rest, the circulation proposed in Fig. 1(a) does not seem unreasonable.

SPINNING SIDEBANDS

As the sample rotates in a linear field gradient, a typical spin passes through a magnetic field which is modulated in intensity, $B_0 + B_m \cos \omega_m t$ and since it follows the instantaneous Larmor precession condition, its rate of precession is frequency-modulated. Viewed in the frequency domain, this corresponds to the introduction of a set of *spinning sidebands* on each side of the main resonance, at distances equal to integral multiples of the spinner frequency,

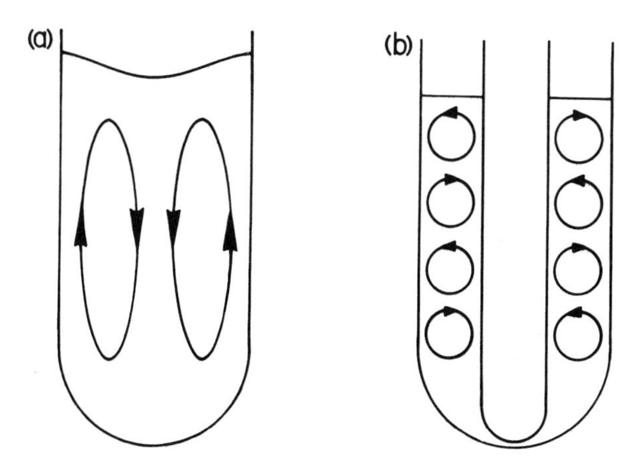

Fig. 1. (a) While the spinner is accelerating the liquid probably undergoes a 'washing machine' motion partially averaging vertical field gradients. (b) In a coaxial sample tube with the central tube rotating at a uniform speed, Taylor vortex flow (2, 3) is initiated in the annular region.

$n\omega_m/2\pi$, normally tens of Hz. The general mathematical expressions for the amplitudes of these sidebands involve Bessel functions, but for the practical cases of interest they can be greatly simplified. This is because the depth of modulation B_m is weak in comparison with the modulation frequency ω_m; in the vocabulary of the electronic engineer, the modulation index is very low. Under these conditions all the sidebands are weak in comparison with the parent resonance line, and only the first-order sidebands ($n = 1$) have appreciable intensity. They may be further reduced in intensity either by reducing the transverse field gradients (by adjusting the currents in the homogeneity coils) or by increasing the spinning rate; both reduce the modulation index. However, there is a practical limit on the spinner speed because of vortexing of the liquid in the sample tube. If this vortex approaches the receiver coil, the discontinuity in magnetic susceptibility degrades the resolution. Plastic 'vortex plugs' are sometimes used to prevent this effect.

Spinning sidebands interfere with the identification of weak signals, for example weak satellite resonances (such as those in a proton spectrum arising from the carbon-13 isotope) and it is important to check the spectrum at two different spinning rates in order to identify spinning sidebands. In time-averaging experiments, spinning sidebands may be broadened out by varying the spinner speed.

While it is the inhomogeneity in B_0 which is the major contribution to the spinning sidebands, other effects are also involved. These include the inhomogeneity of the radiofrequency field B_1, the variation of the coupling of the nuclei to the receiver coil, and the influence of uneven wall thickness in the spinning sample tube. Combinations of these effects introduce one contribution to the spinning sidebands which is independent of the spinning rate (4, 5). All three factors are more important in superconducting solenoids than in iron magnets, partly because of the different receiver coil geometry (saddle-coil) and partly because susceptibility effects are enhanced at high fields. An indication that more than one factor is at work is provided when the low-field and high-field sidebands are observed to have different phases (4, 5). This complex interplay of contributions from different physical phenomena makes it difficult to devise the best strategy for the adjustment of the shim coils for minimum intensity sidebands. For example, it is not necessarily best to optimize the shim coil settings with a stationary sample, since contributions from some transverse gradients may interfere with each other when the sample is spinning (5).

SPINNING SAMPLES AND SPIN ECHOES

Consider a spinning sample in a linear transverse field gradient. Immediately after excitation by a radiofrequency pulse, the signals from all volume elements are in phase, but they begin to lose phase coherence because they precess at different frequencies corresponding to the different regions in the transverse gradient. After one full rotation, each element at a given radius has passed through the same combination of local fields, and the signal components are all in phase again. This may be regarded as a *spinning echo*. As a result the envelope of the free induction decay is modulated at the spinning rate, which transforms in the frequency domain as a central resonance flanked by spinning sidebands.

It is therefore not surprising that spinning can interfere with conventional spin echo* experiments, as the refocusing cycle and the spinning cycle move in and out of coincidence. One solution is to use a stationary sample for spin echo measurements, but this broadens the observed lines and renders the field/frequency regulation less effective as the internal reference line is broadened. It is possible to circumvent this problem by synchronizing the refocusing pulses with the rotation of the sample, a trigger signal being derived from the spinner tachometer. Two complete revolutions of the spinner are allowed to occur between refocusing pulses, so that each volume element passes through the same series of applied fields during defocusing and refocusing periods. This is equivalent to synchronizing the refocusing pulses with the centres of the spinning echoes.

REFERENCES

1. F. Bloch, *Phys. Rev.* **102**, 104 (1956).
2. M. Vera and J. B. Grutzner, *J. Amer. Chem. Soc.* **108**, 1304 (1986).
3. G. I. Taylor, *Phil. Trans. Roy. Soc. (Lond.)* **A223**, 289 (1923).
4. E. R. Malinowski and A. R. Pierpaoli, *J. Magn. Reson.* **1**, 509 (1969).
5. J. Dadok, unpublished report.

Cross-references

Broadband decoupling
Field/frequency regulation
Modulation and lock-in detection
Spin echoes

Saturation Transfer

Most methods of chemical kinetics are only applicable to systems not at thermodynamic equilibrium, but NMR provides an opportunity to measure rates of chemical exchange* between components in a true equilibrium state. The method depends on being able to attach an innocuous label to an atom or group of atoms that is being carried to another site by chemical exchange. These techniques are particularly important in the limit of slow exchange (slow compared with the chemical shift difference between the two sites) when separate resonances are observed for the two exchanging sites. In the early experiments it was the phase of the nuclear precession which acted as the label, exchange causing a loss of phase coherence and a consequent broadening of the lines which increased as the rate of exchange increased. An extra damping term can be introduced into the Bloch equations to represent the effects of exchange. A more direct method was pioneered by Forsén and Hoffman using a population disturbance as the label (1). In the slow exchange limit, sudden saturation of resonance X causes a gradual loss of intensity of resonance A as saturated spins are transferred from site X to site A. This competes with the normal spin–lattice relaxation* acting to restore the thermal equilibrium populations, but if these relaxation times are measured, then the exchange rate can be deduced from the time evolution of the intensity of the A resonance. In this form the technique raises some difficult questions about how fast a given resonance line can be saturated. Nowadays the Forsén–Hoffman experiment is much more conveniently initiated by a selective population inversion pulse, the progress of population transfer being followed by monitoring the free induction decays in a series of experiments with different delays τ between the inversion pulse and the observation pulse. Nevertheless, it is convenient to retain the original term 'saturation transfer' to cover all these experiments.

Consider the simple case of just two exchanging sites with equal populations, so that the forward and backward exchange rates are equal, represented by a rate constant k, and equal spin–lattice relaxation times at the two sites, T_1. Let the X spins be inverted by a selective 180° pulse, followed by a variable delay τ and a 90° monitoring pulse which is non-selective. When τ is infinite both signals reach their equilibrium values $M_\infty(A)$ and $M_\infty(X)$. The solution of the differential equations governing the longitudinal magnetizations gives

$$M_\infty(A) - M_Z(A) = M_\infty(A) [\exp(-\tau/T_1) - \exp(-\tau/T_1 - 2k\tau)] \quad [1]$$

and

$$M_\infty(X) - M_Z(X) = M_\infty(X) [\exp(-\tau/T_1) + \exp(-\tau/T_1 - 2k\tau)]. \quad [2]$$

It is easy to check that these equations have the right form by first setting $k = 0$ so that there is no exchange, and noting that $M_Z(A) = M_\infty(A)$ throughout, while

$$M_Z(X) = M_\infty(X)[1 - 2 \exp(-\tau/T_1)]. \quad [3]$$

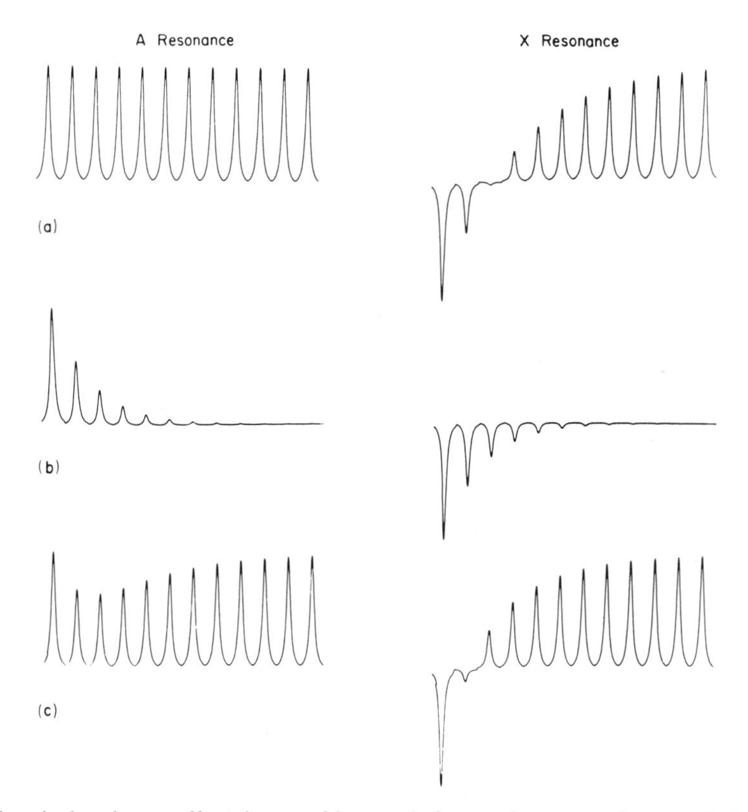

Fig. 1. Chemical exchange effects in a weakly-coupled two-spin system after population inversion of the X resonance. (a) In the case of no chemical exchange ($k = 0$) there is no effect on the A spins, while the X spins recover by spin–lattice relaxation. (b) With chemical exchange but no relaxation ($T_1 = \infty$), transfer of inverted spins to A causes an exponential decay of the A resonance. Both A and X resonances approach a state of saturation. (c) With both relaxation and exchange, the A resonance loses intensity due to transfer but recovers due to relaxation. The X resonance exhibits a biexponential recovery.

This is the expression describing the recovery of the X signal after a population inversion (Fig. 1(a)). Then the case of finite exchange but infinitely long relaxation time may be considered, where $\exp(-\tau/T_1) = 1$. The equations then reduce to

$$M_Z(A) = M_\infty(A) \exp(-2k\tau) \quad [4]$$

and

$$M_Z(X) = -M_\infty(X) \exp(-2k\tau). \quad [5]$$

These describe a situation where the A resonance decays to zero with a time constant $1/(2k)$, while the inverted X resonance grows back to zero with the same time constant (Fig. 1(b)). The factor 2 arises because each exchange 'event' involves the loss of a polarized spin and the gain of an inverted spin (or vice versa). Had this been a *saturation* transfer experiment the factor 2 would be absent.

When both exchange and relaxation occur together, the recovery curves are no longer simple exponentials (Fig. 1(c)). The A signal loses intensity and passes through a minimum before growing back to $M_\infty(A)$ as spin-lattice relaxation effects take over. The X resonance recovers from its initial inversion, quite quickly at first as exchange and relaxation effects reinforce, and then more slowly as the exchange becomes less effective because the transferred spins are less and less perturbed. An analysis of these two curves allows both the exchange rate and the spin-lattice relaxation time to be extracted.

Real examples of chemical exchange are seldom as simple as this. Even when the problem is limited to just two exchanging sites, the relative populations may differ, giving two rate constants k_A and k_X, and the spin–lattice relaxation times are in general different, T_{1A} and T_{1X}. The equations governing the time evolution after a selective population inversion of the X resonance are then (2)

$$M_Z(A)/M_\infty(A) = 1 - (2k_A/\beta)\{\exp[-\tfrac{1}{2}(\alpha - \beta)\tau] - \exp[-\tfrac{1}{2}(\alpha + \beta)\tau]\} \quad [6]$$

and

$$M_Z(X)/M_\infty(X) = 1 - (1/\beta)\{(\beta + R_X - R_A)\exp[-\tfrac{1}{2}(\alpha + \beta)\tau] + (\beta + R_A - R_X)\exp[-\tfrac{1}{2}(\alpha - \beta)\tau]\}, \quad [7]$$

where

$$\alpha = R_A + R_X, \qquad \beta = [(R_A - R_X)^2 + 4k_Ak_X]^{1/2}$$

and

$$R_A = k_A + (1/T_{1A}), \qquad R_X = k_X + (1/T_{1X}).$$

Further complications must be introduced if more than two exchanging sites are involved; these cases are usually treated by a full density matrix calculation.

An illustrative two-site exchange problem is shown in Fig. 2. The 1,2,6-trimethylpiperidinium ion exchanges slowly between two forms, one having the N-methyl group axial and the other equatorial. The experiment studies spin exchange between the 3,5 carbon-13 resonances of the two forms by applying a selective population inversion pulse to the low-field resonance (the equatorial N-methyl isomer). In practice, these experiments are conveniently performed in a difference mode, by subtracting the perturbed free induction decay from the unperturbed decay. All resonances that are unaffected by exchange then cancel, leaving a strong X resonance which decays monotonically to zero and an A resonance which grows from zero, passes through a maximum and then decays asymptotically to zero.

Recent developments in two-dimensional spectroscopy* exploit the same ideas in a different fashion (3). Suppose, for simplicity, that exchange occurs between just two sites A and X. First, the equilibrium longitudinal magnetizations are converted into transverse magnetizations by a 90° pulse, and labelled according to their precession frequencies ω_A and ω_X during a variable evolution period t_1. A part of these vectors is then returned to the Z axis by another 90° pulse and a period τ seconds allowed for chemical exchange between the sites, residual transverse magnetization being dispersed by a field gradient *homospoil* pulse. Then a fixed interval τ is allowed for chemical exchange between the two sites, Z-magnetization being exchanged. At the end of this period, a third 90° pulse generates transverse magnetizations again, and the amount transferred

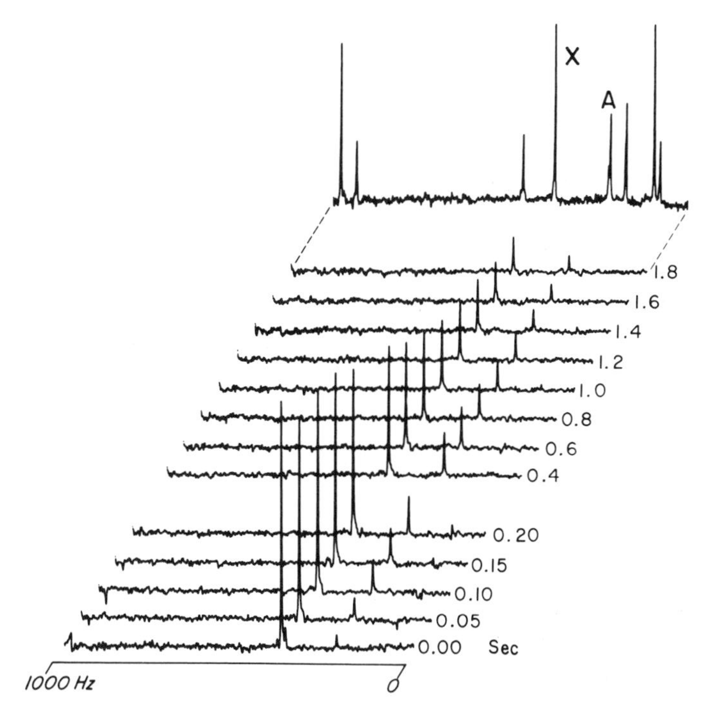

Fig. 2. Chemical exchange studied by difference spectroscopy*. The experiment is initiated by a selective population inversion of resonance X (positive signals in the difference spectrum). Exchange carries inverted spins to site A causing a population disturbance which increases with time and then dies away due to spin–lattice relaxation. Note the break in the timescale where the increments increase from 0.05 seconds to 0.2 seconds. The conventional spectrum (top trace) is from carbon-13 in the two stereoisomers of the 1,2,6-trimethylpiperidinium ion.

from site X to site A is identified by the fact that it now has a new precession frequency ω_A. All the NMR signal components that retain the same precession frequencies throughout the experiment give rise to peaks on the principal diagonal of the two-dimensional spectrum, but those that have suffered a frequency jump due to exchange, give rise to easily identifiable *cross-peaks* lying off the main diagonal. Information about the rate of exchange is contained in the relative intensities of these cross peaks. The beauty of the method lies in the clear picture that it provides of the many possible exchange paths in a complicated system, each pair of cross peaks identifying one of the paths. If the mixing time is short compared with the inverse of the exchange rate only single-step exchange events are observed. In the one-dimensional Forsén–Hoffman experiment each exchange pathway would have to be investigated separately, and the analysis becomes quite difficult for multisite exchange. Further refinements of this experiment (4) make τ a variable by incrementing it in step with the evolution time t_1 in some constant proportion – the 'accordion' experiment.

REFERENCES

1. S. Forsén and R. A. Hoffman, *J. Chem. Phys.* **39**, 2892 (1963).
2. G. A. Morris and R. Freeman, *J. Magn. Reson.* **29**, 433 (1978).
3. J. Jeener, B. H. Meier, P. Bachmann, and R. R. Ernst, *J. Chem. Phys.* **71**, 4546 (1979).
4. G. Bodenhausen and R. R. Ernst, *J. Am. Chem. Soc.* **104**, 1304 (1982).

Cross-references

Chemical exchange
Difference spectroscopy
Selective excitation
Spin–lattice relaxation
Two-dimensional spectroscopy

Selective Decoupling

In the section on coherent decoupling* a pictorial description of double resonance effects is presented where it is seen intuitively that the influence of the B_2 field extends over a frequency range of the order of magnitude $\gamma B_2/2\pi$ Hz. It is therefore possible to contemplate coherent irradiation experiments which strongly perturb one chemically shifted group of spins but have a negligible influence on others. This is called selective decoupling. Consider a very simple case of two protons P and Q with relative chemical shift Δ Hz, coupled with similar direct couplings J_{CH} to carbon-13 spins P′ and Q′, respectively. Each proton resonance is a doublet of splitting J_{CH} but it is assumed that $\Delta \gg J_{CH}$. If proton P is irradiated at its chemical shift frequency with a B_2 field with $\gamma_H B_2/2\pi$ comparable with J_{CH}, then the effective fields E_1 and E_2 in the rotating frame must be equal

$$E_1 = E_2 = [(\tfrac{1}{2}J_{CH})^2 + (\gamma_H B_2/2\pi)^2]^{1/2}. \qquad [1]$$

There is thus no residual splitting, the P′ resonance is completely decoupled. However, the decoupler is Δ Hz off-resonance for the Q proton, giving effective fields E_3 and E_4

$$E_3 = [(\Delta + \tfrac{1}{2}J_{CH})^2 + (\gamma_H B_2/2\pi)^2]^{1/2} \qquad [2]$$

$$E_4 = [(\Delta - \tfrac{1}{2}J_{CH})^2 + (\gamma_H B_2/2\pi)^2]^{1/2}. \qquad [3]$$

Now, because $\Delta \gg J_{CH}$ the residual splitting on the Q′ carbon resonance

$$E_3 - E_4 = J_{CH}\Delta/[\Delta^2 + (\gamma_H B_2/2\pi)^2]^{1/2} \qquad [4]$$

which, since the offset Δ is large compared with $\gamma_H B_2/2\pi$, is approximately equal to J_{CH}. Thus while the P′ spin is decoupled, the Q′ spin is not.

This principle can be used for locating hidden resonances in homonuclear spin systems, or for deciding which multiplets in a complicated spectrum are related by mutual coupling. It is not of course necessary to find the *exact* decoupling condition $\Delta = 0$ since the residual splitting will be small provided that $\gamma B_2/2\pi \gg \Delta$. As outlined above, selective decoupling can be applied to the task of correlating proton shifts with carbon-13 shifts, although the series of different experiments that is required might be quite time-consuming. There are two-dimensional spectroscopy* experiments which accomplish the same result much more quickly. (See Shift correlation*.)

An analogous selective decoupling experiment gives the relative signs of two spin coupling constants in a system of three (or more) coupled spins (1, 2). Suppose these are called A, M and X. One of these, say M, can be regarded as a *passive* spin not directly involved in the decoupling, but splitting the A and X resonances by the appropriate coupling constants J_{AM} and J_{MX}. These are the couplings for which the relative signs will be determined. For simplicity, suppose that $|J_{MX}| > |J_{AX}|$. This makes it feasible to irradiate the low-field doublet of the X resonance with a radiofrequency field $\gamma B_2/2\pi$ comparable with J_{AX}, without appreciably affecting the high-field doublet, a distance J_{MX} away. In this experiment J_{MX} takes on the function of Δ in eqns [3] and [4]. Consequently, only the low-field A doublet or the high-field A doublet is decoupled, but not both. If the former, then J_{AM} and J_{MX} must have like signs, if the latter, they have opposite signs. It is just as if a Maxwell demon had separated the NMR sample into two equal parts, one containing only nuclei with the M spin in an α state, the other with the M spin in the β state, and then performed a coherent decoupling experiment on the first sample. Figure 1

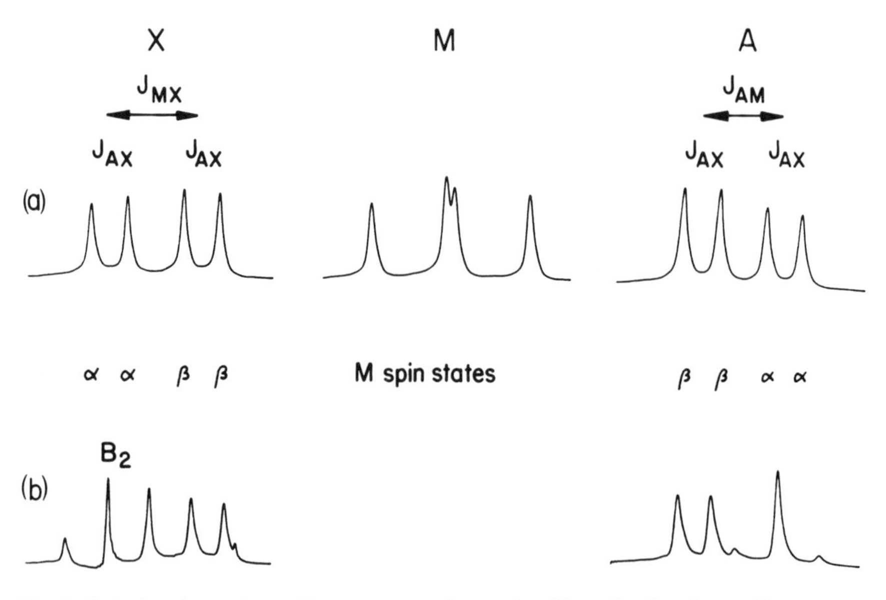

Fig. 1. Relative signs of coupling constants determined by selective decoupling. (a) Conventional spectrum of the AMX spin system of protons in 2,3-dibromopropionic acid. (b) By irradiation of the *left-hand* doublet of the X multiplet (where the M spin has an α state) we observe the coalescence of the *right-hand* doublet of the A multiplet (the result would be the same if α and β are interchanged). This indicates that J_{AM} and J_{MX} have opposite signs. In addition to the decoupled line in the A spectrum, there are two weak satellite lines where both B_1 and B_2 contribute a photon.

shows a selective decoupling experiment on the protons of 2,3-dibromopropionic acid which establishes that the vicinal coupling J_{AM} and the geminal coupling J_{MX} have opposite signs. Coupling constants for a large variety of bond configurations and many different combinations of nuclear species have had their signs determined in this way, based on an absolute sign determination for some suitable case, for example the single-bond CH coupling, believed to be positive. There are now methods based on two-dimensional spectroscopy which provide the same information about relative signs of coupling constants.

For relative sign determinations in heteronuclear spin systems it should be noted that it is strictly the reduced coupling constants $J_{AM}/(\gamma_A\gamma_M)$ and $J_{MX}/(\gamma_M\gamma_X)$ that are being related, so that if one of the magnetogyric ratios γ_A or γ_X happens to be negative, then the argument presented above is reversed. Then if irradiation of a high-field doublet decouples the corresponding high-field doublet, the coupling constants J_{AM} and J_{MX} would have *opposite* signs. This caveat applies to proton-nitrogen-15 decoupling, and to proton-silicon-29 experiments.

REFERENCES

1. D. F. Evans and J. P. Maher, *Proc. Chem. Soc. (Lond.)* 208 (1961).
2. R. Freeman and D. H. Whiffen, *Mol. Phys.* **4**, 321 (1961).

Cross-references

Coherent decoupling
Rotating frame
Shift correlation
Two-dimensional spectroscopy

Selective Excitation

Once a high resolution NMR spectrum has been recorded, the spectroscopist has to decide whether this provides sufficient information to solve the chemical problem in hand, or whether he should go on to perform some supplementary experiments. The latter might be double resonance, relaxation studies, two-dimensional spectroscopy* or selective excitation; these techniques provide assignments or additional identification of the raw NMR data.

In the early NMR spectrometers which employed slow-passage continuous-wave methods, selective excitation could be taken for granted. For example the important saturation transfer* experiment (1) simply required a double irradiation field B_2 to be adjusted to the frequency where it caused saturation of the resonance of a chosen nuclear site X while a second resonance A was examined by the usual frequency-sweep. It was only when Fourier transform spectrometers (see Fourier transformation*) were generally adopted that special techniques were needed for selective excitation. This was partly because *pulsed* selective excitation was often required, and partly because facilities for fine tuning the transmitter frequency are not necessary for non-selective pulse excitation and hence often not available.

Various degrees of selectivity can be contemplated. Usually in high resolution work a single resonance line needs to be excited while adjacent lines are left essentially unperturbed. Some other applications call for an entire spin multiplet to be excited reasonably uniformly while adjacent chemically shifted multiplets are spared. Some spectroscopists call the latter a *semiselective* experiment. For applications in NMR imaging and high resolution *in vivo* NMR, a wide range of selectivities may be required.

The simplest form of selective pulse is a long rectangular pulse of low radiofrequency intensity often called a *soft* pulse. The flip angle α of the pulse is of course fixed by the type of experiment to be performed ($\alpha = 90°$ or $180°$). But whereas a hard pulse is of short duration t_p and high B_1 intensity, a soft pulse has a relatively long duration and B_1 is reduced in proportion, maintaining the same flip angle

$$\alpha = \gamma B_1 t_p. \qquad [1]$$

As a rough rule of thumb, B_1 influences nuclear spins within a frequency range of the order of $\gamma B_1/2\pi$ Hz from the exact resonance condition but has a diminishing influence at larger resonance offsets. Thus the longer the pulse duration the more frequency selective the excitation.

The Bloch equations may be used to calculate the form of the excitation as a function of resonance offset. For a 90° rectangular pulse the absorption-mode component follows the curve

$$A = \sin(\gamma B_{eff} t_p/\pi)/(\gamma B_{eff} t_p/\pi), \qquad [2]$$

where the effective field in the rotating frame is given by

$$B_{eff} = [\Delta B^2 + B_1^2]^{1/2}. \qquad [3]$$

This excitation function does not follow quite the same curve as

$$B = \sin(\gamma \Delta B t_p/\pi)/(\gamma \Delta B t_p/\pi), \qquad [4]$$

which is the curve predicted on the assumption that the frequency-domain excitation is the Fourier transform of the time-domain pulse envelope (Fig. 1).

To be effective, a soft pulse must be finely adjusted to the chosen resonance frequency. If the radiofrequency transmitter is only adjustable in rather coarse steps (as in some Fourier spectrometers) and if it is desirable to switch from normal ('hard') pulses to soft pulses without changing the B_1 intensity, then a different approach can be useful, the 'DANTE' sequence (2). This is a regular sequence of N extremely short duration intense radiofrequency pulses at a specific pulse repetition rate, $f = 1/\tau$. The behaviour is most easily visualized using the vector model*. Each pulse flips the spins through a very small angle α/N so that at exact resonance the net effect is a rotation through α radians. For

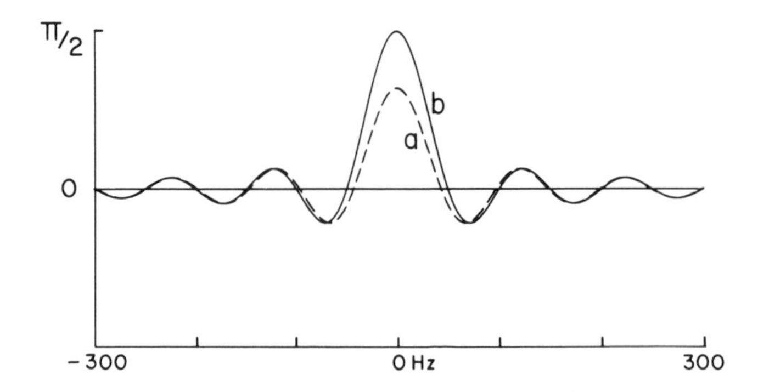

Fig. 1. The frequency-domain excitation function for a 10 ms rectangular selective 90° pulse calculated (a) according to the Bloch equations (b) assuming a Fourier transform relationship (sinc function). The two curves become indistinguishable at large resonance offsets.

nuclei at a resonance offset $\Delta\nu$ the behaviour is more complex. Between each pulse these nuclei precess through a small angle

$$\theta = 2\pi\Delta\nu\tau, \qquad [5]$$

so that the next pulse catches them at a different phase of their precession and the tipping effect of the pulses is no longer cumulative. The trajectories are the family of zig-zag curves illustrated in Fig. 2. For appreciable offsets from resonance, where $\theta \gg \alpha/N$, these trajectories amount to small cyclic excursions near to the north pole of this diagram and very little transverse magnetization is generated at the end of the sequence. In the limit of a very large number N of pulses, the zig-zag component may be neglected and these trajectories approximate those of a single soft pulse of flip angle α, and radiofrequency intensity B_1t_p/τ. The frequency selectivity is just the same as for a soft pulse. The sinc function wiggles of Fig. 1 can be explained by the cyclic excursions near the north pole.

However, there is one important difference. The precession angle θ could have been written $(2n\pi + \theta)$ and the result would have been the same. There is therefore a set of 'sideband' responses corresponding to resonance offsets where, between each pulse, the spins accomplish a small number n of complete

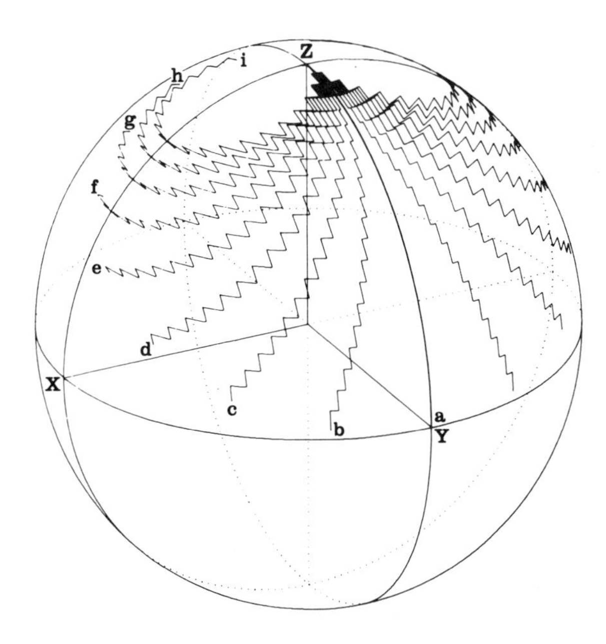

Fig. 2. Selective excitation by the 'DANTE' sequence of short pulses separated by periods of free precession. For exact resonance (a) the magnetization vector goes directly to the Y axis. For increasing resonance offsets, the trajectories (b), (c), . . . , (i) follow zig-zag paths that carry the vectors further and further from the Y axis. For large resonance offsets the vectors execute small, nearly cyclic excursions close to the Z axis. There are also sideband conditions for offsets such that the magnetization vector accomplishes a whole number of complete revolutions about the Z axis between each pulse.

revolutions in addition to the excess rotation θ. If we choose to work at such a sideband condition (usually the first sideband where n = 1) then the fine tuning of the irradiation frequency can be accomplished by setting the pulse repetition rate f which is computer-controlled and hence finely adjustable. There is then no need to change either the transmitter frequency or the radiofrequency level in this experiment; instead the pulse repetition rate, the pulsewidth and number of pulses are the adjustable parameters. The net flip angle at resonance is given by

$$\alpha = \gamma B_1 N t_p, \quad [6]$$

where N is the number of pulses in the sequence. The selectivity in the frequency domain is the same as that of a soft pulse of the same overall duration; that is to say, the *effective* radiofrequency intensity is B_1 reduced by the duty cycle of the pulse sequence, t_p/τ. The main practical restriction on the implementation of this sequence is to avoid pulse imperfections when the pulsewidth t_p is very short (of the order of 1 microsecond). Where there is a problem, the transmitter level B_1 is reduced by switching in a suitable attenuator. Typically, a sequence of pulses with N between 10 and 50 might be used. Care must of course be taken to avoid any excitation by the unused sidebands on each side of the transmitter frequency.

In Dante's Purgatory, souls progress toward Heaven by negotiating a series of seven circular ledges running around a mountain. Progress from one ledge to the next is only permitted after one complete circumnavigation. This complex trajectory clearly anticipates the motion of a nuclear magnetization vector in the rotating frame experiencing a sequence of seven pulses at the first sideband condition. This is the origin of the name DANTE (Delays Alternating with Nutation for Tailored Excitation).

SHAPED SELECTIVE PULSES

The sin x/x function which describes the excitation pattern of a rectangular selective pulse has sidelobes which extend for a surprisingly large distance from exact resonance. This is an undesirable feature for many applications of selective pulses since a distant sidelobe could induce a weak excitation in a second resonance, perhaps the one under observation. The observation of weak nuclear Overhauser enhancements between protons would be particularly susceptible to this type of interference. (See Nuclear Overhauser effect*.)

If we accept that the excitation spectrum is approximately the Fourier transform of the pulse envelope, sidelobes arise because of the step discontinuities at the beginning and end of the pulse. They should therefore disappear if the rise and fall of the pulse envelope could be suitably smoothed. Many possible shaping functions could be considered for this purpose. We require a

Dante's Purgatory. The idea seems to be not so much to escape but to work one's way up to the more interesting vices.

time-domain envelope of short duration (to minimize relaxation losses during the pulse) but which gives the highest selectivity in the frequency domain and no sidelobes. Now it is one of the tenets of the theory (3) of Fourier transforms that if a time-domain function f(t) can be differentiated k times before the derivative exhibits a discontinuity, then the frequency-domain excitation falls off as $\Delta\nu^{-k}$ at large offsets $\Delta\nu$. Thus a rectangular pulse, which has a discontinuous first derivative, gives an excitation which falls off as $\Delta\nu^{-1}$ in the tails, as does a sin x/x function. The theorem suggests that a Gaussian shaping function would be a good choice in this respect, because all the derivatives of a Gaussian are continuous, so it should give a very compact frequency-domain excitation pattern. To the extent that the NMR response is linear, a Gaussian pulse envelope generates a Gaussian excitation pattern; the longer the duration of the former the more selective is the latter. These arguments are vindicated by an exact calculation of trajectories of nuclear magnetization vectors according to the Bloch equations for Gaussian-shaped pulses at a series of resonance offsets (4). The absolute-value mode NMR signal after a Gaussian excitation pulse follows an approximately Gaussian curve as a function of resonance offset (Fig. 3). A rectangular pulse of the same duration, although it has a narrower central lobe, has a prominent set of sidelobes that extend a large distance from the resonance where the Gaussian excitation is negligible. For practical implementation, the Gaussian pulse envelope is usually approximated by a digital function and the extreme tails of the Gaussian are truncated.

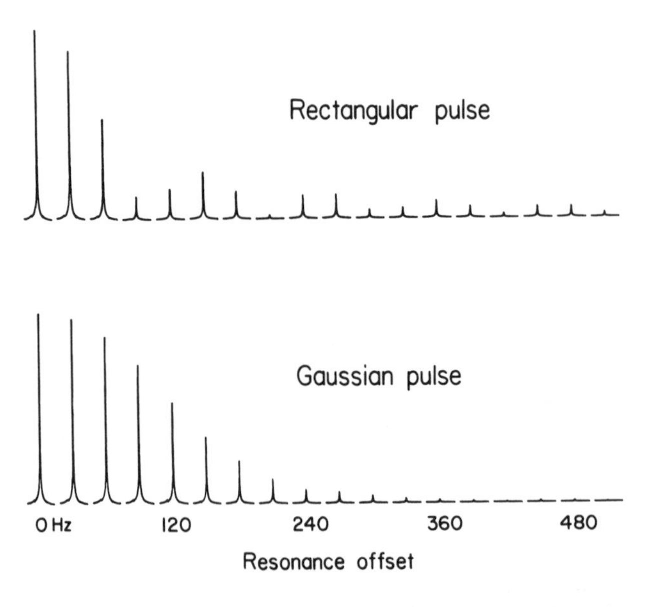

Fig. 3. Frequency-domain excitation patterns of rectangular and Gaussian-shaped selective pulses of the same duration (10 ms). At large resonance offsets, the excitation spectrum of the rectangular pulse still has sidelobes of appreciable intensity, whereas the excitation spectrum of the Gaussian pulse dies away monotonically.

APPLICATIONS OF SELECTIVE PULSES

Simplification of High-Resolution Spectra

Many high-resolution spectra suffer from overcrowding, several adjacent spin multiplet patterns overlapping in a complicated fashion. Selective excitation allows the spectroscopist to begin to disentangle the mess by substituting a subtle approach for the brute force method. For example, the fine structure of an individual carbon-13 site may be examined by exciting the decoupled carbon-13 resonance with a selective pulse followed by acquisition of the free induction decay with the proton decoupler switched off (5). In analogous fashion, individual proton multiplets may be monitored by selective polarization transfer* from a single carbon-13 site, using a modification of the INEPT experiment (6). A more general version of this application picks out proton spin multiplets one at a time without using the coupling to carbon-13 as a criterion. A highly selective spin population inversion pulse is applied to any single proton resonance line that is sufficiently well separated from its neighbours. The perturbation of populations is then spread uniformly over the multiplet by employing an observation pulse which has a 90° flip angle for the entire coupled spin system (usually a composite pulse). The individual proton multiplet is then observed by difference spectroscopy (7). Selective versions of homonuclear chemical shift correlation* experiments (COSY) can also be envisioned, turning a two-dimensional experiment into one-dimensional spectroscopy (4).

*Solvent Suppression**

There are many procedures in use to try to alleviate the problem caused by the extremely intense signal from water interfering with the proper digitization* of weak solute signals. Many of these involve selective presaturation of the solvent peak or selective population inversion or a combination of wideband excitation counterbalanced for the solvent peak by an antiphase selective pulse of the same flip angle. The DANTE sequence or a Gaussian shaped selective pulse may be used for this purpose. Alternatively, the frequency region over which high suppression is achieved may be extended somewhat by employing one of the binomial pulse sequences such as $1\bar{2}1$ or $1\bar{3}3\bar{1}$ (8–10).

*Relaxation and Chemical Exchange**

A whole range of spin–spin and spin–lattice relaxation* studies can be carried out with soft pulses with durations in the millisecond–second timescale (11). On a continuous-wave spectrometer these can be very easily implemented by pulsing modulation sideband responses simply by gating the audiofrequency signal fed to the field modulation coils. A nice feature of these experiments is that the NMR response is observable during the soft pulse, so that pulse flip angles are easily calibrated.

Selective pulses are also used in cross-relaxation studies and in particular for studies of the homonuclear Overhauser effect. Closely related experiments are performed when chemical exchange is studied by saturation transfer (1) or by selective population inversion. In both applications shaped selective pulses would seem to be preferable as it is very important to avoid off-resonance perturbation of the observed resonance since the intensity changes under investigation are so small.

Spin echo* experiments on coupled spin systems can be quite complicated because, in addition to the echo decay due to spin–spin relaxation, there is a modulation of the echo amplitudes due to homonuclear coupling. One way to circumvent the modulation effect and also reduce the echo decay to a single relaxation parameter, is to perform a selective excitation experiment. Each chemical site is then examined one at a time. Another way to suppress the echo modulation is to perform a spin-locking* experiment. Since this requires the radiofrequency field B_1 to be applied for times of the order of seconds, limitations of the radiofrequency circuitry preclude using the full B_1 intensity. A selective spin-locking experiment is accomplished with a much lower B_1 level.

Spatial Selectivity

The rapidly developing science of magnetic resonance imaging and the related field of high-resolution *in vivo* NMR both employ selective excitation in static magnetic field gradients in order to code the NMR signal according to its spatial coordinates. Selective excitation in a field gradient relies on the non-linearity of the NMR response (12) so that a simple pulse-observe sequence gives only a weak response. A better approach is to excite an echo, often accomplished here by reversing the B_0 gradient.

NMR imaging experiments often shape the time-domain pulse envelope according to a sin x/x function in an attempt to achieve a frequency-domain excitation that is close to rectangular, since this defines a slice through the sample more precisely. Pulse durations are very much shorter than in high resolution work since the B_0 gradients can be made very intense. *In vivo* NMR spectroscopy attempts to define an 'active volume' which gives rise to the high resolution spectrum (often phosphorus-31) while suppressing signals from the remainder of the sample volume.

REFERENCES

1. S. Forsén and R. A. Hoffman, *J. Chem. Phys.* **39**, 2892 (1963).
2. G. A. Morris and R. Freeman, *J. Magn. Reson.* **29**, 433 (1978).

3. R. N. Bracewell, *The Fourier Transform and its Applications*. McGraw-Hill: New York, 1978.
4. C. Bauer, R. Freeman, T. Frenkiel, J. Keeler and A. J. Shaka, *J. Magn. Reson.* **58**, 442 (1984).
5. G. Bodenhausen, R. Freeman and G. A. Morris, *J. Magn. Reson.* **23**, 171 (1976).
6. A. J. Shaka and R. Freeman, *J. Magn. Reson.* **50**, 502 (1982).
7. C. Bauer and R. Freeman, *J. Magn. Reson.* **61**, 376 (1985).
8. V. Sklenář and Z. Starčuk, *J. Magn. Reson.* **50**, 495 (1982).
9. D. L. Turner, *J. Magn. Reson.* **54**, 146 (1983).
10. P. J. Hore, *J. Magn. Reson.* **54**, 539 (1983).
11. R. Freeman and S. Wittekoek, *J. Magn. Reson.* **1**, 238 (1969).
12. D. I. Hoult, *J. Magn. Reson.* **35**, 69 (1979).

Cross-references

Chemical exchange
Continuous-wave spectroscopy
Digitization
Fourier transformation
Nuclear Overhauser effect
Polarization transfer
Saturation transfer
Shift correlation
Solvent suppression
Spin echoes
Spin–lattice relaxation
Spin–locking
Spin–spin relaxation
Two-dimensional spectroscopy
Vector model

Sensitivity

Nuclear magnetic resonance spectroscopy has always suffered from poor sensitivity, largely due to the low Boltzmann factors involved. It is therefore important to understand the various factors which limit the attainable signal-to-noise ratio, some of which are out of the control of the operator, others may be optimized. Codrington *et al.* (1) have calculated the sensitivity of a Fourier transform spectrometer by modifying an earlier treatment applicable to continuous-wave spectroscopy (2). Fundamental to this calculation is the assumption that the output of the receiver is made up of two *independent* components, the signal voltage V_s and the noise voltage V_n. Since the Fourier transformation stage is a linear operation, signal and noise components remain independent.

The calculation involves the following parameters

N	Number of nuclear spins per unit volume of the sample
B_0	Applied static magnetic field intensity
γ	Magnetogyric ratio of the nucleus
I	Spin quantum number
T	Absolute temperature
$\hbar$	Planck's constant divided by 2π
k	Boltzmann constant
μ_0	Permeability of free space (SI unit conversion factor)
ω_0	Larmor precession frequency ($\omega_0 = \gamma B_0$)
T_2^*	Decay constant of transverse magnetization
$\Delta\omega$	Offset from exact resonance (drops out)
α	Flip angle of the radiofrequency pulse
ξ	Filling factor of the receiver coil ($\xi < 1$)
ϱ	Ratio of effective inductance to total inductance of coil
Q	Quality factor of the receiver coil (typical value 50)
V_c	Volume enclosed by the receiver coil
n	Number of turns on the receiver coil (cancels out)

a	Length of the receiver coil (cancels out)
F	Noise figure of the preamplifier (F > 1)
G	Overall gain of the receiver (cancels out)
B	Final bandwidth of the receiver (cancels out)
λ	Nagaoka's constant ($\lambda = 1$ for infinitely long solenoid)

THE SIGNAL

The equilibrium nuclear magnetization M_0 is given by

$$M_0 = NB_0\gamma^2\hbar^2 I(I + 1)/3kT. \qquad [1]$$

A radiofrequency pulse of flip angle α produces a Y component of magnetization

$$M_Y = M_0 \sin \alpha. \qquad [2]$$

This is taken as the basis of the NMR signal calculation, relaxation and the attendant steady-state effects in a repeated pulse sequence being disregarded in this treatment. It is assumed that a crossed-coil probe is used and that the transmitter coil is sufficiently large compared with the receiver coil that the transmitter field may be regarded as uniform over the effective sample volume. The signal is only acquired for a limited time (the acquisition time). Otherwise the noise would be unlimited. The free induction signal is weighted with a suitable sensitivity enhancement* function in order to deemphasize the noise components in the tail of the decay. It is assumed that the optimum weighting function is used, having a time constant equal to T_2^*, the equivalent of a matched filter (3). The signal transformed into the frequency domain is

$$S(\omega) = \tfrac{1}{4}GKT_2^*(1 + i\Delta\omega T_2^*)/[1 + \tfrac{1}{4}\Delta\omega^2(T_2^*)^2], \qquad [3]$$

where

$$K = n\varrho\xi QM_0 \sin \alpha\ \omega_0 V_c/a.$$

This is the expression for a Lorentzian line of full width $2/(\pi T_2^*)$ Hz, twice the width of the line before sensitivity enhancement. The real part of this expression represents the absorption-mode signal, and the peak height is given by setting $\Delta\omega = 0$.

$$S_p(\omega) = \tfrac{1}{4}GKT_2^*. \qquad [4]$$

THE NOISE

Most of the noise arises from the thermal agitation of the electrons in the receiver coil, the 'Johnson noise', and depends on kT. Noise from the transmitter (a potential problem in continuous wave spectrometers) may be neglected in a pulse-modulated Fourier transform spectrometer. Noise introduced in the amplification stages, principally by the preamplifier, is accounted for through the noise figure F, a number greater than unity.

Plancherel's theorem states that the total noise energy is the same in the time and frequency domains (4). It is assumed that, within the bandwidth of interest, the noise is 'white', that is to say its distribution function is independent of frequency and time. This allows us to calculate the noise voltage in the frequency domain

$$N(\omega) = G(\tfrac{1}{2}kT\omega_0\lambda\mu_0 V_c \varrho QFT_2^*)^{1/2}(n/a). \quad [5]$$

SIGNAL-TO-NOISE RATIO

The quantity of interest is the ratio of the peak absorption-mode signal to twice the root-mean-square noise. This is the form adopted by the instrument manufacturers. Taking the peak signal amplitude as

$$S_p(\omega) = \tfrac{1}{4}Gn\varrho\xi QM_0 \sin\alpha(V_c/a)\omega_0 T_2^*. \quad [6]$$

this gives

$$R(\omega) = S_p(\omega)/2N(\omega)$$

$$= \frac{1}{24}\left[\frac{2\varrho QV_cT_2^*}{\mu_0 F\lambda}\right]^{1/2}\left[\frac{\omega_0}{kT}\right]^{3/2} N\xi\gamma\hbar^2 I(I+1)\sin\alpha. \quad [7]$$

(We shall see later that the peak-to-peak noise is five times the root-mean-square noise.) Note that the number of turns per centimetre on the receiver coil (n/a) has cancelled out of this expression, as has the receiver gain G and bandwidth B.

This expression may be rewritten in order to group the various factors into three categories: fundamental constants, properties of the NMR sample and receiver parameters.

$$R(\omega) = \left[\frac{\gamma\hbar^2 I(I+1)}{24k^{3/2}\mu_0^{1/2}}\right]\left[\frac{N\omega^{3/2}(T_2^*)^{1/2}}{T^{3/2}}\right]\left[\frac{2\xi^2\varrho QV_c}{\lambda F}\right]^{1/2}\sin\alpha. \quad [8]$$

This is a convenient form for considering the optimization of sensitivity.

SENSITIVITY OPTIMIZATION

There is of course nothing to be done about the fundamental constants, except to recognize that a high gyromagnetic ratio γ is important, and to note in passing that nuclei with $I > 1$ have slightly stronger signals than $I = \frac{1}{2}$ nuclei, although they tend not to be too interesting for high resolution studies.

The next group of parameters is under the control of the investigator. Every effort is made to increase N, the number of nuclei per unit volume. It is proportional to the molar concentration of the solution, the natural abundance of the nuclear species, and the number of equivalent spins contributing to a given resonance line. In certain circumstances there may be no alternative but to enrich a given isotope artificially in order to boost sensitivity.

The operating frequency may or may not be a parameter which the investigator can select. It is not simply a matter of designing magnets at the highest possible field, since there is the parallel requirement that a large enough volume of homogeneous field be accessible. Nevertheless this is one parameter that has seen the most consistent improvement over the years, with the recent rapid increase in the number of superconducting solenoids in use for high resolution NMR. The operating frequency enters the calculation in three separate ways: in the expression for the Boltzmann population difference; as the frequency at which the nuclear magnetization oscillates in the receiver coil; and in the frequency dependence of the noise $\omega_0^{1/2}$. The resulting dependence on $\omega_0^{3/2}$ is not maintained for high γ nuclei such as protons, because at very high operating frequencies the Q factor degrades and the noise figure F increases, while the proportion of the inductance wasted in the receiver coil leads increases. Resonant cavities, similar to those used in electron spin resonance, have been used on a few very high field spectrometers. For nuclei like carbon-13 which require efficient broadband decoupling from protons, higher operating frequencies put higher demands on the bandwidth covered by the decoupler if sensitivity is to be maintained.

Low temperatures favour sensitivity, since the Boltzmann population differences increase and the noise power falls off. The result is a predicted $T^{-3/2}$ dependence. Unfortunately, the temperature range accessible to the operator is limited by the liquid range of the available solvents, and in any case the proportional change in T is quite small. In fact, it may often be more profitable to work at higher temperature since the solubility usually increases (increasing N) and the spin–lattice relaxation time usually decreases, allowing faster repetition of the excitation pulses. Because of the difficulties of precise temperature regulation, most experiments are conducted at the ambient probe temperature (often several degrees above room temperature). However, it is feasible to cool the receiver coil while maintaining the sample near room temperature. This has the advantage of reducing the Johnson noise in the receiver coil and increasing the quality factor Q. Experiments have successfully been carried out with the receiver coil at the temperature of liquid helium (4.2 K) but, although a large improvement in the signal-to-noise ratio was observed for this coil, it suffered from a significant loss of filling factor ξ and it is not yet clear whether an absolute improvement in sensitivity has been achieved compared with a conventional spectrometer under identical conditions (5).

The decay rate of the sensitivity enhancement function affects both the signal intensity (by broadening the lines) and the noise energy (by filtering out rapidly fluctuating noise). Note that the calculation assumed the equivalent of matched filtration. Under these conditions the sensitivity is proportional to $(T_2^*)^{1/2}$. An improvement to the resolving power of the spectrometer is therefore reflected in stronger signals, provided that instrumental broadening still outweighs the natural linewidths. Imperfect heteronuclear decoupling or residual field/frequency instabilities may contribute to the linewidth $(T_2^*)^{-1}$ and therefore degrade the signal-to-noise ratio.

The operator has a rather restricted choice of the receiver coil parameters, essentially limited to the selection of the sample tube diameter. It is nevertheless important to understand how these parameters influence the signal-to-noise ratio. The receiver coil filling factor ξ influences the signal strength directly. Under the assumption that the radiofrequency field of the transmitter coils is essentially uniform when measured over the effective volume of the receiver coil, the filling factor may be approximated by $(d_s/d_c)^2$, where d_s and d_c are the diameters of the sample and the coil respectively. The sample tube wall thickness, any Dewar vessels and the clearance required for a spinning sample tube all contribute to $d_c - d_s$, but the proportional effect on ξ falls off for larger sample diameters. Recently significant improvements in filling factor have been achieved by careful attention to the bulk magnetic susceptibility of the wire used for the receiver coil, making this close to zero by combining metals with susceptibilities of opposite signs. This picks up magnetization from sample regions very close to the coil, normally 'lost' because of distortions of the B_0 field.

The quality factor Q of the coil affects sensitivity as $Q^{1/2}$ since it improves the signal voltage proportional to Q and the noise as $Q^{1/2}$. It represents the ratio of the energy stored in the tuned circuit to the energy dissipated in the series resistance. Hence it is favoured by heavy gauge wire, high-conductivity

Filling factor

copper and low-loss capacitors. The Q factor tends to follow a broad maximum as a function of frequency. Samples which contain ions, for example aqueous solutions of some biological materials, can appreciably degrade the Q factor of the coil and reduce sensitivity. This can be thought of as a capacitive coupling of the tuned circuit to the random Brownian motion of the ions. For this reason, coils constructed of flat strips of copper foil, although they normally possess high quality factors, are not really suitable for the study of ionic solutions since they have a relatively large capacitive coupling; thin wire is preferable. The rate at which a radiofrequency pulse can be switched on or off depends on Q, so single coil spectrometers may require a Q-spoiling device (for example crossed diodes) which operates during the pulse but not during signal acquisition.

The inductance of the leads to the receiver coil is largely wasted for the detection of nuclear magnetic resonance signals, and this loss of efficiency is reflected in the correction factor $\varrho^{1/2}$. This can be quite important at high frequencies when the number of turns on the receiver coil is small (often unity). For a nucleus of intermediate frequency it is often found that increasing the number of turns on the receiver coil improves sensitivity, although this parameter n dropped out of the expression for signal-to-noise ratio; the effect is explained by the change in the ϱ factor.

The simplest way of increasing sensitivity is to use a larger volume sample coil, since the signal-to-noise ratio increases as $V_c^{1/2}$ while Q, ξ and ϱ also tend to increase with V_c. Of course, this does not help if the absolute amount of sample is limited and the concentration has to be reduced, nor is it profitable to use a small diameter sample tube in a large diameter coil, since the filling factor would be very poor. Receiver coils have been specially designed to accept small (2 mm) samples for cases where the sample is limited. If the receiver coil is increased in size too much, problems arise with the homogeneity of the field and with broadband heteronuclear decoupling. A practical limit seems to be 20–25 mm diameter samples.

Nagaoka's constant λ reflects the receiver coil geometry, favouring short fat coils, since $\lambda^{1/2}$ affects noise but not the signal. For a solenoid coil with length equal to its diameter $\lambda = 0.69$. Note that quite different considerations apply to saddle-shaped receiver coils (6) commonly used in superconducting solenoids where the sample tube axis is along the field direction B_0 rather than at right angles. Some probes have been designed for superconducting solenoids which have a 'sideways spinning sample' and a solenoid receiver coil in order to retrieve this loss of sensitivity.

The receiver noise figure enters the calculation as $F^{-3/2}$. With good pre-amplifier design F can be quite close to unity (within 10%), so there is little scope for improvement here for conventional samples in non-aqueous solvents. The picture changes dramatically for proton NMR in aqueous solution because of the dynamic range problem (see Digitization* and Solvent suppression*). Non-linear effects in the radiofrequency mixers, intermodulation distortion in the receiver and the limited vertical resolution of the analogue-to-digital converter all conspire to introduce spurious fluctuations which appear to the spectroscopist as 'noise' and which obscure very weak NMR signals. Note that the specifications for sensitivity quoted by the instrument manufacturers seldom (if ever) apply to biochemical samples in aqueous solution.

For carbon-13 spectroscopy, or for that matter any experiment involving broadband decoupling*, there is another type of 'noisy' artifact. The decoupler radiofrequency applied to the protons is periodically phase-modulated and some spurious modulation is transmitted to the carbon-13 free induction decay. This appears in the spectrum as weak *cycling sidebands*. Under certain conditions the proliferation of these weak sidebands resembles true noise and degrades the spectrometer sensitivity.

It might seem rather surprising that the receiver bandwidth B dropped out of the expression for sensitivity, since it appears in the corresponding expression for a slow-passage continuous-wave spectrometer. In the Fourier transform experiment, the equivalent of the last stage of bandwidth reduction is the sensitivity enhancement weighting function applied to the free induction decay. In the present calculation it has been assumed that this weighting function is an exponential of time constant T_2^*. Hence the counterpart of the receiver bandwidth parameter is the factor $(T_2^*)^{1/2}$ which affects the noise level.

The factor $\sin\alpha$ has been deliberately left until last. For a single isolated pulse acting on a spin system at thermal equilibrium the condition $\alpha = 90°$ obviously gives the maximum detected signal, and for this setting, small deviations in the intensity of the B_1 field over the sample have little effect. Many applications of NMR, however, require that the sensitivity be improved by repetitive excitation, and steady-state effects must be considered. Usually it is the Z-component of magnetization which exhibits appreciable steady-state effects*, and the effects of transverse magnetization can be disregarded. Then the optimum flip angle is the 'Ernst angle' α given by $\cos\alpha = \exp(-T/T_1)$, where T is the interval between pulses.

Perhaps the best hopes for improved sensitivity lie outside the scope of these calculations, relying on increasing M_0 by polarization transfer* or the electron–nuclear Overhauser effect*.

In order to give an appreciation of the kind of sensitivity that can be achieved on present-day (1986) high-field spectrometers, we show in Fig. 1 the proton spectrum of 5 micrograms of penicillin G obtained at 500 MHz. This makes no claim to be a world record.

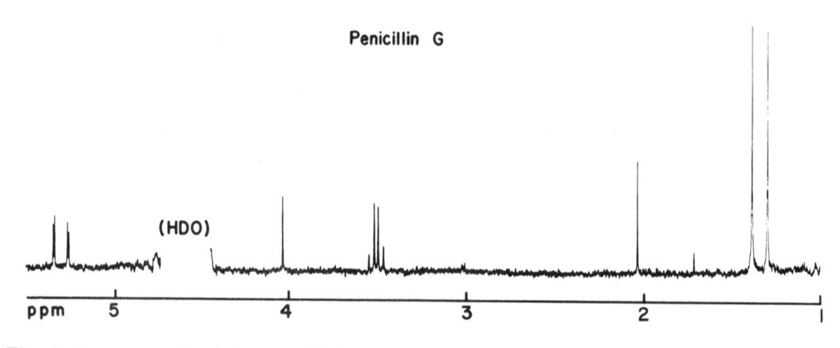

Fig. 1. An example of the sensitivity achievable on present-day high-field spectrometers. A 500 MHz proton spectrum of a 5 μg sample of Penicillin G in D_2O after 90 min of accumulation. The residual HDO peak was irradiated in the 1 s interval between scans. Spectrum, courtesy of Dr A. Derome.

THE PRACTICAL MEASUREMENT OF SIGNAL-TO-NOISE RATIOS

The difficulty lies in measuring the noise level. First, a section of baseline must be selected which does not contain artifacts that might be mistaken for noise, such as weak spinning sidebands. If the noise is to be measured from a chart recording, the gain must be increased by a suitable known factor until the level of noise is sufficiently high. This is not simply to improve the accuracy of the determination, but to avoid the effects of recorder 'dead band' which tends to iron out low-level noise. Recorder plotting programmes often incorporate a variable pen velocity control in the X axis so that (steep-sided) signals are more faithfully reproduced than low-level noise.

The next step is to select a section of noise containing 100 zero-crossings and to measure the peak-to-peak noise in this section; this is taken to be 5.0 times the root-mean-square noise voltage. For spectra that have been properly digitized and stored in the computer, a better method is to calculate the root-mean-square noise directly by means of a suitable computer routine. The 'signal-to-noise ratio' is then the peak signal height divided by *twice* the r.m.s. noise.

Finally, an important caveat. Strictly the term 'sensitivity' is defined in terms of the minimum concentration of test material required to produce a signal that is just detectable above the level of the noise. For most purposes, 'sensitivity' and 'signal-to-noise ratio' may be used interchangeably, but only when all spectrometer operations are *linear*.

REFERENCES

1. R. S. Codrington, H. D. W. Hill and R. Freeman, Varian Research Report, Number 105 (1974).
2. H. D. W. Hill and R. E. Richards, *J. Phys. Soc. (Lond.) E.* **1**, 977 (1968).
3. R. R. Ernst, *Advances in Magnetic Resonance* **2**, 1 (1966).
4. D. I. Hoult and R. E. Richards, *Proc. Roy. Soc. (Lond.)* A**344**, 311 (1975).
5. P. Styles, N. F. Soffe, C. A. Scott, D. A. Cragg, F. Row, D. J. White and P. C. J. White, *J. Magn. Reson.* **60**, 397 (1984).
6. D. I. Hoult, *Progress in NMR Spectroscopy* **12**, 41 (1978).

Cross-references

Broadband decoupling
Digitization

Nuclear Overhauser effect
Polarization transfer
Sensitivity enhancement
Solvent suppression
Steady-state effects

Sensitivity Enhancement

The factors which determine the inherent sensitivity of the NMR spectrometer are set out under Sensitivity*. This section examines methods for enhancing this important parameter, supposing that all the normal precautions have already been taken to optimise the instrumental conditions. These enhancement techniques may be broken down into three categories – methods which increase the signal strength, those which reduce the noise level, and those which separate the two by some type of data processing technique (1).

SIGNAL ENHANCEMENT

It is not always necessary to work with the equilibrium nuclear magnetization M_0. Indeed spectroscopists who perform NMR experiments in the Earth's magnetic field would experience difficulties if they did so (although the sample volume could be several litres since there is a large volume of homogeneous field). Prepolarization in a higher magnetic field is one possibility, the enhanced polarization persisting for times of the order of the spin–lattice relaxation time. The polarizing field need not of course be homogeneous. This method has not as yet had much impact on high resolution spectroscopy.

Signal enhancement through the electron–nuclear Overhauser effect* is a second possibility, promising theoretical improvements as high as two to three orders of magnitude for proton NMR. The practical difficulties stem from the need to supply microwave radiation (as well as radiofrequency excitation) without interfering with the high-resolution probe. Another objection is that the sample has to have some kind of paramagnetic agent added (often a stable free radical like diphenyl picryl hydrazyl). But the most telling disadvantage is that it would not be a viable technique at the high magnetic fields now commonplace in high resolution spectroscopy, because the microwave frequency would be unmanageably high and because the Overhauser enhancement falls off at high frequencies. This leaves the much more modest enhancements available from *nuclear* Overhauser effects, essentially a routine technique for carbon-13 spectroscopy where the maximum enhancement is a factor 3. The price paid for this improvement is the uncertainty in the relative signal intensities, rendering quantitative work unreliable. Mention should also be made of the increase in signal height obtained through coalescence of the carbon-13 multiplets when proton broadband decoupling* is used; a methyl group, for example, increases by 8/3.

Polarization transfer* is an alternative to the nuclear Overhauser effect, either through Hartmann–Hahn rotating frame resonance (see Hartmann–Hahn experiment*) or by the simpler INEPT or DEPT pulse sequences. Unlike the Overhauser effect, these do not require any particular relaxation mechanism so they are more generally applicable. The enhancement factor is simply γ_I/γ_S, where I is the nucleus providing the strong polarization and S is the low-sensitivity nucleus. This gives better enhancement factors than the nuclear Overhauser effect, particularly for large γ_I/γ_S ratios and for systems where γ_I and γ_S are opposite in sign as, for example, the proton–nitrogen-15 system. Often there is a further enhancement due to the more favourable spin–lattice relaxation of the I spins. A prerequisite of the method is a resolved spin–spin interaction, otherwise the transverse I spin magnetization decays through spin–spin relaxation before the transfer can take place.

Another route is by way of indirect detection – using a high-sensitivity spin system to detect the spectrum of a low-sensitivity nucleus. The largest enhancements would be expected through optically detected magnetic resonance (ODMR) or electron–nuclear double resonance (ENDOR) but high resolution NMR has so far been limited to the modest enhancements provided by indirect detection through other nuclei, usually protons. Spin-echo difference spectroscopy* has had some success here. Difficulties arise when the S spin is of low natural abundance since there is then an extremely intense I spin signal from those molecules which do not contain the rare S isotope. The best results have been achieved by schemes which convert proton magnetization into heteronuclear multiple-quantum coherence* which precesses at a frequency which involves the chemical shift of the rare spin, followed by reconversion to observable proton magnetization. This has been very successful for studying nitrogen-15 spectra (2).

Although clearly not a general method, chemically induced nuclear polarization* should be included in this section. It allows the products of a free radical reaction to be detected with enormously enhanced spin populations because of a sorting process which takes place while the radicals are trapped in a solvent cage. The signal enhancements die away through spin–lattice relaxation, usually to an undetectably low level because the concentrations are very low indeed.

An entirely different approach is to increase the *rate* at which repetitive signals can be collected since most experiments involve time averaging. This rate is normally limited by the requirement for spin–lattice relaxation. There are two approaches to this problem; either reduce the flip angle below 90°, or introduce a relaxation delay after signal acquisition. The details of the optimization procedure are set out under Steady-state effects*. Relaxation can be accelerated by adding a suitable relaxation reagent, for example chromium acetyl acetonate. For slowly relaxing carbon-13 nuclei, it may be useful to employ a flow method which replenishes the saturated spins inside the radiofrequency coil with new polarized spins from outside (3). Such devices are only useful if the quantity of sample material is large and the solubility limit has already been reached. One might speculate on the possibility of finding some scheme for *pulsed relaxation* where spin–lattice relaxation is greatly enhanced during the waiting period between signal acquisitions, but somehow turned off while the free induction signal is being detected.

NOISE REDUCTION

There is not really much scope for reducing the thermal noise generated by the random motion of the electrons in the wire of the receiver coil. If all is well with the spectrometer, this is the dominant source of noise, and because the NMR phenomenon is such a weak effect there will always be competition between the two. Somehow the spectroscopist has to learn to live with this noise and not vainly seek to make it go away. Cooling the coil to the temperature of liquid helium does in fact reduce this Johnson noise by a factor of 8.4 (it goes as the square root of the temperature ratio), but the necessary cryogenic apparatus and the need to keep the sample near room temperature make the filling factor poor (4). Actually, this is not entirely a noise reduction experiment since a further advantage accrues from the improved quality factor Q which enhances the signal more than the noise.

Not all 'noise' is true thermal noise. The accepted method for reducing fluctuations from other sources is to modulate the NMR signal and then use synchronous detection, for this should discriminate against the instrumental fluctuations. In fact present-day Fourier transform spectrometers enjoy this benefit already as a by-product of pulsed operation; the fluctuations in transmitter level do not pass into the receiver for the latter is switched off during transmitter pulses.

DISCRIMINATION BETWEEN NOISE AND SIGNALS

Here we are looking for some characteristic property which allows us to separate signals from noise. Often the noise is fluctuating faster than the NMR signal – its Fourier spectrum contains higher-frequency components which may therefore be filtered out. In the frequency domain this may be achieved by a simple low-pass filter (RC circuit) or convolution* with a suitable smoothing function. In the time domain it involves multiplication of the free-induction decay by a *sensitivity enhancement* function, often a decaying exponential, which de-emphasizes the noise components in the tail of the free induction decay. Multiplication by an exponential in the time domain is equivalent to convolution by a Lorentzian in the frequency domain (the two functions form a Fourier transform pair). If the process is carried too far the signal is broadened and the signal-to-noise ratio begins to be degraded. The optimum condition is known as the *matched filter*. In the time domain this corresponds to multiplication with a sensitivity enhancement function which exactly matches the decay of the raw signal. A matched filter broadens a Lorentzian line by a factor 2 and a Gaussian line by a factor $2^{1/2}$. The use of an exponential sensitivity enhancement function is essentially a routine operation in Fourier transform NMR; some two-dimensional experiments (5) are better served by a Gaussian function which gives more suitable lineshapes in two-dimensional spectra*.

A second, more fundamental property of noise that distinguishes it from an NMR signal, is its random character. If an NMR spectrum (or a free induction decay) is recorded a second time, then at any chosen point, the signal component will have the same amplitude whereas the noise will fluctuate in amplitude and perhaps in polarity. Summation of successive measurements accumulates signal components linearly as the number of traces N, whereas noise grows only as $N^{1/2}$ (6). This is basically the extension of the principle of the 'boxcar integrator' to a multichannel device. It is important to remember that this time averaging* method only works properly if the noise is truly random – any coherence in the fluctuations will not 'average out'.

Time averaging is central to the operation of modern Fourier transform spectrometers (7), which themselves have largely superseded continuous-wave instruments because they enjoy the enormous advantage of gathering signals from all frequency regions of the spectrum simultaneously. Unfortunately, the method of time-averaging has inherent practical limits dictated by the dependence of signal-to-noise ratio on the square root of the number of recordings. An increase in the experimental time from an overnight run to an entire weekend only doubles the signal-to-noise ratio at best.

Implicit in time averaging is the requirement for digitization* of the signal. It is important to ensure that this process does not increase the noise content by aliasing high-frequency noise. Just as NMR signals can be aliased by sampling the free induction decay at a rate which violates the Nyquist condition, so too can high frequency noise components. The incoming signal must therefore be band-limited to filter out high-frequency noise before digitization. Similar requirements apply to time averaging of frequency-domain spectra – an integrator or filter must precede the analogue-to-digital converter so that frequency components higher than the sampling rate are filtered out.

REFERENCES

1. R. R. Ernst, *Advances in Magnetic Resonance*, Ed.: J. S. Waugh. Academic Press: New York, 1966, Vol. 2.

2. A. Bax, R. H. Griffey and B. L. Hawkins, *J. Am. Chem. Soc.* **105**, 7188 (1983).

3. D. A. Laude, R. W. K. Lee, and C. L. Wilkins, *J. Magn. Reson.* **60**, 453 (1984).

4. P. Styles, N. F. Soffe, C. A. Scott, D. A. Cragg, F. Row, D. J. White and P. C. J. White, *J. Magn. Reson.* **60**, 397 (1984).

5. R. Freeman, *Proc. Roy. Soc. (Lond.)* A **373**, 149 (1980).

6. M. P. Klein and G. W. Barton, *Rev. Sci. Instr.* **34**, 754 (1963).

7. J. C. Lindon and A. G. Ferrige, *Progress in NMR Spectroscopy* **14**, 27 (1980).

Cross-references

Broadband decoupling
Chemically induced nuclear polarization
Convolution
Difference spectroscopy
Digitization
Hartmann–Hahn experiment
Lineshapes in two-dimensional spectra
Multiple-quantum coherence
Nuclear Overhauser effect
Polarization transfer
Sensitivity
Steady-state effects
Time averaging

Shift Correlation

When NMR spectroscopy is used to elucidate molecular structure, one of the weakest links in the evidence is often that of assignment of the resonances. Comparison with the spectra of similar molecules and the organic chemist's knowledge of trends in chemical shielding commonly provide this information, but where these methods fail, recourse must be made to further NMR experiments such as relaxation measurements or double resonance studies. Recently two-dimensional spectroscopy* has begun to take over from double resonance, and a particularly useful form for assignment purposes is called *chemical shift correlation*.

Consider first for simplicity the heteronuclear case, for example the correlation of proton and carbon-13 chemical shifts. Basically this is a polarization transfer* experiment, usually in the direction protons → carbon-13. Any suitable interaction could be used to implement the transfer of polarization – J-coupling, cross-relaxation or chemical exchange*, but the most common is J-coupling. Since single-bond couplings are very much larger than long-range C–H couplings, the experiment can be set up so that essentially all the transfer takes place via the former. There are two time intervals involved in the two-dimensional experiment. During the first interval, the evolution time t_1, protons are labelled by their natural precession frequencies. Then magnetization is transferred to the appropriate carbon-13 resonances which precess at their natural frequencies during the second interval, the acquisition time t_2. Two-dimensional Fourier transformation converts this picture into the frequency domain. A peak in this two-dimensional spectrum therefore has an abscissa (on the F_1 axis) which is the proton chemical shift, and an ordinate (on the F_2 axis) which is the shift of the carbon-13 atom to which that proton is attached. The proton and carbon shifts are said to be *correlated* and the spectrum is called a shift correlation spectrum. If the proton assignment is known, then the carbon-13 assignment naturally follows and *vice versa*.

The observed carbon-13 signal is restricted to magnetization components transferred from protons, and the 'natural' carbon-13 signal is suppressed by a difference method since it serves no useful purpose in shift correlation. The transferred magnetization is four times stronger than the natural contribution since proton Boltzmann populations are four times larger than carbon-13 populations. Since only transferred magnetization is detected, the cycle time for repeating the experiment can be quite short, because it is only necessary to wait for *proton* spin–lattice relaxation. The experiment is therefore fast and has quite good sensitivity.

The simplest way to visualize the magnetization transfer is in terms of spin population effects, based on the fact that coupled proton and carbon spins share the same energy levels. A perturbation of proton populations necessarily influences carbon-13 intensities. In this sense the experiment is related to the 'selective population transfer' experiments which employ a selective 180° pulse applied to one of the carbon-13 satellite lines in the proton spectrum.

Consider the simplest case of a CH group. The pulse sequence is shown in Fig. 1. The initial 90° pulse about the X axis excites transverse proton magnetization along the Y axis represented by two vectors which then precess at $2\pi(\delta_H \pm \frac{1}{2}J_{CH})$ rad sec^{-1}. At the midpoint of the evolution period a 180° pulse is applied to carbon-13 which reverses the labels on these vectors so that they now precess at $2\pi(\delta_H \mp \frac{1}{2}J_{CH})$ rad sec^{-1}, giving a net precession angle at the end of the evolution period $\theta = 2\pi\delta_H t_1$ radians. The 180° pulse has thus had the effect of removing the J_{CH} term from this expression, effectively decoupling the protons from carbon-13 during the evolution period. As a result, CH splittings do not appear in the F_1 dimension of the shift correlation spectrum. At this point the two proton vectors are parallel (Fig. 2(d)) and a *fixed* interval $\Delta_1 = 1/(2J_{CH})$ is introduced to allow these vectors to precess into antiparallel orientations in preparation for the actual magnetization transfer stage (Fig. 2(e)).

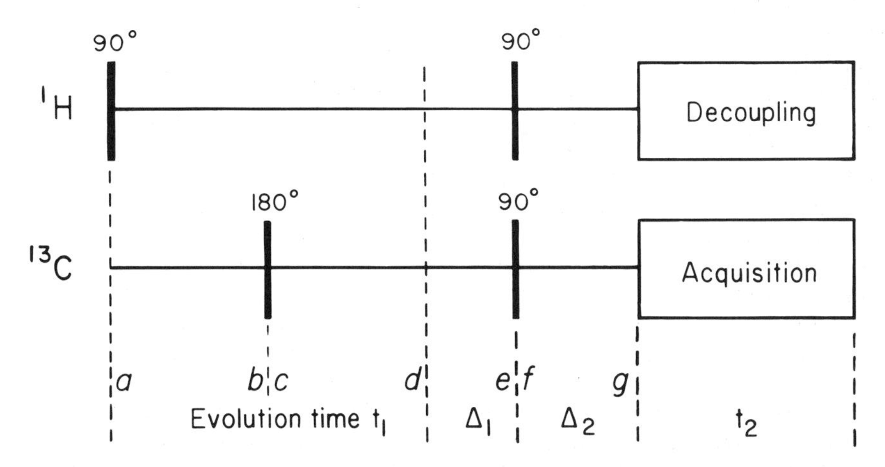

Fig. 1. Pulse sequence for a two-dimensional shift correlation experiment employing polarization transfer from protons to carbon-13. The timing markers (a)–(g) refer to the vector diagram of Fig. 2. The fixed delays Δ_1 and Δ_2 prevent signal cancellation.

It is now convenient to decompose these vectors into their components $\pm M \cos\theta$ along the X axis and $\pm M \sin\theta$ along the Y axis. A proton 90°(X) pulse rotates the latter into $\pm M \sin\theta$ along the Z axis, corresponding to a population disturbance on the appropriate energy levels. The other component ($\pm M \cos\theta$) is converted into multiple-quantum coherence* and is 'lost'. The population disturbance therefore evolves as $\sin\theta$ where $\theta = 2\pi\delta_H t_1$; this is how the information about the proton shift is transferred to the observed carbon-13 signal. There is no perturbation if $\theta = 0$ and a maximum perturbation if $\theta = \pi/2$. For this extreme condition ($\theta = \pi/2$) the spin system behaves as if a selective 180° pulse had been applied to one proton transition to give a population inversion while the other was left undisturbed. Figure 2 shows the corresponding populations on the energy-level diagram (reduced to the simplest numerical representation based on $\gamma_H/\gamma_C = 4$).

These population disturbances influence carbon-13 transitions that share the same energy levels. They may be detected by applying a carbon-13 'read' pulse which converts them into transverse magnetization (Fig. 2(f)). For the case of maximum transfer ($\theta = \pi/2$) this would be observed as an up–down pattern of carbon-13 signals with relative intensities −3 and +5. If the unit intensity natural carbon-13 signal is suppressed, the signals due to magnetization transfer have intensities −4 and +4. This represents the γ_H/γ_C sensitivity enhancement factor.

Now, there is an important simplification of the correlation spectrum if proton broadband decoupling can be used during the acquisition of the

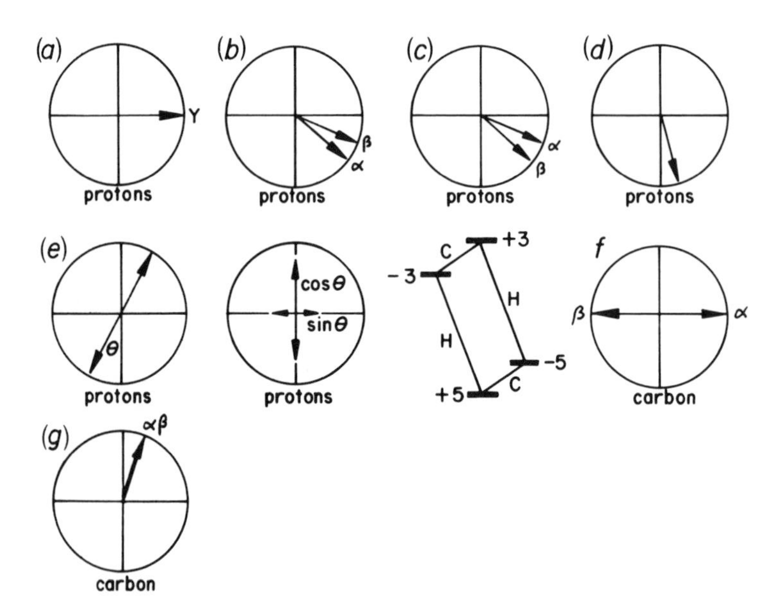

Fig. 2. Vector diagram illustrating the shift correlation sequence set out in Fig. 1. The polarization transferred is proportional to $\sin\theta$ and is thus a function of the proton shift and the evolution time. The mechanism for transfer is the disturbance of the populations on the energy level diagram; for simplicity the maximum perturbation is indicated ($\sin\theta = 1$).

carbon-13 signal in the t_2 interval. However, at point (f) the carbon-13 vectors are antiparallel and would cancel if broadband decoupling were applied. Consequently, a further *fixed* period Δ_2 is introduced to allow these vectors to precess by $\pm\pi/2$ radians into a parallel orientation. For a CH group Δ_2 is set equal to $1/(2J_{CH})$ but in general where there are also CH_2 and CH_3 groups present, a compromise setting $\Delta_2 = 1/(3.3J_{CH})$ should be used. At the end of the Δ_2 delay (Fig. 2(g)), proton broadband decoupling is switched on and the carbon-13 signal is acquired.

HOMONUCLEAR SHIFT CORRELATION (COSY)

Two-dimensional chemical shift correlation spectroscopy can also be carried out on homonuclear systems; indeed this was the first two-dimensional Fourier transform experiment performed by Jeener (1) and which pioneered this new field of research. Proton–proton shift correlation spectra exhibit some important differences with respect to the proton–carbon experiment. The coupling constants lie in a smaller, continuous range so the proton shift correlation spectrum maps out the entire network of couplings. The correlation spectrum is square in shape, both axes span the proton chemical shift range. The technique is often called COSY (for correlation spectroscopy). There is a treatment of this experiment in the section on the product operator formalism*.

The pulse sequence is very simple (Fig. 3) with just two 90° pulses, which accounts for the fact that this was the first form of two-dimensional spectroscopy. We can distinguish two types of proton signal (2). In the first kind the protons do not change frequency between the evolution and detection periods (or at most the frequency changes from one line to another line of the same multiplet). The resulting peaks fall on (or very near) the principal diagonal of the spectrum and are known as *diagonal peaks*. No magnetization is transferred

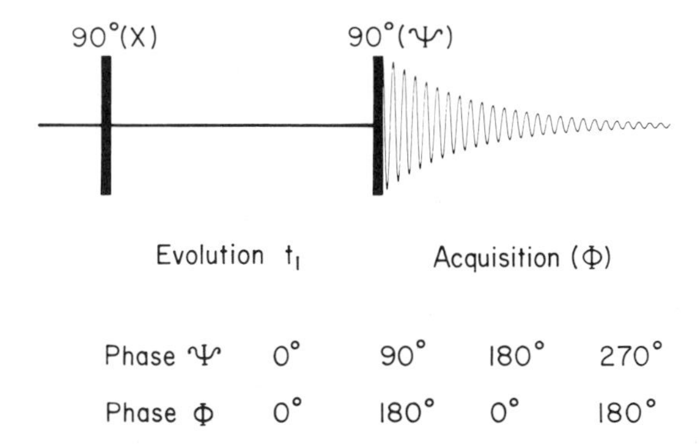

Fig. 3. Pulse sequence for a homonuclear shift correlation (COSY) experiment. The phase of the second pulse (Ψ) and the receiver reference phase (ϕ) are cycled as shown.

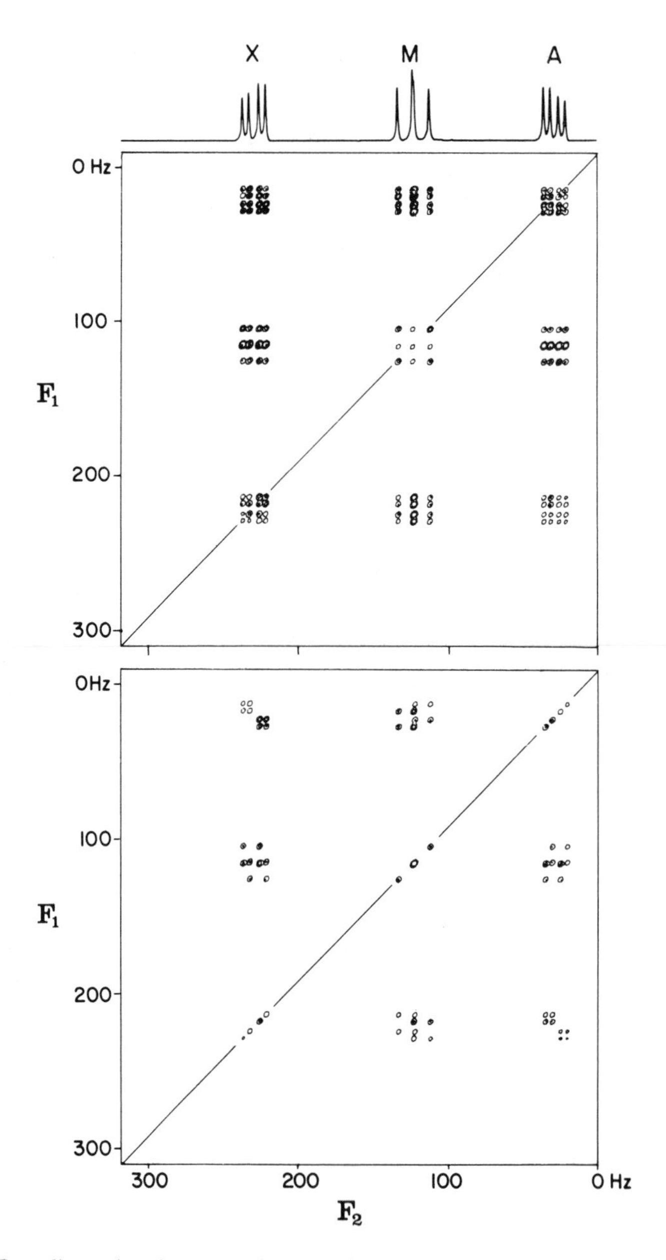

Fig. 4. Two-dimensional spectra of protons in 2,3-dibromopropionic acid. The effect of reducing the flip angle of the second pulse in a homonuclear shift correlation (COSY) experiment from 90° (upper spectrum) to 45° (lower spectrum). The number of multiplet components in each cross-peak is halved (indicating the relative signs of the two passive coupling constants) and the diagonal peaks are confined to those exactly on the diagonal.

from one proton site to another, so they provide no correlation information. For two protons A and X the coordinates would be $(\delta_A \pm \frac{1}{2}J, \delta_A \pm \frac{1}{2}J)$ and $(\delta_X \pm \frac{1}{2}J, \delta_X \pm \frac{1}{2}J)$. Diagonal peaks normally appear as two-dimensional dispersion mode signals.

Our attention is focused on the other type of response – the *cross-peaks* which arise from coherence transfer (3) from site A to site X, a frequency jump approximately equal to the chemical shift difference. These peaks form two square patterns symmetrically disposed with respect to the principal diagonal, with coordinates $(\delta_A \pm \frac{1}{2}J, \delta_X \pm \frac{1}{2}J)$ and $(\delta_X \pm \frac{1}{2}J, \delta_A \pm \frac{1}{2}J)$. Cross-peaks are normally in the pure absorption mode and alternate in sign around the square pattern $[\pm \ \mp]$. This characteristic arrangement of cross-peaks is easy to recognize and provides direct evidence that protons A and X are spin-coupled. If the coupling constant becomes comparable with the natural linewidth, appreciable mutual cancellation occurs between the antiphase components of the cross-peak; eventually this leads to vanishingly low intensity, but surprisingly weak spin–spin couplings can be detected in this experiment. It is possible to 'tune' for a particular long-range coupling by introducing a fixed delay into the evolution period (4).

The coupling J_{AX} which gives rise to the coherence transfer may be designated the *active* coupling. When a further proton M couples to this two-spin system the new couplings are *passive* and simply introduce new splittings into the cross-peaks, that is, they convert δ_A and δ_X into effective chemical shifts $(\delta_A \pm \frac{1}{2}J_{AM})$ and $(\delta_X \pm \frac{1}{2}J_{MX})$. Now, if the second pulse of the sequence is reduced (a convenient setting is 45°) then the passive spin does not change its orientation during the experiment, so the number of multiplet components within a given cross-peak is halved. The relative signs of J_{AM} and J_{MX} may then be obtained by inspection, for the cross-peaks are now made up of two identical square patterns displaced in a direction arctan (J_{AM}/J_{MX}) as illustrated in Fig. 4 for a simple three-spin system.

Recently the COSY experiment has been considerably improved by the introduction of filtration through double-quantum coherence (5). This greatly reduces the relative intensity of the diagonal peaks, making it easier to detect weak cross-peaks close to the diagonal. It is also advantageous to use the pure absorption-mode display (see Lineshapes in two-dimensional spectra*) for this improves resolution and allows pattern recognition techniques (6) to be used more effectively.

REFERENCES

1. J. Jeener, *Ampère International Summer School*, Basko Polje, Yugoslavia (1971).
2. W. P. Aue, E. Bartholdi and R. R. Ernst, *J. Chem. Phys.* **64**, 2229 (1976).
3. A. A. Maudsley, A. Wokaun and R. R. Ernst, *Chem. Phys. Lett.* **55**, 9 (1978).

4. A. Bax and R. Freeman, *J. Magn. Reson.* **44**, 542 (1981).

5. U. Piantini, Ø. W. Sørensen and R. R. Ernst, *J. Amer. Chem. Soc.* **104**, 6800 (1982).

6. B. U. Meier, B. Bodenhausen and R. R. Ernst, *J. Magn. Reson.* **60**, 161 (1984).

Cross-references

Chemical exchange
Lineshapes in two-dimensional spectra
Multiple-quantum coherence
Polarization transfer
Product operator formalism
Two-dimensional spectroscopy
Vector model

Solvent Suppression

Now that slow-passage continuous-wave spectroscopy* has been superseded by pulse excitation and Fourier transformation* it is hard to think of any practical advantages of the old method, except perhaps that it is conceptually simpler and reminiscent of the procedures of optical spectroscopy where the rotation of a prism provided a frequency scan. The one important difference which puts Fourier spectroscopy at a disadvantage is the problem of signals with high dynamic range (see Digitization*). In practice, the worst case is that of dilute aqueous samples, for water is about 110 molar in protons and it is often necessary to study millimolar concentrations of the solute. Use of heavy water is often ruled out because it exchanges with the NH protons of peptides, proteins and nucleic acids.

Signals with a high dynamic range cause four kinds of practical problems in the spectrometer. Unless the entire receiver system is perfectly linear there will be spurious harmonics of the solvent frequency and intermodulation distortion (sums and difference frequencies) wherever there are two or more strong signal components. Fourier spectroscopy requires a digitized free induction decay and hence an analogue-to-digital converter (ADC). These have practical limitations on the dynamic range they can handle (typically 12 or 15 bits) and no part of the free induction signal can be allowed to overflow. This means that the very weak solute signals may be considerably smaller than the least-significant bit of the converter. One might imagine that this means that such weak signal components are therefore lost and do not get passed to the Fourier transform stage, but this is not the case since they are riding on top of the intense solvent signal. However, the ADC makes a digitization error of the order of the least-significant bit at each sampling operation, so the digitized free induction decay can be thought of as the superposition of the true signal and a 'digitization error' signal. The Fourier transform of this digitization error introduces 'digitization noise' into the frequency-domain spectrum which obscures the weak solute lines. This is not, of course, true noise but in practice it resembles random noise and is spread throughout the spectrum.

The remaining two problems are round-off errors in the Fourier transform program, since these can obscure the weak solute signals, and the limited word-length of the computer storage. Both these problems can be solved by using double-precision arithmetic. Hence, since the spectroscopist can learn to live with the weak harmonics induced by receiver non-linearity, the dynamic range problem is essentially localized at the analogue-to-digital conversion stage.

Many different approaches have been examined to solve the dynamic range problem. In continuous-wave spectroscopy, the strong solvent peak was simply clipped, so that if digitization were required (for time averaging) the weak signals could be properly converted. The only loss would be any signals directly underneath the solvent peak. In Fourier spectroscopy such clipping is not permissible. Efforts have therefore been directed almost exclusively at schemes which operate on the nuclear spin system to ensure that the solvent signal is greatly reduced in amplitude while the rest of the spectrum remains (as far as possible) unchanged. This may be achieved in three ways (a) by examining the free induction decay while the solvent signal is passing through a null condition after population inversion or (b) by selectively saturating the solvent or (c) by arranging not to excite the solvent peak.

SOLVENT SIGNAL PASSING THROUGH NULL CONDITION

This modification (1) of an inversion-recovery spin–lattice relaxation* experiment has been called *Water-eliminated Fourier transformation* (WEFT). It relies on the fact that of all the components in the sample, water probably has the longest spin–lattice relaxation time, since the solute tends to be a slowly tumbling macromolecule. Consequently, the signals of interest are well on their way to recovery towards their equilibrium magnetization when the water signal is passing through its null condition. The advantage of this technique is that it provides one of the best chances for observing solute signals very close to the water resonance. The disadvantage is that relative intensities are distorted by differential relaxation effects. There is a modification where the initial spin inversion pulse is made frequency selective; this largely avoids the intensity distortions and does not require differential relaxation rates between water and solute.

An experiment which is conceptually related to WEFT is one that exploits differences in spin–spin relaxation* times. A paramagnetic relaxation reagent is added to the solution in low concentration (<0.2 millimolar). By penetrating the inner hydration sphere of the ion, water is efficiently relaxed whereas solute molecules are much less influenced by the relaxation agent. A spin-echo* experiment is performed, the final half-echo being acquired at a time when the water signal has greatly diminished amplitude. Suppression ratios of 10^4 are claimed for this method (2). Echo modulation through scalar coupling is suppressed by using a high pulse repetition rate. Similar experiments have been performed in which chemical exchange is used to shorten the spin–spin relaxation time.

SELECTIVE SATURATION OF THE SOLVENT PEAK

This method has the advantage of simplicity. A soft radiofrequency pulse or a DANTE sequence (see Selective excitation*) is used to cause intense saturation of the water signal, then a non-selective 90° pulse is applied to elicit the free induction signal. Large suppression ratios can be achieved by saturation, but it is impossible to avoid some saturation of signals near to the water peak. Since the water peak is inhomogeneously broadened, it does not necessarily saturate uniformly across the line profile. This problem can be alleviated somewhat by making small changes in the irradiation frequency.

One practical disadvantage of saturation methods, particularly important if the duration of the saturating pulse is quite long, is the possibility of saturation transfer* by cross relaxation or chemical exchange with other nuclei. This gives rise to anomalous intensities for the lines affected or even the loss of some lines altogether.

AVOIDING EXCITATION OF THE SOLVENT PEAK

One of the principal advantages claimed for rapid-scan correlation spectroscopy (3) is that it can be set up to avoid scanning through the solvent peak. This would usually require two separate scans. In practice, the scan must be stopped well clear of the solvent resonance since the B_1 field in this experiment is quite intense and would excite the solvent at some distance from resonance. The method is thus unsuitable for observing signals close to the solvent.

Pulse schemes for selective excitation have been more successful, the solvent signal falling at a null in the excitation pattern. Pioneered by Alexander (4), this has evolved into a whole family of methods involving soft pulses, composite pulses (5) and the more recent binomial pulse sequences (6–10). Implicit in this approach is the acceptance of intensity and phase distortions in the spectrum near to the solvent line and the inability to observe lines very close to the solvent resonance. The phase distortion (which may not be linearly dependent on frequency) can normally be corrected quite effectively.

One of the most widely used schemes is the 214 sequence introduced by Redfield (5), probably the first example of a composite pulse*. It is a sandwich of five pulses of constant B_1 amplitude and relative widths $2\ \overline{1}\ 4\ \overline{1}\ 2$, where the overbars indicate reversal of the radiofrequency phase. There are no significant intervals between the pulses. It can be thought of as a single long (soft) pulse of duration τ superimposed on a negative-going pulse pair with separation $\tau/2$ and an amplitude $2B_1$. It produces a null in the excitation at a resonance offset at about $0.97/\tau$ (hertz). A relatively broad null condition is achieved by arranging that the first derivative with respect to resonance offset is also near to a null condition. In practice, the relative widths of the pulses are finely adjusted for the best solvent suppression; two orders of magnitude suppression is readily achieved. The phase shifts introduced by this technique follow a quite complicated function of offset, partially corrected by a computer-controlled phase correction scheme.

In an endeavour to find alternative sequences that are easier to implement in practice and that are less sensitive to imperfections such as the spatial inhomogeneity of B_1, a new family of pulse sequences has been introduced based on the binomial coefficients (6–10). Here the guiding principle has been the approximate Fourier transform relationship between the time-domain pulse envelope and the frequency-domain excitation pattern. Although this is only strictly followed for a linear system, the NMR excitation involved here is not grossly non-linear so the transform acts as a good guide. We consider here the $1\ \bar{3}\ 3\ \bar{1}$ sequence as a representative example since it combines good performance with relative simplicity. (The first member of the family, the $1\ \bar{1}$ *jump and return* sequence (9) was not initially regarded as a binomial sequence.)

Ideally, there would be four infinitely narrow pulses (delta functions) with relative intensities $1\ \bar{3}\ 3\ \bar{1}$ and alternating radiofrequency phases (represented by the overbars) separated by equal intervals of time τ. The Fourier transform of this function is an excitation function which follows $\sin^3 (\Delta\omega\tau/2)$, where $\Delta\omega$ is the offset from the transmitter frequency; this has a null in the excitation at the transmitter frequency. The essential feature is that the null condition is relatively broad because of the $\sin^3$ dependence (the alternative $1\ \bar{2}\ 1$ sequence (10) follows only a $\sin^2$ dependence). This is important because the flanks of the intense water resonance normally extend over an appreciable frequency range.

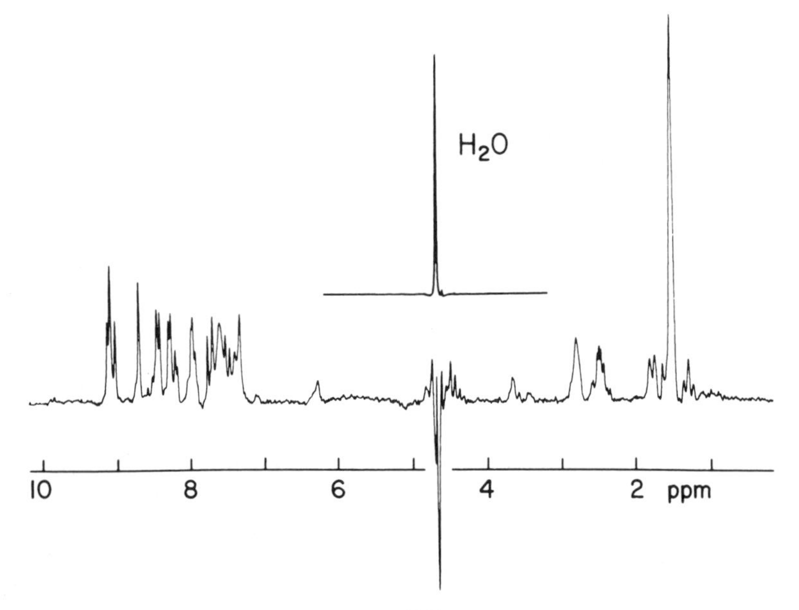

Fig. 1. Solvent peak suppression using the $1\bar{3}3\bar{1}$ sequence. 200 MHz proton spectrum of 50 mM viomycin in 90% H_2O and 10% D_2O. Inset is the H_2O signal obtained with conventional excitation scaled down by a factor one thousand; the suppression ratio is thus more than three orders of magnitude. Reproduced by permission from P. J. Hore (7).

It also helps to avoid exciting the spinning sidebands. A broad maximum in the excitation pattern occurs at offsets near $\pm 1/2\tau$ (Hz); this is used to examine the solute spectrum.

In practice, of course, the pulses have finite durations and it is the pulsewidths (rather than intensities) that are in the ratio 1:3:3:1. The efficiency of this sequence is not particularly sensitive to the setting of B_1 and is therefore not much affected by spatial inhomogeneity in the radiofrequency field. The sequence also tolerates missetting of the 3:1 ratio of pulsewidths. Time must be allowed for spin–lattice relaxation of water between scans so that no steady-state effects* occur. Phase distortion within the spectrum is readily corrected with a term linearly dependent on offset, with a 180° phase discontinuity at the transmitter frequency. The $1\,\bar{3}\,3\,\bar{1}$ sequence achieves suppression ratios of the order of 10^3 without the need for painstaking adjustment of the instrumental parameters (see Figure 1).

There are now many competing proposals for water suppression sequences, all of which appear to work well *in principle* but in practice there are so many instrumental imperfections involved that the suppression ratios are seldom as good as the predictions. Some sequences are very sensitive to errors in the radiofrequency phase or to radiation damping* effects. It may be necessary to combine more than one water suppression technique to achieve the desired result.

THE ANALOGUE-TO-DIGITAL CONVERTER (ADC)

All these spin gymnastics would be unnecessary if an analogue-to-digital converter could be designed with a much better dynamic range. Unfortunately, NMR spectroscopy demands faster and faster sampling rates in order to cover wide spectral widths. Because the ADC operates by a method of successive approximations, the time required for a conversion increases in proportion to the number of bits. It therefore seems unlikely that we can have both very high dynamic range and very fast sampling.

Another approach is to compress the dynamic range of the incoming signal by some suitable trick. One possibility is to examine only the *change* in signal intensity between one sampling operation and the next (11). Suppose that the first point on the free induction decay is not fed to the ADC but stored in a sample-and-hold circuit and used to measure the change in voltage for the second sample point, using a differential amplifier. This second ordinate is then used as reference for the third sampling point and so on. If the water signal is reasonably close to the transmitter frequency and if the sampling rate is high, then the intense water signal only changes a very small amount between adjacent sampling operations. The spectrometer gain may then be set quite high so as to avoid digitization errors on the weak solute signals without causing overflow of the water signal. Once the free induction decay has been digitized in this difference mode, dynamic range is no longer a problem because double-precision computer storage can be used. Consequently, the original free induction decay can be reconstructed by adding the ordinates cumulatively. It remains to be seen how effective this scheme will turn out to be in practice.

REFERENCES

1. S. L. Patt and B. D. Sykes, *J. Chem. Phys.* **56**, 3182 (1972).
2. R. G. Bryant and T. M. Eads, *J. Magn. Reson.* **64**, 312 (1985).
3. J. Dadok and R. F. Sprecher, *J. Magn. Reson.* **13**, 243 (1974).
4. S. Alexander, *Rev. Sci. Instr.* **32**, 1066 (1961).
5. A. G. Redfield, S. D. Kunz and E. K. Ralph, *J. Magn. Reson.* **19**, 114 (1975).
6. D. L. Turner, *J. Magn. Reson.* **54**, 146 (1983).
7. P. J. Hore, *J. Magn. Reson.* **54**, 539 (1983).
8. P. J. Hore, *J. Magn. Reson.* **55**, 283 (1983).
9. P. Plateau and M. Guéron, *J. Am. Chem. Soc.* **104**, 7310 (1982).
10. V. Sklenář and Z. Starčuk, *J. Magn. Reson.* **50**, 495 (1982).
11. S. Davies, C. Bauer, P. Barker and R. Freeman, *J. Magn. Reson.* **64**, 155 (1985).

Cross-references

Composite pulses
Continuous-wave spectroscopy
Digitization
Fourier transformation
Radiation damping
Saturation transfer
Selective excitation
Spin echoes
Spin–lattice relaxation
Spin–spin relaxation
Steady-state effects

Spin Echoes

When the sample is in an inhomogeneous applied magnetic field B_0, it is convenient to visualize it as a three-dimensional mosaic of small volume elements, each small enough that field gradients across that element can be neglected, but with each element in a slightly different applied field. The macroscopic nuclear magnetization from one such element is called an *isochromat* since it has a fixed precession frequency (provided that diffusion and convection can be neglected). The magnetization from each isochromat may be represented by a small vector, the resultant of all such vectors giving the total macroscopic magnetization M_0. When excited by a strong radiofrequency pulse, all these isochromatic vectors are aligned along the Y axis of the rotating frame* of reference, that is to say they all start their free precession in phase. As they evolve with time they fan out in the X–Y plane and their resultant decays monotonically, accounting for the decay of the observed free precession signal. For simplicity, this is often represented by the decay constant T_2^* although this should not be confused with a true relaxation time. For smallish molecules in non-viscous media, T_2 will normally be much longer than T_2^* and this will be the situation we consider in what follows.

After a delay of (say) $3T_2^*$ the observed NMR signal has decayed to a negligible level. Yet if it were possible to confine the observation to only a small region of the sample, the spins there would still be precessing in a concerted fashion and there would be a significant macroscopic NMR signal which would persist for a time of the order of the spin–spin relaxation* time T_2. The T_2^* decay is thus a spurious instrumental effect. Hahn (1) was the first to realize that this 'lost' signal could be retrieved in a two-pulse experiment. After the initial free induction signal had disappeared, a second pulse at time τ induced a new response which peaked at time 2τ. By analogy with the reflection of sound waves, this new response was called a 'spin echo'.

Hahn's experiment used a 90° pulse followed by another 90° pulse, but the picture is much clearer in the method devised by Carr and Purcell

(2) where the second pulse rotates all vectors by 180°. Consider a simplified case where the array of isochromats is represented by just three vectors a, b, and c, with c the furthest from resonance, and thus the fastest vector in the frame rotating at the transmitter frequency. All three vectors are aligned along the +Y axis by the first 90° pulse (Fig. 1(a)) and after a time τ disperse into a fan of vectors in the X–Y plane. A 180° pulse about the X axis turns these vectors into mirror image positions with respect to the X axis, so arranged with the fast vectors behind the slow vectors that all move into coincidence along the −Y axis after a further period τ of free precession. In terms of a 'phase evolution diagram' (Fig. 1(b)) it is clear that the 180° pulse has a refocusing effect.

If the echo peak amplitude is monitored as a function of 2τ in a series of experiments, it is observed to decay with a time constant T_2 which may be much longer than T_2^* for the type of sample we are considering here. In spectra with many lines, Fourier transformation of the echoes allows the individual spin–spin relaxation times to be determined. It is as if the linewidths had been measured in a perfectly uniform applied magnetic field B_0. This, however, presupposes that all isochromats retain the same precession frequency throughout the experiment. In practice, some diffusion across the field gradients occurs. Carr and Purcell showed that this effect could be minimized by keeping 2τ short and by following the time evolution by repeatedly refocusing with 180° pulses, giving a 'train' of spin echoes. Only the diffusion between refocusing pulses matters, and this can be kept to a negligible level. Multiple refocusing introduces the possibility of cumulative errors if the 180° pulses are imperfect, but by applying these pulses about the Y axis rather than the X axis, these errors are compensated for the even-numbered echoes, provided there is no modulation on the echo. This modification, suggested by Meiboom and Gill (3), is widely used in spin echo work. An alternative is to use a composite 180° pulse (4) designed to be self-compensating.

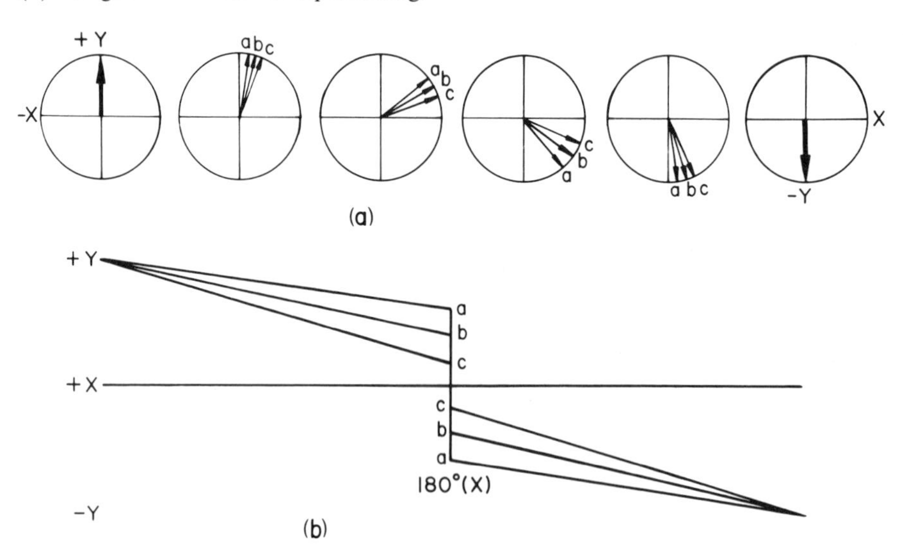

Fig. 1. (a) Vector diagram illustrating the refocusing of isochromats in a spin echo experiment, together with (b) the corresponding phase evolution diagram.

Spin echo experiments provide the most direct method of measuring spin–spin relaxation times in the common situation where the natural linewidths are obscured by field inhomogeneity broadening. Figure 2 shows a sketch of a typical spin echo train.

When a Carr–Purcell spin echo train with multiple refocusing is used to study the effect of chemical exchange on spin–spin relaxation times, an interesting new phenomenon is observed (5). A chemical exchange 'event' shortens the phase memory of the spins by replacing a precessing spin that has good phase coherence with another spin of random phase. If the exchange rate is fast compared with the rate of application of refocusing pulses, then exchange causes fast decay of the observed echoes. In contrast, if the exchange rate is slow compared with the pulse repetition rate, the probability of an event occurring between refocusing pulses is negligibly small, and the echo decay becomes essentially independent of chemical exchange effects.

Spin echoes can be used to monitor molecular diffusion. Usually, this is carried out by applying a known intense field gradient during one of the τ delays, thus interfering with the refocusing if molecular motion occurs. A variation of this experiment can be used to study concerted flow of the sample along an applied magnetic field gradient. Needless to say, these experiments are carried out on a non-spinning sample. For high-resolution spectroscopy involving spin echoes, a spinning sample is normally used since this provides narrower lines in the observed spectrum and a narrower deuterium signal for field/frequency regulation*. Occasionally, there can be serious interactions between the spinning and spin-echo formation. It may then be necessary to adopt a scheme where the pulse repetition rate is synchronized with the sample spinning rate, or to switch the spinner off altogether.

It is possible to engineer gross differences in spin–spin relaxation times of solvent water and dissolved macromolecules by addition of low concentrations (~0.2 millimolar) of paramagnetic ions. Each ion is surrounded by a hydration sphere which inhibits access by the macromolecules but which is in rapid

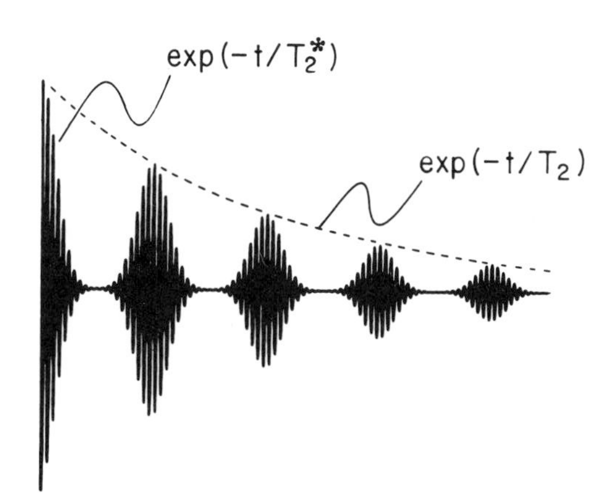

Fig. 2. Simulation of a train of spin echoes showing the rapid signal decay due to instrumental effects (T_2^*) and the slower decay of the echo envelope due to spin–spin relaxation (T_2).

exchange with bulk water thus shortening its spin–spin relaxation time. This has been made the basis of several techniques for solvent suppression* in which a spin echo experiment is used to discriminate between water and the slowly relaxing solute.

If we neglect spin–spin relaxation on this timescale, a spin echo is exactly symmetrical with respect to its midpoint. Thus the Fourier transform of an entire spin echo can contain no antisymmetrical components (dispersion mode signals) but only symmetrical components (absorption mode).

ECHO MODULATION

In a spin echo experiment on a homonuclear spin system, chemical shifts have the same effect as an offset from the transmitter frequency, being refocused at the time of the echo. In contrast, spin–spin coupling is not refocused but modulates the echoes at a frequency of $\frac{1}{2}$J Hz (6). (We consider here only the simple case where the coupling is first-order.) The echo modulation arises

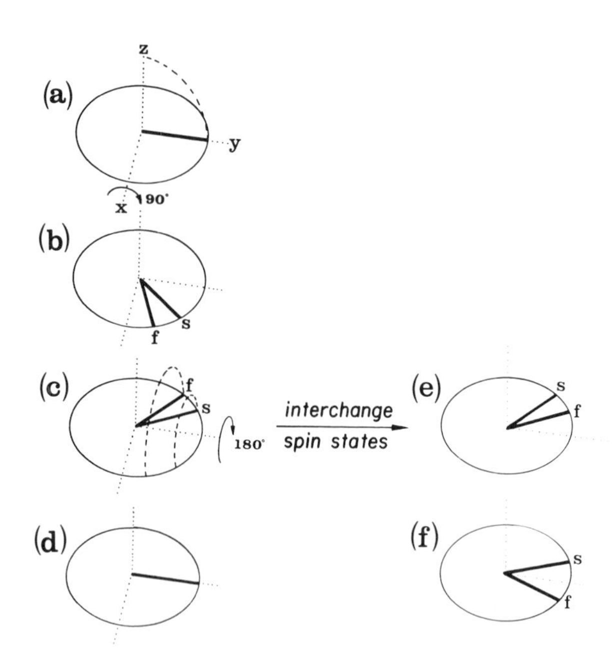

Fig. 3. Vector diagram illustrating the origin of echo modulation. In the sequence a–b–c–d, both chemical shift and spin coupling effects are refocused and there is no echo modulation. By contrast, in the sequence a–b–c–e–f, where the spin state labels (f and s) are interchanged, the two vectors continue to diverge due to spin coupling and the echo therefore carries J-modulation.

because the 180° refocusing pulse of the spin echo experiment flips both coupled spins I and S, leaving the coupling term J I_ZS_Z unchanged.

The origin of this J-modulation of the echo is readily visualized on a simple vector picture, taking the case of a two-spin IS system as an example. The I spin resonance is made up of two lines corresponding to the α and β states of the S spin, and may be represented by two magnetization vectors f (fast) and s (slow) which have natural precession frequencies in the rotating frame of $\Delta f \pm \frac{1}{2}J$ Hz, where Δf is the offset of the I shift from the transmitter frequency. The initial 90° pulse aligns both vectors along the +Y axis (Fig. 3) and they then precess freely, building up a relative phase difference of $2\pi J\tau$ rad by the end of the first interval τ seconds. (At the same time field inhomogeneity spreads out each vector into a fan of isochromats, but since this is reversed in the second τ interval, it may be disregarded for the purposes of this discussion.) At this point the 180° pulse is applied and has two quite distinct effects. First, it rotates f and s into mirror-image positions with respect to the Y axis of the rotating frame. Had this been the only effect, f and s would come to an exact focus at time 2τ (Fig. 3(d)). However, the 180° pulse also flips the S spin, interchanging the α and β spin states and thus interchanging the labels f and s in the diagram. Now, f is further from the transmitter and precesses faster than s in the first τ interval, but becomes the slower vector in the second τ interval. The divergence between the two vectors does not therefore cancel but is cumulative, building up into a phase difference $4\pi J\tau$ radians at the time of the echo (Fig. 3(f)). The echo is thus made up of two components, and their resultant is $M_0 \cos(\pi Jt) \exp(-t/T_2)$ rather than the 'expected' echo amplitude $M_0 \exp(-t/T_2)$. If the experiment is repeated for a series of different settings of 2τ, or if an echo train is generated by multiple refocusing, the echo peaks are observed to be amplitude modulated according to $M_0 \cos(\pi Jt) \exp(-t/T_2)$.

Inversion of *both* coupled spins is thus the key to echo modulation. For a heteronuclear spin system, the I spin echoes are not normally modulated by the

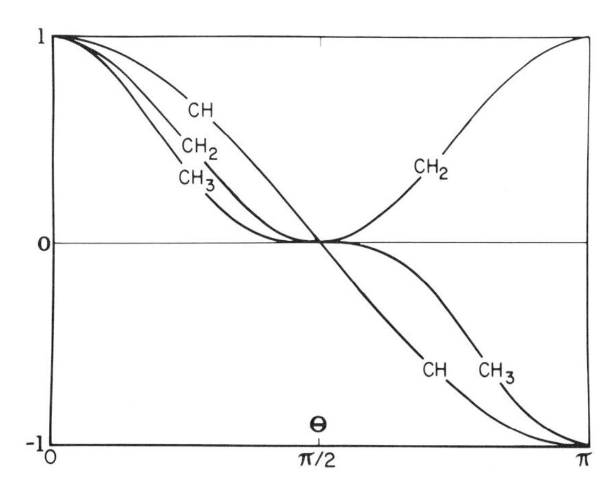

Fig. 4. Echo modulation patterns for CH, CH_2 and CH_3 groups as a function of $\theta = 2\pi J_{CH}\tau$. This can be used as a technique for 'editing' carbon-13 spectra. Quaternary sites have such small CH couplings as to give virtually no modulation on this timescale.

coupling J_{IS}, and it is only if another 180° pulse is deliberately applied to the S spins that the modulation is observed. In homonuclear spin systems the 180° is normally non-selective and affects both the I and the S spins, generating modulated echoes, but if the 180° pulse is made frequency-selective so that it flips the I spins without influencing the S spins, then the I spin echoes remain unmodulated. More complicated rules apply in the case of strong coupling or situations where the second pulse is not exactly 180°. For very high pulse repetition rates (comparable with chemical shift differences) the echo modulation disappears. This can be important when using spin echoes to suppress broad lines in a J-coupled spectrum.

Echo modulation can be exploited for all kinds of applications in high-resolution spectroscopy. Suppose we are interested in detecting the weak satellites which appear in proton spectra due to the presence of low-abundance carbon-13 or nitrogen-15 spins. Under normal circumstances, these weak signals are swamped by the much larger proton signals from molecules containing carbon-12 or nitrogen-14 spins. If a spin echo experiment is performed on the proton system, using a relatively short timing interval 2τ, the Fourier transform of the second part of the echo has the same form as the Fourier transform of the free induction decay*; there is only a small loss of intensity due to spin–spin relaxation. By contrast, if the same experiment is repeated with a 180° pulse applied to carbon-13 (or nitrogen-15) simultaneous with the proton 180° refocusing pulse, then the proton echo is changed in amplitude by a factor $\cos(2\pi J_{IS}\tau)$, where J_{IS} is the coupling constant to the heteronucleus. By setting $\tau = 1/(2J_{IS})$ the echo is simply inverted. Difference spectroscopy* then eliminates the 'parent' proton spectrum, leaving only the satellites (7). Suppression ratios can be quite high, particularly for the nitrogen-15 experiment because the protons on nitrogen-14 are preferentially broadened, giving a cleaner difference spectrum. Related techniques are used for multiplicity determination* in carbon-13 spectroscopy; the form of the echo modulation for the various groups (methine, methylene and methyl) is shown in Fig. 4.

In the general case where several spin–spin couplings are involved, the corresponding echo modulation can be quite complicated and then it proves convenient to analyse it by Fourier transformation. This is J-spectroscopy*. A J-spectrum may be thought of as a conventional high-resolution spectrum that has been collapsed by removing all chemical shift effects, leaving only J-couplings.

REFERENCES

1. E. L. Hahn, *Phys. Rev.* **80**, 580 (1950).
2. H. Y. Carr and E. M. Purcell, *Phys. Rev.* **94**, 630 (1954).
3. S. Meiboom and D. Gill, *Rev. Sci. Instr.* **29**, 688 (1958).
4. M. H. Levitt and R. Freeman, *J. Magn. Reson.* **43**, 65 (1981).

5. Z. Luz and S. Meiboom, *J. Chem. Phys.* **39**, 366 (1963).
6. E. L. Hahn and D. E. Maxwell, *Phys. Rev.* **84**, 1246 (1951).
7. R. Freeman, T. H. Mareci and G. A. Morris, *J. Magn. Reson.* **42**, 341 (1981).

Cross-references

Chemical exchange
Composite pulses
Difference spectroscopy
Field/frequency regulation
Free induction decay
J-spectroscopy
Multiplicity determination
Rotating frame
Solvent suppression
Spin–spin relaxation
Two-dimensional spectroscopy

Spin–Lattice Relaxation

Suppose it were possible to drop the NMR sample into the probe instantaneously and then to apply an excitation pulse immediately. Would we observe an NMR signal? The simple answer is no. The nuclear polarization would be negligibly small in such an experiment for the nuclei were only polarized in the very weak Earth's field, giving a negligible population difference between the lower and upper energy levels. When the field is suddenly changed to the intense field B_0 of the spectrometer, the process of building up the appropriate Boltzmann population distribution can be quite slow. Initially, the nuclear spins are 'hot' (essentially equal populations) and they cool down by transferring some magnetic energy to their surroundings, a process called relaxation. If we were to plot the variation of the population difference (Z-magnetization) as a function of time, we would find an exponential curve starting near zero and rising asymptotically towards the Boltzmann equilibrium condition. It would have a time constant T_1, the spin–lattice relaxation time. 'Lattice' here is used as a general term for the nuclear environment, dating from the early theories of spin–lattice relaxation which were all concerned with the solid state.

In practice, the high resolution NMR spectroscopist is seldom concerned with the delay which ensues after the insertion of the sample and before the detection of the first free induction signal (although this is a problem for studies on nuclei like silicon-29 in solids). Spin–lattice relaxation is involved in a much more fundamental way. The Boltzmann equilibrium nuclear magnetization is usually represented by M_0. Any deviation of the actual longitudinal magnetization M_Z away from its equilibrium value M_0 is followed by an exponential recovery according to the equation

$$(M_Z - M_0)_t = (M_Z - M_0)_0 \exp(-t/T_1). \qquad [1]$$

This expression is simply derived from one of the Bloch equations in the rotating frame, and in the absence of any B_1 field

$$dM_Z/dt = -(M_Z - M_0)/T_1, \qquad [2]$$

which defines T_1 phenomenologically as a *longitudinal* relaxation time.

In continuous-wave spectroscopy, irradiation with too high a level of the radiofrequency field B_1 causes 'saturation' of the spins by pumping them from lower to upper level at a rate so fast that spin–lattice relaxation cannot compete effectively. The detected signal M_Y becomes weak and may disappear in the limit. This is reflected in the expression for the steady-state solution of the Bloch equations

$$M_Y = M_0\gamma B_1 T_2/[1 + (\Delta\omega T_2)^2 + \gamma^2 B_1^2 T_1 T_2]. \qquad [3]$$

At exact resonance ($\Delta\omega = 0$) this expression shows that M_Y increases with B_1 provided that $\gamma^2 B_1^2 T_1 T_2 < 1$, reaches a maximum where this term is unity and then decreases when $\gamma^2 B_1^2 T_1 T_2 > 1$. This saturation behaviour is given a geometrical interpretation in the section on continuous-wave spectroscopy*. Since the wings of the line profile saturate less readily than the peak, saturation also broadens the line in a slow-passage continuous-wave spectrum.

A process equivalent to saturation can occur in pulsed NMR if a train of pulses is applied. Suppose that perfect 90° pulses are employed, converting M_Z into M_Y, which corresponds to an equalization of the spin populations of the two energy levels, that is to say saturation. If the interval between pulses is comparable with the spin–lattice relaxation time T_1, then the recovery of M_Z towards M_0 is incomplete. A steady-state* regime is set up where the observed signal amplitude M_Y is a function of T_1

$$M_Y = M_0[1 - \exp(-t/T_1)]. \qquad [4]$$

This is an important consideration when setting up a pulsed NMR experiment for optimum sensitivity. There are two schools of thought. One advocates the use of 90° pulses with suitable delays (t_d) after signal acquisition (t_a) so that the pulse interval ($t_d + t_a$) allows adequate spin–lattice relaxation. The recommended ratio is $(t_d + t_a) \simeq 1.3\,T_1$ (1). The second school sets $t_d = 0$ and balances the effects of saturation by reducing the flip angle α of the pulses. Provided that there is no transverse magnetization just before the pulse (a difficult condition to fulfil in practice) the optimum flip angle can be shown to be given by

$$\cos\alpha_{OPT} = \exp(-t_a/T_1), \qquad [5]$$

where t_a is now the interval between pulses (2).

These relaxation phenomena can also be used to measure spin–lattice relaxation times by monitoring the recovery of the Z-magnetization after a suitable perturbation, using eqn [1]. Through Fourier transformation this permits the determination of the T_1 values of individual lines in spectra of many resonances. These *inversion recovery*, *saturation recovery* and *progressive saturation* techniques are examined in detail in a later section.

MECHANISMS FOR SPIN–LATTICE RELAXATION

Nuclei are almost completely isolated from the violent rotations of the molecular framework. As if they were gyroscopes mounted on perfectly frictionless bearings, they align themselves with the direction of the magnetic field whatever the orientation of the molecule. They cannot exchange their magnetic energy with their surroundings by direct mechanical coupling but only indirectly by magnetic dipole or electric quadrupole interactions. For nuclei with $I = \frac{1}{2}$ the only interactions with the lattice are magnetic in origin. Transfer of energy from the spins to the lattice requires that there be a magnetic field at the nucleus fluctuating at the Larmor precession frequency in order to induce an NMR transition. This field originates from magnetic dipoles in the lattice which is in thermal agitation, and since the frequency spectrum of these motions is very wide, the component actually in tune with the Larmor frequency is very weak. This is why NMR spin–lattice relaxation is a slow process compared with the rates at which populations come to equilibrium in, say, infra-red or optical spectroscopy.

Compared with the frequencies of molecular motion in a liquid (translation, rotation and vibration), nuclear precession frequencies are slow, so it is only the slowest molecular motions (rotation) that are even approximately matched to the Larmor frequency. The distribution function or *spectral density* function for reorientational motion is not known in detail, but it is usually taken to be a Lorentzian function of frequency, since the time-domain correlation function is assumed to be exponential. It is customary to plot the spectral density function on a logarithmic frequency scale, giving it the appearance of a rather flat curve which falls off as it approaches the critical frequency $1/\tau_c$, where τ_c is the *correlation time* for reorientation. Inertial and viscosity effects tend to determine τ_c.

Figure 1(a) sketches typical spectral density functions for fast, intermediate and slow molecular tumbling, the area under these curves remaining constant. In the fast limit the component at the Larmor frequency is small, but increases as the molecular reorientation rate is slowed down. Most small organic molecules in non-viscous solvents are in this category. This means that any factor which slows the molecular motion shortens the spin–lattice relaxation time. A minimum is reached near the condition sketched for intermediate molecular tumbling rates, and any further slowing of the motion reduces the component at the Larmor frequency and therefore increases the relaxation time (Fig. 1(b)). Large biological molecules fall into this category. For high resolution NMR spectroscopy, a short T_1 favours sensitivity but too short a T_1 begins to broaden the lines and degrades resolution (since T_2 cannot be longer than T_1).

There are three principal types of magnetic interaction which contribute to spin–lattice relaxation of $I = \frac{1}{2}$ nuclei. The most important is the *dipole–dipole interaction*, where the nucleus experiences a fluctuating field due to the motion of neighbouring magnetic dipoles (unpaired electrons or other nuclei). The field due to a dipole of strength μ at a distance r and subtending an angle θ with respect to the B_0 direction is given by the simple formula

$$B_{DD} = \pm(\mu_0/4\pi)\mu(3\cos^2\theta - 1)/r^3. \qquad [6]$$

(The term $(\mu_0/4\pi)$ is the conversion factor for SI units.) The sign of this field depends on the orientation of the dipole (with or against the magnetic field). Unpaired electrons have much stronger magnetic dipoles than nuclei, and samples intended for high-resolution work usually exclude paramagnetic transition metal compounds, free radicals and triplet molecules in order to avoid excessive spin–lattice relaxation. Even paramagnetic oxygen gas dissolved in the solvent can have a significant effect on long spin–lattice relaxation times (of the order of 1–10 seconds) and is usually removed by several freeze–pump–thaw cycles for very high-resolution work or for determinations of long spin–lattice relaxation times. The remaining dipolar fields originate from other nuclear spins in the immediate environment and can therefore provide information about molecular structure or molecular motion.

A second type of magnetic interaction arises because the chemical shielding of the nucleus depends on the orientation of the molecule with respect to the B_0 field direction. It is therefore known as the *chemical shift anisotropy* mechanism. Shielding can be represented as a weak secondary magnetic field set up when the electrons surrounding the nucleus precess in the applied field B_0. This field is thus proportional to B_0 and fluctuates as the molecule tumbles. For protons, which have small chemical shifts, this relaxation mechanism is not important, but it is often significant for nuclei with large chemical shift ranges, for example fluorine-19, carbon-13 and many metals. Since the relaxation rate depends on the square of the appropriate field at the nucleus, the rate of spin–lattice relaxation by chemical shift anisotropy increases as the square of the applied field strength B_0. There is an ingenious test for this mechanism if the

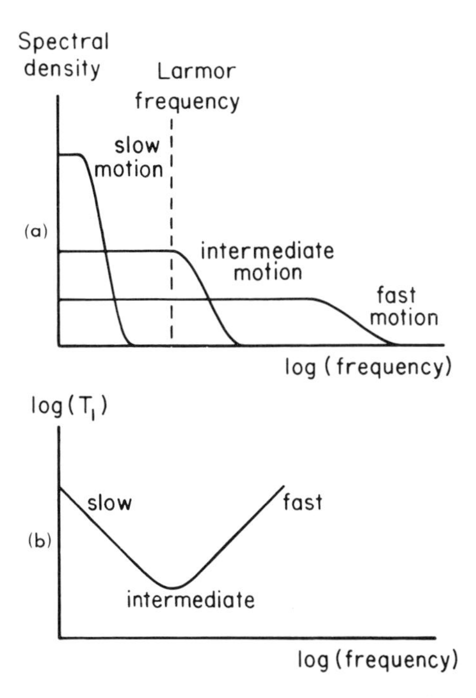

Fig. 1. Spectral density functions for slow, intermediate and fast molecular reorientation (the curves are Lorentzian, distorted by the logarithmic frequency scale). Molecular motion at an intermediate rate generates the largest component at the Larmor frequency, giving a minimum in the curve of T_1 against frequency (lower graph).

spin–lattice relaxation time is sufficiently long – the sample is temporarily removed from the B_0 field during an inversion recovery experiment, thus eliminating the shift anisotropy effect for this period (3). Relaxation by chemical shift anisotropy becomes relatively more important as spectrometers are constructed at higher and higher fields. It may then dominate all other mechanisms and there is some concern that this may set an upper limit on the resolving power and sensitivity of phosphorus-31 spectroscopy at high fields by broadening the lines faster than the increase in chemical shifts.

The third mechanism is known as *spin–rotation interaction*. As a molecule rotates, the moving charges of electrons and nuclei create small electric currents which in turn induce a small magnetic field at the nucleus. Although the charges of electrons and nuclei balance, the resulting fields do not quite cancel because in general the 'centre of gravity' of the electron charge does not coincide with that of the corresponding nucleus. This leaves a net field at the nucleus which fluctuates because the reorientational motion is not uniform but proceeds by a series of random jumps. This mechanism increases in effectiveness as the tumbling motion speeds up, so that, unlike the dipolar mechanism, spin–rotational relaxation is more effective for small molecules in low-viscosity solutions at high temperature. This opposite temperature dependence may be used to distinguish the two mechanisms.

A fourth, extremely rare relaxation mechanism relies on the weak magnetic field induced at the site of a spin I by spin–spin coupling J_{IS} to a neighbouring spin S. This is only an effective relaxation mechanism if the induced field is rapidly fluctuating because S is a rapidly relaxing quadrupolar nucleus, and also has a Larmor frequency very close to the Larmor frequency of nucleus I. This occurs for carbon-13 coupled to bromine-81, which has such a broad resonance that its tail overlaps the carbon-13 resonance (4).

If the nucleus possesses a quadrupole moment there is a much more effective mechanism coupling the spin to the motions of the environment. The distribution of the nuclear charge is no longer spherical, but is of the form of an ellipsoid, and this can interact strongly with an electrostatic field gradient. Most chemical bonds generate such a gradient of the electrostatic field at the nucleus and the interaction fluctuations as the molecule reorients in solution. Spin–lattice relaxation thus tends to be rapid and the spin–spin relaxation* time is similarly shortened, giving broad lines and making these nuclei unsuitable for high-resolution work, since any small chemical shift effects are swamped by the excessive line broadening. At sufficiently high fields, chemical shifts may nevertheless be measurable. In special cases, notably the deuterium nucleus ($I = 1$) in small molecules, the quadrupole moment is so small that high-resolution spectra may be recorded with only rather slight line broadening. In general, however, nuclei which do not have $I = \frac{1}{2}$ prove of little interest for high-resolution work. This includes all the isotopes of the halogens of significant natural abundance.

Rapid relaxation of quadrupolar nuclei also has a beneficial effect in that it leads to some simplification of the spectra of protons and carbon-13. Although quadrupolar nuclei such as chlorine-35 have significant coupling to protons and carbon-13, spin–lattice relaxation washes out the splitting if the relaxation rate is fast compared with the coupling constant J. Thus it is possible for most purposes to ignore the presence of halogen atoms in the molecule, much as we ignore non-magnetic nuclei like oxygen-16. Nitrogen-14 may split protons into a 1 : 1 : 1 triplet under conditions of slow spin–lattice relaxation. It is, however, important to remember that coupling to a quadrupolar nucleus may shorten the spin–spin relaxation time of the observed nucleus (which may or may not result in a detectable broadening), and that the quadrupolar nucleus may have a significant magnetic dipole moment and thus contribute to spin–lattice relaxation of near neighbours.

The time dependence of the nuclear magnetization during spin–lattice relaxation is not necessarily a simple exponential curve. In a coupled spin system, for example, cross-relaxation causes multiexponential behaviour, and in the case of chemical exchange* where the experiment is initiated with a selective inversion pulse, some lines will first lose intensity and then recover.

DETERMINATION OF SPIN–LATTICE RELAXATION TIMES

Nowadays spin–lattice relaxation measurements are usually carried out by pulse methods (5), and the T_1 values of individual lines in a multiline spectrum are readily extracted by separating the appropriate frequency components by Fourier transformation (6). The decay time constant for transverse magnetization (T_2) can often be shorter than T_1, so in all these experiments it is important to eliminate transverse components of magnetization while the longitudinal magnetization is being monitored.

The most widely used technique is that of *inversion recovery* (7). This requires a preparation period long compared with T_1 for the spins to reach thermal equilibrium before the pulse sequence is applied. A 180° pulse then inverts the Z component of magnetization and after a variable delay t during which the inverted spin population recovers back towards equilibrium, the component M_Z is examined by applying a 90° pulse. If S_t is the amplitude of the detected NMR signal, and S_∞ is the corresponding asymptotic signal observed when $t \gg T_1$, then if perfect inversion can be assumed

$$S_\infty - S_t = 2S_\infty \exp(-t/T_1). \qquad [7]$$

Figure 2 shows a typical inversion-recovery experiment on the carbon-13 spectrum of *meta*-xylene. The spin–lattice relaxation time is then evaluated either by making a graphical plot on semilogarithmic paper or through a suitable computer routine which accepts digital peak height data. Since timing errors are usually negligible in computer-controlled spectrometers, the errors are essentially only in the measurement of intensity. Note that the statistical errors on S_∞ are really just as significant as those on S_t, so that it can be argued that an equal number of measurements of S_∞ and S_t should be made. If the spectrometer resolution is not reproducible, then the normalized intensity difference $(S_\infty - S_t)/S_\infty$ should be computed. If the initial inversion pulse is imperfect, the residual transverse magnetization should be cancelled by phase alternation of the monitoring pulse (8) or eliminated by means of a strong field gradient *homospoil* pulse. The recovery is *still* exponential

$$S_\infty - S_t = S_\infty(1 - k) \exp(-t/T_1), \qquad [8]$$

where k measures the efficiency of the inversion (ideally $k = -1$), depending on the error in the flip angle and the extent of any resonance offset effects. An alternative is to employ a composite 180° pulse* such that offset effects or flip angle errors are compensated (9).

Most relaxation experiments are time-consuming, and this may be the critical consideration for low-sensitivity nuclei which require extensive time-averaging. One problem is that the determination of S_∞ requires a rough estimate of the longest T_1 to be measured, so that an interval of 4–5 times T_1 can be programmed. Some workers leave the unknown value of S_∞ as a variable parameter in a non-linear least-squares computer program, along with the time constant T_1. It should, however, be remembered that a two-parameter fit seriously reduces the significance of a given piece of experimental data compared with a single-parameter fit, and there are objections in principle against deriving a quantity by a fitting procedure when it can in fact be measured directly. If the measurement of S_∞ is abandoned, the experiment may be further accelerated by reducing the number of different values of t employed, and in the limit, approximate values of the relaxation times may be obtained with just two settings of t (10). This procedure requires estimates of the relaxation times in order to choose suitable values of t, so it would be applicable to situations where T_1 is measured in a series of closely related compounds, or for a given molecule under different physical conditions.

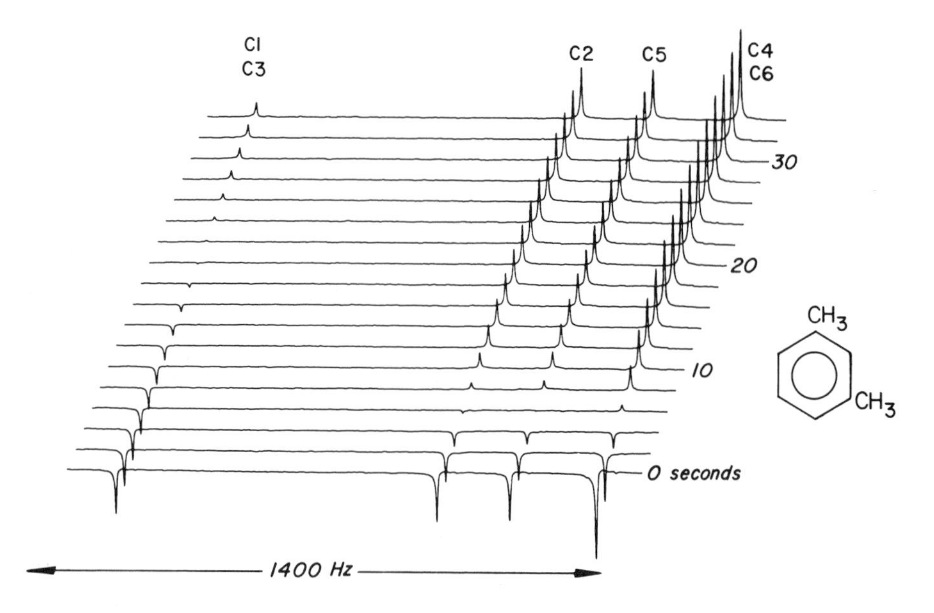

Fig. 2. A typical inversion–recovery experiment to determine individual spin–lattice relaxation times. Carbon-13 nuclei 1 and 3, having no close proton neighbours, have a very long relaxation time. The slightly different relaxation rates of carbons 2 and 5 compared with carbons 4 and 6 suggest that the molecular reorientation is slightly anisotropic.

The other time-consuming aspect of the inversion-recovery method is the waiting period 4–5 T_1 required to allow the establishment of thermal equilibrium before each pulse sequence is initiated. Two alternative techniques *progressive saturation* and *saturation recovery* have been devised to circumvent this problem. Progressive saturation is the analogue in pulsed spectroscopy of the method used in continuous-wave instruments, where the spectrum is examined at different levels of the B_1 field and the degree of saturation monitored. A regular sequence of 90° pulses is applied at a rate fast compared with the rate of spin–lattice relaxation so that a steady-state regime is established for the longitudinal magnetization. Immediately after each 90° pulse the spin populations in upper and lower energy levels are equal ($M_Z = 0$) and in the short interval before the next pulse, spin–lattice processes return some spins to the lower level and M_Z grows towards M_0. The observed NMR signal is thus proportional to the growth in M_Z, and if the experiment is repeated for several different pulse intervals t, one of which is long compared with all spin–lattice relaxation times, then

$$S_\infty - S_t = S_\infty \exp(-t/T_1). \qquad [9]$$

Instead of a waiting period of 4–5 T_1, this experiment requires only one or two pulse intervals to settle into a steady state. Free induction decays are therefore acquired more rapidly, giving improved signal-to-noise in a given time; however, each ordinate on the final graph ($S_\infty - S_t$) is only one half that of the inversion–recovery method, as can be seen by comparing eqn [7] with eqn [9]. Progressive saturation measurements presuppose that some technique is used to prevent the establishment of steady-state components of transverse magnetization immediately before each pulse (11). Systematic errors can be introduced by imperfections in the 90° pulse, particularly errors in the pulse length.

The saturation recovery technique is related to progressive saturation but has the advantage that the minimum setting of the relaxation interval t is not limited by the acquisition time. Again there is no preparation period required, the spins are simply saturated by a sequence of radiofrequency pulses* and any transverse magnetization is dispersed by a field gradient pulse (12). After a variable interval t (which allows partial recovery of Z-magnetization), this is monitored by a 90° pulse. The time evolution of the signal intensity is again described by eqn [9].

In all these experiments there remains the difficulty of estimating an approximate value for the spin–lattice relaxation time in order to set up the experimental parameters. In multiline spectra this difficulty is compounded because there is usually a distribution of T_1 values. The most satisfactory remedy to this problem has been suggested by Canet (13) and called the *fast inversion recovery* method. It recognizes that sensitivity requirements will not permit a sufficiently long preparation period for the slowly relaxing spins and therefore accepts a compromise setting of this interval. For the rapidly relaxing resonances the experiment is thus a conventional inversion–recovery sequence, but for the slowly relaxing resonances the Z-magnetization just before the 180° pulse is considerably less than M_0, and in the limit the experiment resembles a saturation recovery sequence. Spin–lattice relaxation times can be extracted for all the resonance lines, the only penalty being that the sensitivity is reduced for the slowly relaxing species.

REFERENCES

1. J. S. Waugh, *J. Mol. Spectr.* **35**, 298 (1970).
2. R. R. Ernst, Sensitivity Enhancement in Magnetic Resonance. In *Advances in Magnetic Resonance*, Ed.: J. S. Waugh. Academic Press: New York, Vol. 2, 1966.
3. G. C. Levy, D. M. White and F. A. L. Anet, *J. Magn. Reson.* **6**, 453 (1972).
4. J. R. Lyerla, D. M. Grant and G. D. Bertrand, *J. Phys. Chem.* **75**, 3967 (1971).
5. H. Y. Carr and E. M. Purcell, *Phys. Rev.* **94**, 630 (1954).
6. R. L. Vold, J. S. Waugh, M. P. Klein and D. E. Phelps, *J. Chem. Phys.* **43**, 3831 (1968).
7. E. L. Hahn, *Phys. Rev.* **76**, 145 (1949).
8. D. E. Demco, P. van Hecke and J. S. Waugh, *J. Magn. Reson.* **16**, 467 (1974).
9. M. H. Levitt and R. Freeman, *J. Magn. Reson.* **33**, 473 (1979).
10. R. Freeman, H. D. W. Hill and R. Kaptein, *J. Magn. Reson.* **7**, 82 (1972).
11. R. Freeman and H. D. W. Hill, *J. Chem. Phys.* **54**, 3367 (1971).
12. J. L. Markley, W. H. Horsley and M. P. Klein, *J. Chem. Phys.* **55**, 3604 (1971).
13. D. Canet, G. C. Levy and I. R. Peat, *J. Magn. Reson.* **18**, 199 (1975).

Cross-references

Chemical exchange
Composite pulses
Continuous-wave spectroscopy
Radiofrequency pulses
Spin–spin relaxation
Steady-state effects

Spin Locking

In a spin-echo* experiment, the amplitude of the echo observed at time t is given by $M_0 \exp(-t/T_2)$, where T_2 is the spin–spin relaxation* time. Certain other factors may influence this amplitude, for example diffusion of the spins in a B_0 gradient, slow chemical exchange* or echo modulation by homonuclear spin–spin coupling. When a repeated train of 180° refocusing pulses is used, all three effects get weaker and finally disappear as the pulse repetition rate is increased. In this limit the refocusing pulses are applied in such rapid succession that the amount of free precession is negligible and all magnetization vectors remain essentially 'locked' along the Y axis of the rotating frame.

The same effect is achieved if a continuous B_1 field is applied along the Y axis in lieu of the sequence of 180° pulses. The experiment consists of a 90° pulse about the X axis followed by a rapid 90° phase shift of the radiofrequency field, which is then left on for some very long time of the order of milliseconds to seconds. This is called *spin locking* or *forced transitory precession* (1). Alternatively, the same condition can be reached with an adiabatic fast passage* interrupted when the sweep reaches the exact resonance condition, but with the radiofrequency field B_1 left on.

If we are dealing with a single-line spectrum, B_1 will normally be so intense compared with off-resonance effects attributable to static field inhomogeneity that the latter may be neglected. Each small volume element of the sample (*isochromat*) experiences almost exactly the same effective field B_{eff}, essentially equal to B_1 and aligned along the Y axis of the rotating frame. The NMR signal consequently decays only by spin–spin relaxation and not by field inhomogeneity effects (as in a free induction decay). Since this relaxation takes place longitudinal to the effective field, the decay time constant is more correctly called the *spin–lattice relaxation time in the rotating frame**, $T_{1\varrho}$ but for a liquid sample this is essentially the spin–spin relaxation time T_2.

In certain practical situations the spin-locking experiment might be preferred over the better-known spin-echo method for determining spin–spin relaxation times in liquids. For example it neatly avoids the problem of cumulative errors in the 180° refocusing pulses. As mentioned above, it also avoids the complications associated with echo modulation in coupled spin systems. Furthermore, when it is feasible to perform a selective spin-locking experiment (where B_1 influences an entire spin-multiplet but leaves adjacent spin multiplets essentially unaffected) the individual spin–spin relaxation times can be measured one at a time. In practice, limitations imposed by the electronics and sample heating effects often preclude the use of a B_1 field so intense as to cover the entire spectral width. Selective spin locking, with a much reduced B_1 intensity, would therefore be the preferred approach.

A spin–spin relaxation study simply involves a series of experiments with different lengths of the spin-locking time t_L, the free induction decay being acquired after the extinction of the B_1 field. Intensities of individual resonance lines are then followed as they decay as a function of t_L. Figure 1 shows the example of the carbon-13 resonances of ortho-dichlorobenzene, where the carbon-13 chemical shifts are sufficiently small that the entire spectrum could be covered by a relatively weak spin-locking field $\gamma B_1/2\pi = 2.6$ kHz. It illustrates (2) that the carbon sites with directly attached chlorine atoms exhibit a short spin–spin relaxation time attributable to scalar coupling to chlorine-35

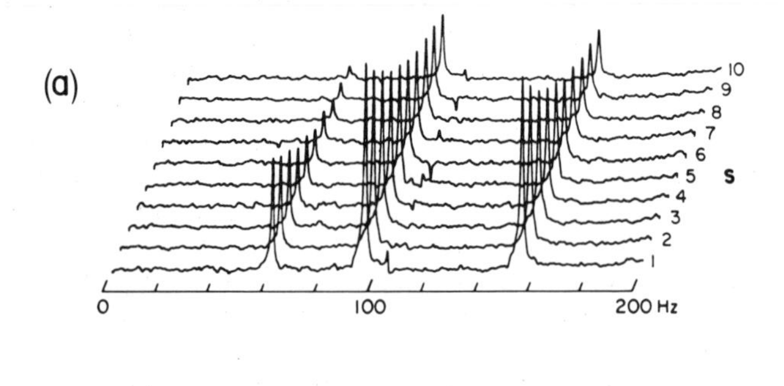

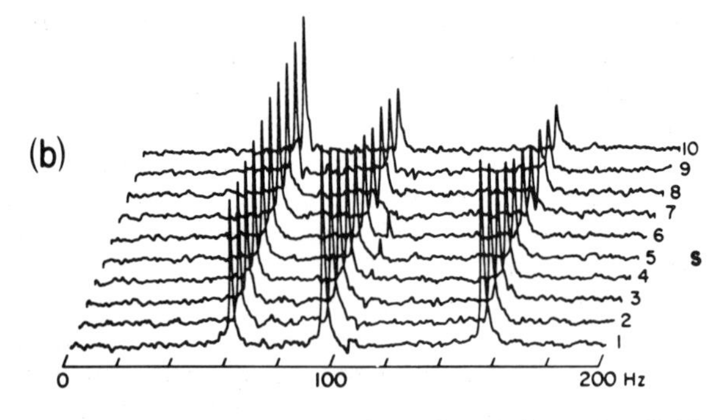

Fig. 1. (a) Spin–spin relaxation of the carbon-13 nuclei in ortho-dichlorobenzene studied in a spin-locking experiment. The quadrupolar chlorine nuclei cause a shortening of T_2 for the low-field line ($T_2 = 4.2$ s). (b) Spin–lattice relaxation curves for comparison. Now, the chlorine-bearing sites are seen to have a long T_1 since there are no nearby protons ($T_1 = 66$ s). Otherwise T_1 and T_2 are essentially equal.

and chlorine-37 nuclei. These experiments employed coherent proton decoupling and could only be carried out successfully by specifically avoiding the Hartmann–Hahn* condition

$$\gamma_H B_2/2\pi = \gamma_C B_1/2\pi, \qquad [1]$$

otherwise spurious decay rates were observed (3). Spin–lattice relaxation measurements on the same sample are shown for comparison; in these, the carbon sites directly attached to chlorine showed extremely *long* spin–lattice relaxation times because there were no close proton spins.

While spin locked, the nuclei possess a very large magnetization ($\approx M_0$) while experiencing only a weak effective field B_{eff}, much less intense than the polarizing field B_0 initially used to create M_0. The spins may therefore be thought of as very 'cold' and they return to thermal equilibrium by losing magnetization, eventually reaching the extremely small value M_0B_1/B_0. This explains why the relaxation, although *longitudinal*, nevertheless corresponds to a decay rather than a recovery.

Whereas the spin–spin relaxation time is sensitive to components of the spectral density of molecular motions near to zero frequency, spin-locking experiments probe motions around the frequency $\gamma B_1/2\pi$. For the majority of situations in liquids, the spectral density function does not change significantly between these two frequencies, but occasionally there might be very slow motions that influence T_2 more than $T_{1\varrho}$. Spin locking is thus a more versatile method for examining slow molecular motion than the spin-echo experiment, since B_1 becomes a useful variable parameter.

REFERENCES

1. I. Solomon, *Compt. Rend. Acad. Sci. Paris* **248**, 92 (1959).

2. R. Freeman and H. D. W. Hill, *J. Chem. Phys.* **55**, 1985 (1971).

3. R. Freeman and H. D. W. Hill, *Dynamic Nuclear Magnetic Resonance Spectroscopy*, Ed.: L. M. Jackman and F. A. Cotton. Academic Press: New York, 1975, Ch. 5.

Cross-references

Adiabatic fast passage
Chemical exchange
Hartmann–Hahn experiment
Rotating frame
Selective excitation
Spin echoes
Spin–spin relaxation

Spin–Spin Relaxation

A client was trying to negotiate a loan from the pawnbroker and offered him his wristwatch. The pawnbroker declined, saying that he had an entire drawer full of wristwatches already. Well, said the client, What time is it? Now even if we suppose that the pawnbroker had very carefully set all the watches to the correct time at 9 a.m. on Monday morning, there would of course be some difficulty in deciding on the exact time some days later, in fact we might expect some kind of Gaussian distribution function of the readings on the large ensemble of watches in the drawer. A similar problem arises with the timekeeping of precessing nuclear spins because of small magnetic interactions between the nuclei which interfere with the exact nuclear precession frequencies. If the spins are set to precess all in phase at time zero (by the application of a short radiofrequency pulse) this phase coherence is gradually lost with a time constant called the spin–spin relaxation time T_2. In order to observe this in practice, we would need to have a perfectly stable spectrometer and a perfectly uniform magnetic field B_0; then a free induction signal would decay with time constant T_2. In effect this relaxation time is defined by the Bloch equations as the decay constant of transverse nuclear magnetization, neglecting instrumental shortcomings. In the steady-state solutions of the Bloch equations, T_2 may be identified as the inverse of the natural linewidth, and this would be observed in a slow-passage continuous wave spectrometer with a perfectly stable and uniform magnetic field.

It is important to make a clear distinction between the spin–spin relaxation time T_2 and the lifetime of the spins in a given excited energy state, the spin–lattice relaxation* time T_1. In many cases of high resolution NMR in the liquid phase, the same physical mechanisms determine both T_1 and T_2, so they are equal. The cases of interest are those where there are *additional* mechanisms for spin–spin relaxation such that T_2 is shorter than T_1. This situation can be emphasized by defining a *phase memory time*, accepting that certain interactions between spins may destroy their phase coherence faster than the spin-state lifetime (T_1). On the vector model* this means that transverse (XY) magnetization decays faster than the regrowth of longitudinal (Z) magnetization. In what follows we shall concentrate attention on these *additional* mechanisms for spin–spin relaxation, taking the spin–lattice relaxation processes for granted.

The two relaxation times may be distinguished in another way. The spin–lattice relaxation time measures the time constant for transfer of energy from the spin system to the environment (the lattice). We could say that it measures the time taken for 'hot' spins to cool down. There is no energy change in the spin system due to spin–spin relaxation. Interactions between spins, characterized by T_2^{-1}, maintain a true thermal equilibrium within the spin system. In solids, where $T_2 \ll T_1$, this allows us to define a spin temperature, which may be quite different from the lattice temperature.

In practical spectrometers, the transverse nuclear magnetization M_Y usually decays with a time constant T_2^* which, for small molecules, can be significantly shorter than T_2 as *isochromats* from different regions of the sample interfere destructively because of the non-uniformity of B_0. For this reason T_2 is commonly determined by a spin-echo* experiment in which field inhomogeneity effects are refocused. The envelope of the spin echo peaks decays exponentially with time constant T_2. Spin locking* may also be used.

In the frequency-domain it is possible to distinguish broadening by B_0 field inhomogeneities from broadening by spin–spin relaxation effects by the fact that in the former case it is possible to 'burn a hole' in the line by frequency-selective irradiation just prior to a non-selective monitoring pulse. Provided there is good field/frequency regulation*, the saturating radiation affects only a restricted volume of the sample which has a B_0 field very close to exact resonance. By contrast, when the linewidth is determined purely by spin–spin relaxation, irradiation of any part of the line saturates the entire line profile – it is said to be *homogeneously broadened.*

THE FREQUENCY SPECTRUM OF THE FLUCTUATIONS (1)

The *spectral density function* which describes the reorientational motion of the molecules has been discussed in the section on spin–lattice relaxation*. We saw that the spin–lattice relaxation time was determined by the component of this spectral density at the Larmor frequency of the nuclei, for this induces NMR transitions. Spin–spin relaxation is also sensitive to fluctuating fields at the Larmor frequency, but it is also sensitive to fluctuations near to zero frequency. If molecular tumbling is fast compared with the Larmor frequency then both relaxation times are influenced in the same manner and $T_1 = T_2$.

When we consider interactions that fluctuate more slowly that this, T_2 becomes shorter than T_1. When T_1 passes through its minimum, it is about $1.6\,T_2$. At lower frequencies still, T_1 starts to rise again while T_2 continues to fall (Fig. 1). Although fluctuations near the Larmor frequency are still involved,

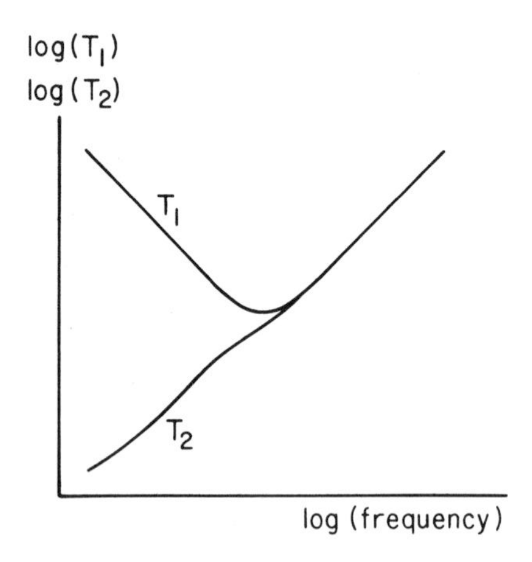

Fig. 1. The dependence of the spin–spin and spin–lattice relaxation times on the correlation frequency for molecular reorientation. For motion fast compared with the Larmor frequency the two relaxation times are essentially equal, but for slow motions the spin–spin relaxation time continues to decrease while the spin–lattice relaxation time increases.

spin–spin relaxation is now dominated by the very low frequency components of the spectral density function. Large biological molecules and polymers often fall into this category where the tumbling frequency is slower than the Larmor precession frequency; linewidths are then determined by spin–spin relaxation rather than the instrumental broadening and the nuclear Overhauser effect* becomes negative. For the investigation of these macromolecules an improvement in the field inhomogeneity has no effect, and the power to resolve chemically shifted resonances must be obtained by increasing the intensity of the applied field B_0.

When the liquid freezes there is an abrupt increase in the broadening due to spin–spin relaxation since there are now static dipole–dipole interactions. As the temperature of the solid is reduced, any remaining reorientational motions eventually freeze out. The corresponding change in linewidth may be used to investigate this kind of transition. At sufficiently low temperatures an asymptotic value of the linewidth is attained, known as the *rigid-lattice linewidth*. Any given spin usually has a large number of near-neighbour spins, and since each of these can be in either the α or β state, the function describing the distribution of local fields is close to a Gaussian in shape. In general, resonances from solids exhibit very short spin–spin relaxation times and long spin–lattice relaxation times (Fig. 1). This is an unfavourable situation from the point of view of sensitivity, but is usually compensated by the high density of spins in the solid state.

MECHANISMS OF SPIN–SPIN RELAXATION

Since spin–lattice relaxation limits the lifetime of the spins in the upper energy state and introduces a Heisenberg broadening on the NMR lines, mechanisms of spin–lattice relaxation should also be considered mechanisms of spin–spin relaxation. However, our interest is focused on any *additional* interactions which fluctuate at low enough frequency that T_2 is shortened without any appreciable effect on T_1. In NMR parlance they are loosely referred to as 'T_2 mechanisms'.

Chemical exchange* is probably the most important of these. An atom moves from one chemically distinguishable site to another, carrying with it a nuclear spin, and is replaced by a similar atom with a similar spin, except that the new precession phase is random. The details of what is observed depend on whether the exchange rate is slow, intermediate or fast compared with the chemical shift difference between the two exchanging sites, but from the point of view of spin–spin relaxation the fluctuation is slow compared with the Larmor precession frequency and there is a perceptible change in T_2. Where the effect is large it may be monitored as a broadening of the frequency-domain spectrum; otherwise it is detected by way of a spin echo experiment (2, 3).

The scalar spin–spin coupling provides another possible source of spin–spin relaxation if it is modulated at a suitable rate. The small magnetic field at a spin I due to coupling J_{IS} to a second spin S may be interrupted either if S exchanges chemically or if it exhibits rapid spin–lattice relaxation (this is a common occurrence if S is a quadrupolar nucleus). These two mechanisms have been called, respectively, *scalar relaxation of the first and second kinds* (1). When the rate of fluctuation is fast compared with J_{IS}, the I spin has only a single resonance line with no multiplet structure. The contribution to the broadening of the I resonance may be written

$$T_2^{-1} = \tfrac{1}{3}(2\pi J_{IS})^2 S(S+1)\left[\tau + \frac{\tau}{1 + (2\pi f_I - 2\pi f_S)^2}\right], \qquad [1]$$

where τ is the correlation time due to chemical exchange or the spin–lattice relaxation time of the S spin. Appreciable T_2 effects are detectable on proton or carbon-13 resonances that are coupled to nitrogen-14, chlorine-35 or chlorine-37 nuclei which have short T_1 values due to their quadrupole moments. Occasionally, this may be used as structural information.

REFERENCES

1. A. Abragam, *The Principles of Nuclear Magnetism.* Oxford University Press, 1961.
2. E. L. Hahn, *Phys. Rev.* **80**, 580 (1950).
3. H. Y. Carr and E. M. Purcell, *Phys. Rev.* **94**, 630 (1954).

Cross-references

Chemical exchange
Field/frequency regulation
Nuclear Overhauser effect
Spin echoes
Spin–lattice relaxation
Spin-locking
Vector model

Spin Tickling

While decoupling effects require irradiation fields $\gamma B_2/2\pi$ comparable with the coupling constant J_{IS}, certain double-resonance effects persist at much lower levels of irradiation. The phenomenon is based on a non-crossing rule for energy levels transformed into the rotating frame* of reference so it is to be expected intuitively that the maximum displacement of resonance lines would be of the order of $\gamma B_2/2\pi$ Hz. Observable changes in the spectrum occur provided that $\gamma B_2/2\pi$ is comparable with the linewidth, and provided that the frequency f_2 of this irradiation is located close to exact resonance of one of the lines of a spin multiplet (not necessarily the chemical shift frequency). The two requirements – precise location and weak perturbation – have given rise to the term 'spin tickling' for this kind of double resonance experiment (1).

In order to observe tickling effects in an NMR spectrum, two nuclei must be spin-coupled, but the magnitude of the coupling is irrelevant provided it is large

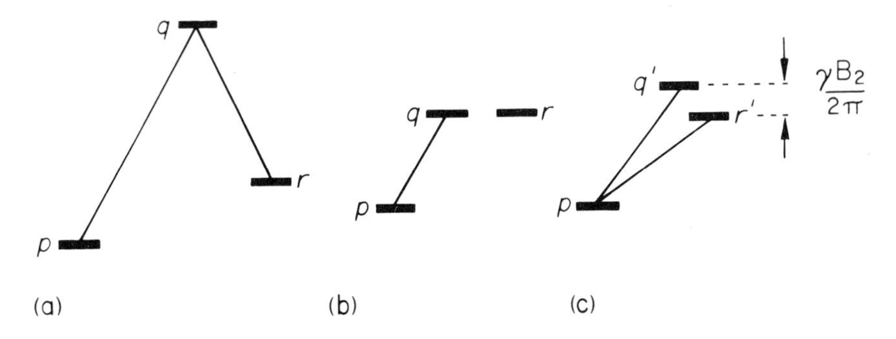

Fig. 1. Transformation of the laboratory-frame energy level diagram (a) into a frame (b) rotating at the frequency of the transition q–r (not drawn to scale). When a B_2 field is applied (c) at this frequency it lifts the degeneracy of levels q and r, introducing a splitting $\gamma B_2/2\pi$ Hz and making the transitions pq′ and pr′ equally allowed.

enough to give a resolved splitting so that one line of the multiplet can be irradiated without significantly perturbing the others. In fact, the method is most useful for complicated networks of spin coupling between large numbers of spins, as often encountered in proton NMR systems. Suppose we isolate just three energy levels p, q and r from the complex network of energy levels appropriate to such a system, and focus attention on the allowed transition at frequencies f_{pq} and f_{qr}. Consider what would happen if f_{qr} were to be irradiated with a weak field B_2 exactly at resonance, $f_2 = f_{qr}$. In order to remove the time-dependence of the problem, the energy-level diagram is transformed into a rotating reference frame at the frequency f_2, where B_2 becomes a static field (Fig. 1(a)). In the first step of this process, levels q and r become degenerate (Fig. 1(b)) and then as B_2 is introduced, their degeneracy is lifted and they are forced apart by an amount $\gamma B_2/2\pi$ Hz.

The selection rules are now changed so that there are two transitions from level p to the perturbed levels q′ and r′. After transformation back from the rotating frame to the laboratory frame, the transition f_{pq} becomes a doublet $f_{pq} \pm \frac{1}{2}\gamma B_2/2\pi$. Thus in this simple case, tickling has given rise to a doublet where only a single line existed initially. We can see that the irradiated line and the split line are related by the fact that they share an energy level in common (q). They are said to be *connected* transitions. Irradiation of a connected transition at exact resonance gives a doublet of splitting $\gamma B_2/2\pi$ Hz.

Consider now the more general case where f_2 is not at exact resonance for f_{qr} but is offset a small amount Δf_2 Hz. Transformation to the rotating frame at frequency f_2 leaves levels q and r not quite degenerate, separated by Δf_2 Hz, and the introduction of the static field B_2 increases this splitting to S Hz, where

$$S = [(\Delta f_2)^2 + (\gamma B_2/2\pi)^2]^{1/2}. \quad [1]$$

One of the new transitions (say $f_{pq'}$) takes on more the character of an allowed transition and gains intensity at the expense of $f_{pr'}$, which can now be said to take on the character of a forbidden transition. In the limit where Δf_2 becomes large compared with $\gamma B_2/2\pi$, then one component of the doublet is fully allowed and falls at the unperturbed frequency f_{pq}, while the other is forbidden and has vanishing intensity. By monitoring the imbalance in the intensities of the two doublet components, it is possible to search for the exact resonance condition $f_2 = f_{qr}$ and thus locate a hidden or unobservable line with high

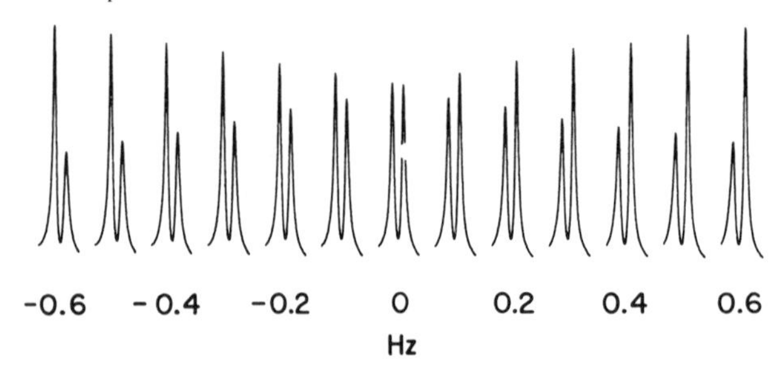

Fig. 2. Illustration of the frequency-dependence of the spin-tickling experiment. The low-field carbon-13 satellite in the proton spectrum of formic acid is monitored while the B_2 field irradiates close to the high-field carbon-13 transition. At exact resonance (0 Hz) the observed line is a symmetrical doublet.

precision. Figure 2 shows a heteronuclear spin tickling experiment to determine the frequency of a carbon-13 transition indirectly by observation of the splitting of a satellite line in the proton spectrum. At the exact tickling condition the two lines of the observed doublet have equal intensity; as the B_2 frequency goes off resonance the doublet intensities become asymmetrical. The frequency of the carbon-13 transition can be determined with an accuracy that is better than ± 0.1 Hz.

The treatment has so far considered only a simple AX system of coupled spins. For an AX_2 system where one of the X lines is irradiated, three levels become degenerate in the rotating frame and the introduction of the B_2 field splits this degeneracy, leaving a triplet for each A resonance. Similarly, for the AX_3 system, the A lines become 1:3:3:1 subquartets because of the tickling condition.

While spin tickling gives more precise information about the frequencies of hidden transitions, it is more tedious to carry out than coherent decoupling since the experiment must be repeated many times with quite small increments of f_2. For complicated spectra, Fourier transform difference spectroscopy* may be used to clarify the results.

In continuous-wave spectroscopy*, the arrangement of the spin tickling experiment can be reversed with the B_1 field monitoring the peak of one resonance line while the frequency f_2 is swept through the rest of the spectrum to locate connected transitions. This is called the 'INDOR' experiment (2) by analogy with electron–nuclear double resonance (ENDOR). Unfortunately, two physical processes contribute to the observed effect – the splitting of the monitored line due to spin tickling, and a population disturbance caused by saturation effects of the B_2 field. For very weak levels of irradiation, the population rearrangement effects appear to predominate. In order to understand the *direction* of the intensity changes, we must make a distinction between two kinds of connected transitions – the *progressive* configuration and the *regressive* configuration (Fig. 3). In the progressive case the monitored line

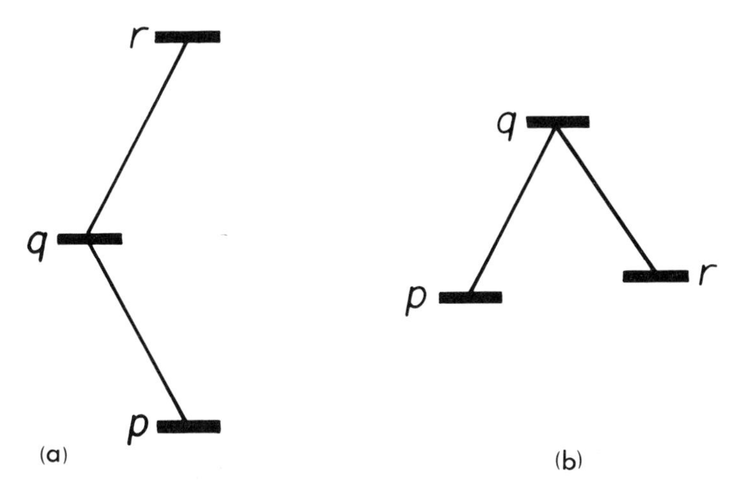

Fig. 3. Two transitions which share a common energy level are said to be *connected*. They may have (a) the *progressive* configuration or (b) the *regressive* configuration.

increases in intensity as spins are pumped into the lower level or alternatively pumped out of the upper level. By contrast, the regressive configuration always shows a decrease in intensity of the monitored line because the population disturbances have the opposite sense.

The progressive and regressive cases are distinguishable in another interesting way when spin tickling occurs. In the progressive case the action of the B_2 field mixes in some of the double-quantum transition, rendering the doublet *more* sensitive to static field inhomogeneity effects, whereas in the regressive case it mixes in some of the zero-quantum transition and the observed doublet is unusually well-resolved (1). This is a situation where we cannot simply work out the allowed transition frequencies and then convolute with a field broadening function to obtain the observed lineshapes. Tickling takes place within a molecule, over such a short distance that field inhomogeneity may be completely neglected. If that molecule is situated in a region where the B_0 field is (say) slightly higher than the mean, then the resonance offsets for both the B_1 and B_2 fields must be in the same sense. For the progressive case this makes inhomogeneity broadening worse, but for the regressive case it is reduced. Similar effects occur, of course, in experiments with multiple-quantum coherence*.

REFERENCES

1. R. Freeman and W. A. Anderson, *J. Chem. Phys.* **37**, 2053 (1962).
2. E. B. Baker, *J. Chem. Phys.* **37**, 911 (1962).

Cross-references

Continuous-wave spectroscopy
Difference spectroscopy
Multiple-quantum coherence
Rotating frame

Steady-State Effects

It is tempting to analyse the Fourier transform NMR experiment in terms of a single pulse followed by a single free induction decay, for this avoids a host of complicating factors. However, the demands of time averaging* and optimized data acquisition compel us to repeat the excitation rapidly, consistent with the resolution requirements which determine the acquisition time t_a. The inter-pulse interval t_p is set equal to $t_a + t_d$, where t_d is a delay to allow for relaxation; in one common mode of operation $t_d = 0$. The choice of t_a is determined by the instrumental decay constant T_2^*; if t_a is less than T_2^* the free induction decay is prematurely truncated, and if t_a is several times T_2^*, time is wasted gathering negligible signals in the tail. In many applications the decay rate is accelerated by imposing a sensitivity enhancement* function, often the equivalent of a matched filter.

These considerations mean that the pulse excitation is repeated at a rate which does not allow complete spin–spin or spin–lattice relaxation* ($t_p < T_2$, T_1). As a result, the radiofrequency pulse is not applied to a magnetization vector M_0 at Boltzmann equilibrium, but to a system still suffering the effects of previous pulses. A steady-state is set up where there are finite transverse components M_X^- and M_Y^- immediately before the pulse and a partially saturated longitudinal component M_Z^-, weaker than M_0. The steady-state conditions are readily calculated (1–4) giving for the transverse magnetization immediately after the pulse

$$M_X^+ = M_0(1 - E_1)(E_2 \sin \alpha \sin \theta)/D \qquad [1]$$

$$M_Y^+ = M_0(1 - E_1)[(1 - E_2 \cos \theta) \sin \alpha]/D, \qquad [2]$$

where

$$E_1 = \exp(-t_p/T_1)$$

$$E_2 = \exp(-t_p/T_2)$$

$$D = (1 - E_1 \cos \alpha)(1 - E_2 \cos \theta) - (E_1 - \cos \alpha)(E_2 - \cos \theta)E_2.$$

Here α is the pulse flip angle and θ is the excess precession angle of the spins during t_p, a function of the resonance offset.

SPIN–LATTICE RELAXATION

In situations where spin–spin relaxation is very much faster than spin–lattice relaxation (large molecules) we may set the pulse repetition rate such that $E_2 \ll 1$ and all transverse magnetization will decay between pulses. A steady-state is then established for longitudinal magnetization only, and the flip angle for optimum signal-to-noise is given by the 'Ernst angle' α_E

$$\cos \alpha_E = \exp(-t_p/T_1). \quad [3]$$

Since, generally, $t_p < T_1$, the Ernst angle is small in comparison with 90°. If an estimate of the spin–lattice relaxation time is available, this suggests a suitable setting for the flip angle.

SPIN–SPIN AND SPIN–LATTICE RELAXATION

For the small-to-medium sized molecules of interest to the organic chemist, T_2 is not very different from T_1 and steady-state transverse magnetization must be considered. The question then arises why these effects are not eliminated by instrumental decay processes such as B_0 inhomogeneity, where T_2^* is short compared with t_p. The answer is that there is a refocusing effect analogous to the Hahn spin echo*, attributable to the repeated radiofrequency pulses. Although the envelope of the free induction signal decays with time after the pulse, it grows again before the next pulse as a consequence of the steady-state conditions.

DETERMINATION OF SPIN–SPIN RELAXATION TIMES

Under suitable conditions (5) the establishment of a steady-state condition with respect to transverse magnetization can be used to measure the spin–spin relaxation time T_2. The steady-state equations simplify considerably if the broadening due to static field inhomogeneity is very strong ($T_2^* \ll t_p$) and where the spin–lattice relaxation time is long ($T_1 \gg t_p$). Then the signal just before the radiofrequency pulse is readily separable from the signal just after the pulse; comparison of these intensities gives a measure of the spin–spin relaxation time T_2. The experiment is thus the analogue of the progressive saturation method for measuring T_1. It is sometimes known as *steady-state free precession* (1, 5). It shows promise for measuring spin–spin relaxation times in biological systems (5) since it is rather insensitive to spatial inhomogeneity of the radiofrequency field B_1.

INFLUENCE ON HIGH-RESOLUTION SPECTROSCOPY

Steady-state effects on transverse magnetization have three principal effects. The signals are rather stronger than expected, the relative intensities are perturbed, sometimes drastically, and individual resonances have varying amounts of dispersion mode signal so that the spectrum cannot be correctly phased. These phase and intensity anomalies reflect the fact that the steady-state conditions [1] and [2] are strongly dependent on θ which is a rapidly oscillating function of the resonance offset. Rather surprisingly, if the pulse flip angle is set equal to the Ernst angle according to eqn [3], then the intensity variations disappear because θ drops out of the steady-state equations.

These phase and intensity anomalies are most easily observed in a homogeneous magnetic field; strong instrumental broadening tends to obscure them by averaging them over a range of θ values. An important result emerges for the case where T_2^* is short compared with t_p: the curve relating signal-to-noise to pulse flip angle has a broad flat maximum virtually independent of θ. The optimum flip angle is then near 70° (much larger than the Ernst angle) and it is by no means critical. A variation of α between 60° and 90° only changes the signal intensity by a few percent. A sensitivity enhancement weighting function would have a similar influence to T_2^* in this respect. Consequently, a Fourier spectrometer can be adjusted for near-optimum sensitivity under these conditions by setting $\alpha \approx 70°$ irrespective of the spin–spin and spin–lattice relaxation times. This is an important simplification.

The formation of spin echoes in a repetitive pulse sequence is inhibited by instrumental instabilities, for example poor field/frequency regulation*, and in severe cases the phase and intensity anomalies disappear. For carbon-13 spectroscopy incoherence effects attributable to proton noise decoupling also interfere with the formation of echoes. Finally, the phase and intensity anomalies can be artificially suppressed by introducing a small timing jitter (of the order of 10 milliseconds) into the t_d delay, randomizing the phase of the transverse magnetization at the time of the next pulse.

RECOMMENDATIONS

We may therefore conclude that in a straightforward Fourier transform experiment with repeated pulse excitation, the sensitivity should be near optimum if the following steps are taken. Decide on the resolution required in a particular experiment and choose the acquisition time t_a accordingly. Set $t_d = 0$. Multiply the free induction signal by a suitable sensitivity enhancement function so that it decays to a low level during t_a, without unduly broadening the lines. For work with small-to-medium sized molecules where T_1 and T_2 are expected to be comparable, set the pulse flip angle to 70°; it is not critical. For large molecules, where T_2 is considerably shorter than T_1, set the pulse flip angle to the Ernst angle, using an estimate of T_1. This will normally be significantly less than 70°. Alternatively use 90° pulses and a suitable value of the relaxation delay t_d; again an estimate of T_1 is required in order to set $t_a + t_d$ equal to about $3T_1$. In experiments that require accurate integration these sensitivity optimization procedures have to be sacrificed in favour of slow pulsing or small pulse flip angles.

REFERENCES

1. H. Y. Carr, *Phys. Rev.* **112**, 1693 (1958).
2. R. R. Ernst and W. A. Anderson, *Rev. Sci. Instr.* **37**, 93 (1966).
3. R. R. Ernst, *Advances in Magnetic Resonance*, Ed.: J. S. Waugh. Academic Press: New York, 1966, Vol. 2.
4. R. Freeman and H. D. W. Hill, *J. Magn. Reson.* **4**, 366 (1971).
5. S. Matsui, M. Kuroda and H. Kohno, *J. Magn. Reson.* **62**, 12 (1985).

Cross-references

Field/frequency regulation
Sensitivity enhancement
Spin echoes
Spin–lattice relaxation
Spin–spin relaxation
Time averaging

Stochastic Excitation

The natural desire for a simple picture has led the spectroscopist to concentrate on continuous-wave slow-passage spectroscopy or the pulse-excited Fourier transform method, since these phenomena are readily described in terms of the steady-state or transient solutions of the Bloch equations. There are, however, more complicated modes of excitation which may have some important advantages; one of these is *stochastic excitation* (1–3). There is a certain theoretical elegance associated with an experiment which employs random noise as the excitation, extracting the desired NMR information by sophisticated data processing techniques. If the excitation is set up so as to drive the spin system in an appreciably non-linear fashion, then the stochastic excitation method can provide, along with the conventional spectrum, all the kinds of information normally extracted by multiple resonance or multipulse techniques. This includes chemical shift correlation*, chemical exchange* and the nuclear Overhauser effect*. In fact, stochastic excitation can be thought of as the superposition of a large number of different multipulse experiments.

Strictly speaking the experiment should be carried out with white noise excitation (rms level independent of frequency) having a Gaussian distribution of amplitudes. In practice, it turns out to be feasible to use binary noise (pseudo-random noise) generated by a shift register. This is used to control the pulse phases (0° or 180°) of a regular sequence of radiofrequency pulses, with a duty cycle typically 1%. The NMR response is sampled (once) in the interval between consecutive radiofrequency pulses. The pulse repetition rate is thus the sampling rate, and must be set slightly higher than the highest resonance offset in the spectrum (for a spectrometer with quadrature detection). Along with the NMR response, it is necessary to store a record of that section of the pseudo-random excitation sequence used to obtain this response, since the data processing involves a *cross-correlation* between excitation function and signal response.

Once the excitation has been switched on, the spin system comes into a steady-state condition in a time of the order of the spin–lattice relaxation time. The power level of the excitation is then adjusted by setting the pulsewidths until the oscillations of transverse nuclear magnetization caused by the pulses are near their maximum excursions. This ensures near-optimum signal-to-noise ratio in the final spectra; in principle, this optimization of the power level could be achieved under computer control. Signal acquisition can then be initiated at any time, together with a record of the appropriate section of the excitation function. Normally, something of the order of 10^6 samples (excitation and response) would be taken.

Analogue-to-digital converters sample the excitation record x(t) and the corresponding spin system response y(t), the former requiring only one bit while the latter would typically use 7 bits. Data reduction is considerably more complicated than in conventional Fourier transform spectroscopy and only an outline of it is presented here; detailed descriptions appear in the literature (4, 5). Cross-correlation of excitation and response functions can be performed either in the time domain or in the frequency domain. For practical reasons of speed of computation and data storage requirements, the calculation is almost always carried out in the frequency domain. This carries the important advantage that it is possible to examine just those selected regions of the frequency domain which contain the lines of interest.

The total of N acquired data points is first divided into M equal sections each with a duration T, where 1/T determines the attainable digital resolution. Each section is then Fourier transformed, the excitation function becoming an excitation spectrum X(F) which has the appearance of random noise. The system response becomes the response spectrum Y(F) which contains recognizable resonance lines grossly distorted by the 'excitation noise'. This systematic noise is reduced by two processes. First, X(F) and Y(F) are cross correlated (by complex multiplication and averaging) and then the results are averaged with those from the remaining M − 1 sections of the record. Since each section is statistically independent of the others (a very long pseudo-random sequence is used), the signal-to-noise ratio is further improved since both excitation noise and random (Johnson) noise are reduced. The result is a one-dimensional cross-correlation spectrum

$$S_1(F) = \langle Y(F)X^*(F)\rangle / \langle |X(F)|^2 \rangle. \qquad [1]$$

Here, the asterisk denotes the complex conjugate. This has the same general form as the conventional NMR spectrum, although the lines may be slightly broader due to saturation.

The really interesting information appears in the corresponding three-dimensional cross-correlation (the two-dimensional cross-correlation spectrum does not exist). The three-dimensional cross-correlation is normally evaluated from the formula

$$S_3(F_1, F_2, F_3) = \frac{\langle Y_3(F_1 + F_2 + F_3)X^*(F_1)X^*(F_2)X^*(F_3)\rangle}{6\langle |X(F_1)|^2|X(F_2)|^2|X(F_3)|^2\rangle}, \qquad [2]$$

where $Y_3(F) = Y(F) - S_1(F)X(F)$. There are of course problems associated with displaying a three-dimensional spectrum, but it turns out that a specific two-dimensional section through the diagram assumes particular importance. This is the *sub-diagonal section* $S_3(F_1, F_2, F_3)$, where F_3 is constrained to equal

$-F_2$. This corresponds to examining transitions involving three quanta where the net change in quantum number $|\Delta m| = 1$ rather than 3. (In strongly coupled three-spin systems these can be observed in conventional NMR; they are usually called the *combination lines*.)

This sub-diagonal section resembles the form of the two-dimensional spectra obtained in homonuclear shift correlation spectroscopy (COSY). There are diagonal peaks and cross-peaks indicating a particular spin–spin interaction. The cross peaks show the usual square pattern of lines with alternating sense ($\pm \mp$). Unfortunately, one region of the spectrum is obscured by strong noisy ridges passing through the intense diagonal peaks. These arise mainly from cross-talk from the one-dimensional cross-correlation spectrum.

Figure 1 shows a two-dimensional shift correlation spectrum for the protons of ethanol obtained as a subdiagonal section through the three-dimensional cross-correlation spectrum (4). The cross-peaks indicating coupling between the methyl and methylene protons are clearly visible with the expected alternation of positive and negative intensities.

Chemical exchange can also be studied by this method and (with special precautions) nuclear Overhauser effects can also be investigated. Although

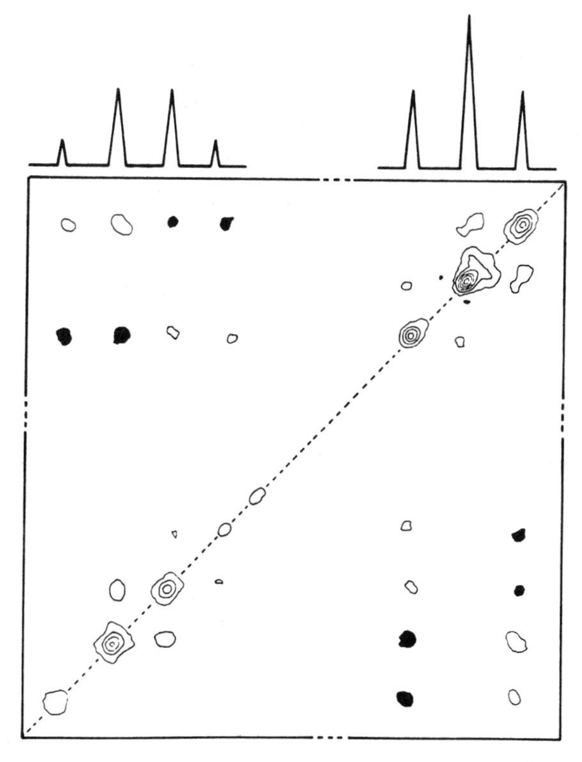

Fig. 1. A two-dimensional homonuclear shift correlation spectrum of ethanol obtained by stochastic excitation. Negative intensity contours have been filled in in black. This represents the real part of a subdiagonal cross section through a three-dimensional data matrix and shows selected regions of the 1024 × 1024 two-dimensional spectrum. Reproduced by permission from B. Blümich and R. Kaiser (6).

these are currently studied by two-dimensional spectroscopy* using the usual coherent pulse excitation schemes, stochastic excitation has a potentially important advantage. *In a single experiment* all the information is collected and transferred to a computer, freeing the spectrometer for other work. The various types of spectra are then derived simply by data processing, giving the conventional one-dimensional spectrum or two-dimensional correlations involving spin–spin coupling, chemical exchange or cross-relaxation. This should turn out to be a more efficient use of spectrometer time than the current procedures where several quite different one- and two-dimensional experiments have to be set up and run consecutively.

REFERENCES

1. R. R. Ernst, *J. Magn. Reson.* **3**, 10 (1970).
2. R. Kaiser, *J. Magn. Reson.* **3**, 28 (1970).
3. B. Blümich and D. Ziessow, *J. Magn. Reson.* **52**, 42 (1983).
4. B. Blümich and D. Ziessow, *J. Chem. Phys.* **78**, 1059 (1983).
5. B. Blümich and D. Ziessow, *Mol. Phys.* **48**, 955, 969 (1983).
6. B. Blümich and R. Kaiser, *J. Magn. Reson.* **58**, 149 (1984).

Cross-references

Chemical exchange
Nuclear Overhauser effect
Shift correlation
Two-dimensional spectroscopy

t_1 Noise

Two-dimensional NMR spectra suffer from a persistent artifact usually known as *t_1 noise*. This is a ridge of noise that runs parallel to the F_1 axis wherever there is a strong resonance peak in the two-dimensional spectrum. The problem is particularly serious for aqueous solution since the noisy ridge may remain even though the parent water line has been largely cancelled by difference spectroscopy. The artifact is actually in two parts – an apparently random fluctuation of the baseline, significantly larger than the true thermal noise, and a d.c. offset of the baseline too large to be accounted for by the tails of the strong resonance peak (even allowing for aliasing at the edges of the spectrum). The presence of this ridge tends to make the noisy component more obtrusive since it raises the mean baseline of the t_1 noise.

We consider first of all the noisy component. It should be clearly distinguished from true thermal noise originating in the Johnson noise of the receiver coil. It arises because various types of instrumental instability cause fluctuations in the amplitude of the free induction decay from one increment of t_1 to the next (1). These instabilities have no effect unless there is a strong NMR response at that particular F_2 ordinate. These are exactly the same types of long-term instability that cause problems in difference spectroscopy* as often employed in studies of the nuclear Overhauser effect or in experiments using multiple-quantum filters.

A common and often serious source of instability arises in the field/frequency regulation* scheme if there is an inadequate signal-to-noise ratio for the deuterium reference signal. In one-dimensional spectroscopy this shows up as a line broadening and noisy fluctuations in the skirts of the resonance lines since the bandwidth of these variations tends to be quite narrow. By contrast, in two-dimensional spectroscopy* the t_1 noise is distributed along the entire F_1 frequency dimension. The remedy is to employ a sufficiently strong deuterium reference material and to restrict the bandwidth of the field/frequency regulation loop.

Another source of t_1 noise arises if the phase or amplitude of the transmitter pulses are not properly reproducible. This type of instability has little impact on one-dimensional spectra, since individual free induction decays may vary in amplitude without impairing resolution when subject to time-averaging. In two-dimensional spectroscopy, however, this can be an important noisy contribution to horizontal (F_1) traces that carry appreciable resonance lines. Similar long-term fluctuations in signal intensity (and linewidth) also come about because of variations in the spatial inhomogeneity of the B_0 field. These slow variations are sampled at a very low repetition rate as t_1 is incremented and thus appear as noisy contributions over the entire F_1 axis.

Instabilities in the receiver gain or in the receiver phase shift usually are so small over the short-term that they may be neglected in one-dimensional spectroscopy. On the long timescale appropriate to the evolution time of a two-dimensional experiment, this is a source of baseline noise wherever there is a strong signal component.

The limited dynamic range of the analogue-to-digital converter used to digitize the time-domain data is known to generate *digitization noise* in one-dimensional spectra containing a strong solvent peak (see Solvent suppression*). The process occurs in both dimensions in two-dimensional spectroscopy and is relatively more important in the F_1 dimension if the spectral width in F_1 is smaller than the spectral width in F_2.

In conventional NMR spectroscopy it is usually a good approximation to assume that the sampling of the free induction signal is perfectly regular in the time-domain. This may not always be the case for the t_1 dimension of two-dimensional spectroscopy because of poor programming or variable timing delays due to disc transfers. Although these are strictly deterministic errors they appear to be random after Fourier transformation. The signal $S(t_1)$ actually recorded may be thought of as the sum of the ideal signal $S_0(t_1)$ (one that would have been observed for a perfectly regular sampling operation) plus a superimposed error signal $S_N(t_1)$ which is only present when $S_0(t_1)$ is present. It is the Fourier transform of $S_N(t_1)$ which provides a contribution to the noisy ridge parallel to the F_1 axis.

Another consequence of irregular timing of the sampling in the t_1 domain is that even a *coherent* perturbation such as a sinusoidal modulation of the magnetic field B_0 may generate apparently random fluctuations of the baseline in the F_1 dimension. This is because the sampling rate is both irregular and very slow in comparison with the modulation frequency. Consequently, spinning modulation and other spurious modulations of the NMR response can contribute to t_1 noise in two-dimensional spectroscopy.

Spectra recorded with small spectral widths in the F_1 dimension are particularly susceptible to t_1 noise. In conventional spectroscopy precautions are taken to avoid aliasing of high-frequency noise components by filtration of the signals before analogue-to-digital conversion. It is not feasible to carry out the equivalent operation in the t_1 dimension, with the result that noisy artifacts may be aliased several times, increasing the amplitude of the observed t_1 noise. A particularly serious case occurs in two-dimensional J-spectroscopy* where the spectral width in the F_1 dimension is very small.

Since t_1 noise is not true random noise it may be feasible to suppress it by data processing techniques without harming true signals. For example, in homonuclear chemical shift correlation* spectroscopy (COSY) the spectrum is known to be symmetrical with respect to the principal diagonal $F_1 = F_2$. When there is a cross-peak with coordinates (a,b) there must also be another cross-peak with coordinates (b,a). It is thus possible to perform a symmetrization procedure by comparing points in the data matrix $S(F_1,F_2)$ that are symmetrically related with respect to the principal diagonal, replacing both with their mean value (for example). Since signals necessarily appear at both locations they remain essentially unchanged, but t_1 noise and the underlying ridges are attenuated (2). A more drastic procedure replaces the two symmetrically related ordinates by the lower value of the two; this largely eliminates t_1 noise. Such non-linear processes should not be used where the cross-peaks are comparable in amplitude with the thermal noise background.

F_1 RIDGES

It is difficult to disentangle the t_1 noise from the baseline ridge running beneath it, but it is clear that there is usually an appreciable d.c. component which we may call the *F_1 ridge*. This implies that the first data point on the t_1 interferogram has an anomalously large intensity whenever there is an NMR signal present. One possible explanation is that the repetition rate in the t_1 domain is too high to allow adequate spin–lattice relaxation so that a steady-state is established where all resonances are partially saturated. The experiment is, however, usually initiated with $t_1 = 0$ and this first data acquisition may occur on an unsaturated spin system, giving an abnormally intense first data point. The usual remedy is to program a few 'dummy runs' through the pulse sequence in order to establish a partially saturated steady-state condition before any actual data acquisition. Another expedient is to run the t_1 evolution in reverse, starting with the longest value t_1(max) and ending with $t_1 = 0$.

In some spectrometer systems there is a more insidious problem which amounts to the improper preparation of the data fed to the Fourier transform program (3). This is treated in detail in the section on Baseline correction*. The resulting baseline offset can go unnoticed in one-dimensional spectroscopy, but in two-dimensional spectra it shows up as an F_1 ridge which makes the t_1 noise more prominent.

REFERENCES

1. A. F. Mehlkopf, D. Korbee, T. A. Tiggleman and R. Freeman, *J. Magn. Reson.* **58**, 315 (1984).

2. R. Baumann, A. Kumar, R. R. Ernst and K. Wüthrich, *J. Magn. Reson.* **44**, 76 (1981).

3. R. Bracewell, *The Fourier Transform and its Applications.* McGraw-Hill: New York, 1965.

Cross-references

Baseline correction
Difference spectroscopy
Digitization
Field/frequency regulation
J-spectroscopy
Shift correlation
Solvent suppression
Two-dimensional spectroscopy

Time Averaging

The NMR phenomenon is such a tiny effect that the thermal fluctuations of the electrons in the receiver coil compete with it. This is called *Johnson noise*. This irreducible minimum of thermal noise masks the weaker NMR signals and thus limits the sensitivity* of the technique. Proper design of the preamplifier and receiver avoids the introduction of additional noise components from the electronic amplifiers and normally ensures that Johnson noise remains the dominant source of noise in the final NMR spectrum.

It then seems natural to enquire whether there is any characteristic property that might distinguish noise from true NMR signals when they both have comparable intensities. The first such criterion is that the noise commonly includes higher-frequency fluctuations than the signals; viewed in the form of a free induction decay, the NMR signals are seen to decay with a time constant T_2^*, whereas the noise remains at a constant level. Sensitivity is therefore improved by the application of a suitable weighting function which de-emphasises the tail of the free induction decay, usually by multiplying with a decaying exponential; the optimum being the *matched filter* which falls off at the same rate as the NMR signals. Any more drastic filtration reduces the signals faster than the noise and there is no further advantage to be gained.

The second criterion is that signals are reproducible if the experiment is repeated, whereas noise should be uncorrelated from one measurement to the next. (Certain other undesirable artifacts may be reproducible, for example some of the effects responsible for 't_1-noise'* in two-dimensional spectroscopy*). In the early 1960s it was realized that this could be made the basis of an important new method of sensitivity improvement known as *time averaging*. The instrumentation was developed from the multichannel analyser used in nuclear physics. Klein (1, 2) seems to have been the first to bring this important technique to the attention of NMR spectroscopists. That was in the days of continuous-wave spectroscopy* but the same principles apply to Fourier transform methods. Indeed time averaging is essential if the Fourier method is to have any sensitivity advantage.

The spectrum or the free induction decay is sampled at regular intervals and the voltage converted into digital form for storage in a computer. At any given sampling point one expects the same intensity signal each time the experiment is repeated, so the total signal builds up linearly with the number of passes. Noise, being a random phenomenon, fluctuates in sign and intensity and builds up only as the square-root of the number of passes. Consequently the signal-to-noise ratio improves as the square-root of the number of free induction decays that are time averaged.

Clearly the method presupposes that the spectrometer has good field/frequency regulation* otherwise the signal component at a given sampling point would also fluctuate and thus the signal-to-noise ratio would not improve as rapidly as expected. As the number of accumulations increases, the dynamic range of the stored signal increases and the number of bits of data storage (the computer word-length) eventually becomes a limitation, and double-precision arithmetic may be required. In some circumstances it can be advantageous to carry out Fourier transformation* after a certain 'block' of time averaging has been completed, continuing the averaging of the blocks in the frequency domain. This allows very intense solvent peaks to be truncated so that weak solute spectra can be stored with a shorter wordlength. Care must be taken with the round-off errors made by the Fourier transform algorithm since, although these may give the appearance of noise, they are to some extent reproducible and therefore not reduced by time averaging.

Ernst (3) has pointed out that the principle of time averaging still applies if the NMR signal, instead of being identical in each experiment, varies in a periodic manner, for example if it is amplitude modulated as in certain two-dimensional spectroscopy* experiments. Consequently, the necessary repetition of the measurements for a large number of different values of the evolution period (t_1) gives rise to a sensitivity improvement similar to that enjoyed by one-dimensional spectroscopy. This accounts for the surprisingly low sensitivity penalty paid by two-dimensional NMR.

The dependence on the square-root of the number of recorded free induction decays imposes practical limitations on the extent of time averaging. For many laboratories the operational maximum is the overnight accumulation ($\approx$15 hours); a weekend run only improves the sensitivity by another factor of two. Beyond this, sensitivity improvements must be sought, not through noise reduction, but through sensitivity enhancement* or more efficient spin–lattice relaxation*.

REFERENCES

1. M. P. Klein, *American Chemical Society Meeting in Miniature*, University of California at Berkeley, 1962.

2. M. P. Klein and G. W. Barton, *Rev. Sci. Instr.* **34**, 754 (1963).

3. R. R. Ernst, Sensitivity Enhancement in Magnetic Resonance. In *Advances in Magnetic Resonance*, Ed.: J. S. Waugh. Academic Press: New York, 1966, Vol. 2.

Cross-references

Continuous-wave spectroscopy
Field/frequency regulation
Fourier transformation
t_1 noise
Sensitivity
Sensitivity enhancement
Spin–lattice relaxation
Two-dimensional spectroscopy

Time-Share Modulation

In constructing a spectrometer for high resolution NMR, particular attention must be paid to noise and interference at the front end of the receiver, otherwise they may compete with or even dominate the NMR signal. The thermal (or Johnson) noise from the receiver coil is inevitable and constitutes a fundamental limitation on the smallest signal that can be detected. In continuous-wave spectroscopy*, fluctuations and slow drifts of the transmitter level are a separate concern, and unless suitable precautions are taken, they can interfere with the detection of very weak NMR signals. There are basically three ways to circumvent the problem of transmitter noise and drift – electronic balancing in a bridge circuit, geometrical balancing in a crossed-coil probe, and separation in the time domain or *time-share modulation*. Fourier spectrometers implicitly use this time-domain separation, but a higher-frequency time-share modulation scheme can also be useful for certain applications (1).

Continuous-wave spectrometers suffer from the effects of slow drifts of the transmitter level or slow changes in the balance of the bridge circuit or of the crossed-coil probe. This leads to corresponding drifts of the baseline of the high resolution spectrum, which is particularly troublesome when integration is used for this is very sensitive to small variations in the baseline. Much better baseline stability is achieved if a modulation and lock-in detection* scheme is employed, and one method to accomplish this is by time-share modulation. Rapid pulsing (10 Hz to 10 kHz) with pulses of small flip angle has an effect very similar to that of low-level continuous-wave irradiation. A radiofrequency source, amplitude modulated by short rectangular pulses, behaves like a single-frequency continuous-wave transmitter (the centreband) flanked by a set of sidebands spaced at intervals equal to the pulse repetition rate. Normally the centreband would be set at resonance, the sidebands being too far from resonance to have any significant effect. The NMR signal is then carried by modulation at the pulse repetition rate and detected in a lock-in detector with this as a reference frequency. The beauty of this method is that all the drifts and spurious variations of the transmitter leakage are rejected by the lock-in detector, which is only sensitive to signals modulated at the pulse repetition rate.

Modern Fourier transform spectrometers still tend to employ a time-share modulation scheme for the deuterium field/frequency regulation* scheme. Typically the pulse repetition rate would be of the order of 20 Hz, so the response is not so much a free induction decay as a steady-state response to time-share modulation. The deuterium receiver is gated off while the transmitter is on. The dispersion-mode deuterium signal is used as the error signal in the field/frequency control loop, while the absorption mode is displayed on a meter or oscilloscope to aid in adjusting the homogeneity. To all intents and purposes the time-share modulation system appears to behave like a continuous-wave device.

There is one occasion where even higher radiofrequency levels are present in the probe, threatening to saturate the preamplifier; this is when homonuclear double resonance experiments are being carried out. (For heteronuclear double resonance, radiofrequency traps and filters can be employed.) It is then necessary to impose a time-share modulation on the B_2 field at a frequency equal to the sampling rate of the free induction decay. Thus the receiver is gated off whenever B_2 is pulsed on, sampling taking place only in the intervals when B_2 is switched off.

Similar concepts can be used to explain a remarkable effect which occurs with spin echoes* in the presence of J-coupling. In a homonuclear Carr-Purcell spin echo experiment, where both coupled spins (A and X) experience the effects of the 180° pulses, the echoes are seen to be amplitude-modulated at a frequency J_{AX}. If, however, the pulse repetition rate is increased to a point where it greatly exceeds the chemical shift difference between A and X, the modulation disappears. Under these conditions the train of 180° pulses can be thought of as a time-share modulation scheme where the repetition rate is so high that only the centreband has any influence on the A and X spins, the sidebands being too far away. This approaches the spin-locking* situation where both A and X are aligned along the centreband field. The echo modulation then disappears.

REFERENCES

1. J. T. Arnold, *Phys. Rev.* **102**, 136 (1956).

Cross-references

Continuous-wave spectroscopy
Field/frequency regulation
Modulation and lock-in detection
Spin echoes
Spin-locking

Transient Nutations

The earth spins about its axis and this axis itself moves on the surface of a cone, that is, it *precesses*. In addition, the angle of this cone is not constant, there is an additional wobbling of the precession axis, called a *nutation*. Nuclei have the same types of motion – spin, precession and nutation, the latter being taken now to mean any periodic variation of the precession axis. In order to stay with classical motion (as opposed to quantum-mechanical) we must consider the net macroscopic magnetization vector M, the resultant of all the individual nuclear magnetizations in the sample. The motion takes on its simplest form in the rotating frame* of reference.

At thermal equilibrium M is aligned along the +Z axis and not much of interest is happening. Suppose we suddenly apply a radiofrequency field B_1 along the X axis, with B_1 strong compared with any resonance offset ΔB. The subsequent motion of M is called a *transient nutation*. In this simple case M describes a circular path in the Z–Y plane; in the laboratory frame this would appear as a rather drastic nutational motion. This induces a sinusoidally modulated NMR response in the receiver, peak signals being observed as M passes through the +Y and −Y axes.

Torrey (1) has analysed this experiment and has shown that the magnetization decays with a time constant T given by

$$\frac{1}{T} = \frac{1}{2}\left[\frac{1}{T_1} + \frac{1}{T_2}\right]. \qquad [1]$$

That is to say both spin–spin and spin–lattice relaxation influence the decay of the observed NMR signal. In the majority of NMR experiments B_1 is switched off after a very short time and we have (say) a 90° or 180° pulse; during such short intervals relaxation is usually neglected.

In situations where the transient nutations are allowed to continue for many rotations, it is observed in practice that the signal decays appreciably faster than predicted by eqn [1]. This is a result of the spatial inhomogeneity of the radiofrequency field B_1, magnetization vectors from different volume elements of the sample (*isochromats*) nutating at different rates and fanning out in the Y–Z plane. Solomon (2) discovered that this effect could be refocused by suddenly reversing the phase of the radiofrequency field B_1 giving a *rotary echo* analogous to the spin echoes* generated by refocusing the effect of B_0 inhomogeneities.

Transient nutations are also observed if the radiofrequency field B_1 is applied continuously but the nuclear magnetization is generated abruptly, and this is

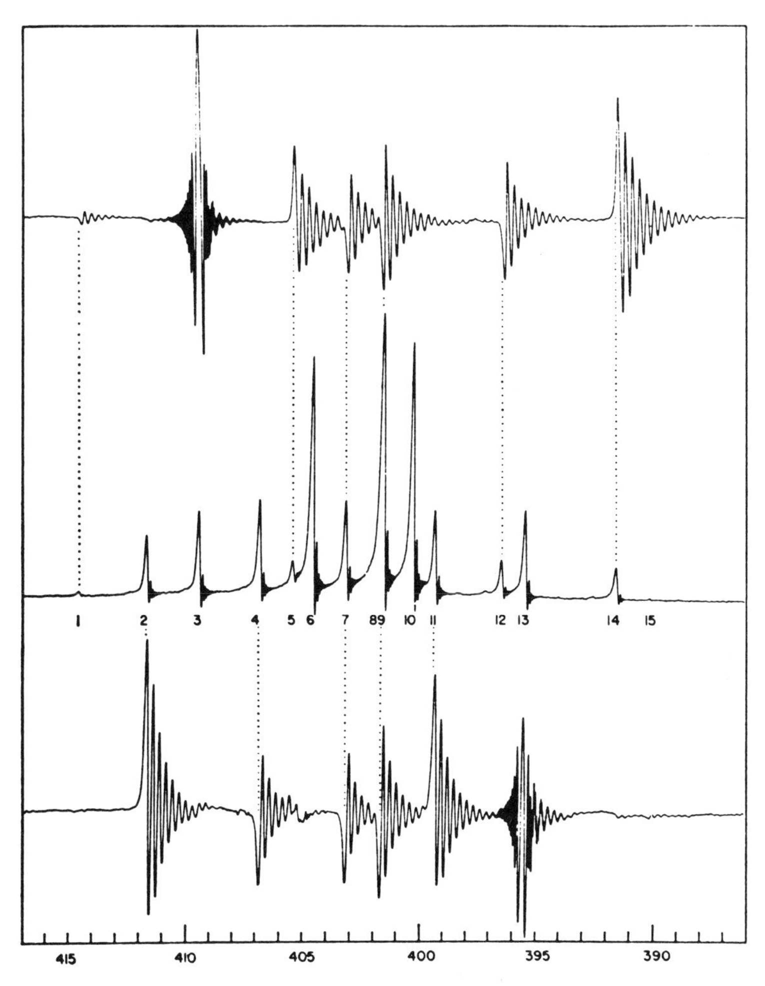

Fig. 1. Transient nutations excited by monitoring one of the lines of the proton spectrum of 2-chlorothiophene (centre) with a saturating continuous-wave field B_1 while sweeping a strong perturbing field B_2 through the entire spectrum. Line 3 is monitored in the upper trace and line 13 in the lower trace. A progressively connected transition gives a positive-going response and a regressively connected transition a negative-going response. The nutation frequency is $\gamma B_1/2\pi$ multiplied by the transition probability of the monitored line.

sometimes a useful mode of detection because the constant-frequency nutation signal provides a characteristic, easily recognized pattern. For example, transient nutations are observed in pulse-modulated double resonance experiments (3) and in chemically induced electron spin polarization (4) where magnetization is suddenly created by a laser pulse. The method provides a convenient practical scheme for monitoring polarization transfer* effects. Instead of sitting on the peak of a line with a very weak B_1 field, subjected to the vicissitudes of poor field/frequency regulation*, a strong (saturating) B_1 field may be used since it is not the natural NMR signal that is of interest but the sudden polarization transfer from another site.

Figure 1 illustrates the transient nutations observed for the three-spin ABC system of protons in 2-chlorothiophene (5). This is a continuous-wave experiment where the B_1 field is adjusted to exact resonance for a chosen line and has sufficient intensity to saturate that line. The second radiofrequency field B_2 is swept through the same spectrum under conditions of adiabatic fast passage*. Whenever B_2 passes through a transition it inverts the appropriate spin population; if this transition has an energy level in common with the transition monitored by B_1 a transient nutation is excited. Progressively connected transitions show a positive-going nutation whereas regressively connected transitions show a negative-going nutation.

REFERENCES

1. H. C. Torrey, *Phys. Rev.* **76**, 1059 (1949).
2. I. Solomon, *Phys. Rev. Lett.* **2**, 301 (1959).
3. R. Freeman, *J. Chem. Phys.* **43**, 3087 (1965).
4. P. J. Hore and K. A. McLauchlan, *Mol. Phys.* **42**, 533 (1981).
5. J. A. Ferretti and R. Freeman, *J. Chem. Phys.* **44**, 2054 (1966).

Cross-references

Adiabatic fast passage
Field/frequency regulation
Polarization transfer
Rotating frame
Spin echoes

Two-Dimensional Spectroscopy

The advent of Fourier transform methods in NMR achieved two important advantages that were immediately exploited. The first was the dramatic improvement in sensitivity brought about because all the resonance frequencies in the spectrum are monitored simultaneously rather than one at a time. The second was the possibility of studying time-dependent phenomena such as spin–spin and spin–lattice relaxation, chemical exchange, spin-echo modulation and chemically induced polarization. Prior to the Fourier transform revolution, the techniques available for monitoring these effects had been very cumbersome indeed. Imagine sweeping through a spectrum in a time of the order of minutes when intensities are varying due to relaxation on a time scale of seconds.

A third development, of comparable importance to the other two, passed almost unnoticed at the time. At a summer school in Yugoslavia, Jeener (1) described a novel experiment in which a coupled spin system was excited by a sequence of two pulses separated by a variable time interval t_1. Jeener realized that if t_1 were to be varied in small steps in a series of experiments, this would constitute a new time dimension and a Fourier transformation could be performed as a function of t_1. This is now known as the *evolution time*. There are then two independent time dimensions t_1 and t_2. The latter is the running variable for sampling the free induction decay. It is convenient to think of the experimental data (2–5) in terms of a matrix $S(t_1,t_2)$ which, after two-dimensional Fourier transformation, becomes a frequency-domain matrix $S(F_1,F_2)$. This is a two-dimensional spectrum where the intensity S is represented by a surface in three-dimensional space; a conventional spectrum is a curve in two-dimensional space. The two frequency dimensions are independent and are therefore represented by orthogonal axes F_1 and F_2.

The receiver is switched off during the evolution period but is operative during the detection period t_2. Information about the behaviour of the spin system is only obtained indirectly, by observing the influence on a set of free induction signals $S(t_2)$. This relies on the fact that nuclear spins possess a memory of what happened to them in the past, a memory with a time constant T_2, the spin–spin relaxation* time. The evolution of the spins is thus mapped out point-by-point by incrementing t_1 in small steps and by monitoring the corresponding free induction decays. Note that no signal is observed during t_1; this therefore provides a method of indirect detection of multiple-quantum coherence, which cannot be observed directly in the spectrometer.

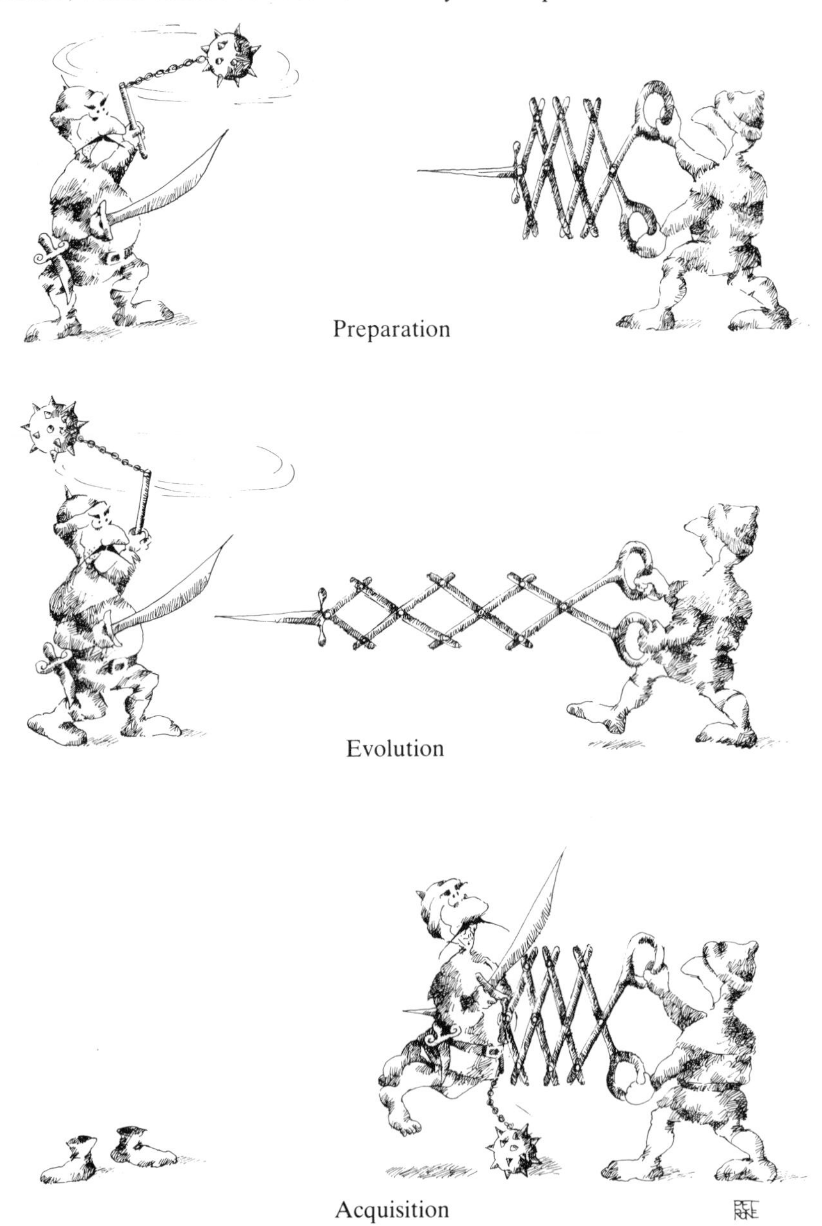

While it is possible to imagine converting $S(t_1,t_2)$ directly into $S(F_1,F_2)$ in a single stage of two-dimensional Fourier transformation, in practice it is usually carried out in separate stages, first with respect to t_2, then with respect to t_1. To see how this is accomplished, consider the simplest case of a single isolated chemical species of resonance frequency δ, acted on by a pulse sequence 45°-t_1-45°. Initially, $t_1 = 0$; and a strong free induction decay is detected in the form of a damped cosine wave. As t_1 is incremented, the nuclear magnetization excited by the first pulse has time to precess during t_1 and the final free induction decay is reduced in amplitude. In this way the signal observed in t_2 is modulated as a function of t_1. The complete set of observed signals $S(t_1,t_2)$ might look something like Fig. 1 where the signals $S(t_2)$ are modulated with another damped cosine wave.

In general this data matrix is so large that it must be stored on a disc unit. The individual free induction decays are retrieved one at a time and Fourier transformed, the resulting spectra being replaced on the disc. This intermediate data matrix $S(t_1,F_2)$ might look something like Fig. 2, where an absorption-mode signal is modulated in amplitude as a function of t_1, decaying in time due to spin–spin relaxation during the evolution period.

The next step is to re-examine this matrix not as traces parallel to the F_2 axis, but as sections parallel to t_1. Since the matrix $S(t_1,F_2)$ is stored on the disc as a series of sequential spectra, it would be an excruciatingly slow process to

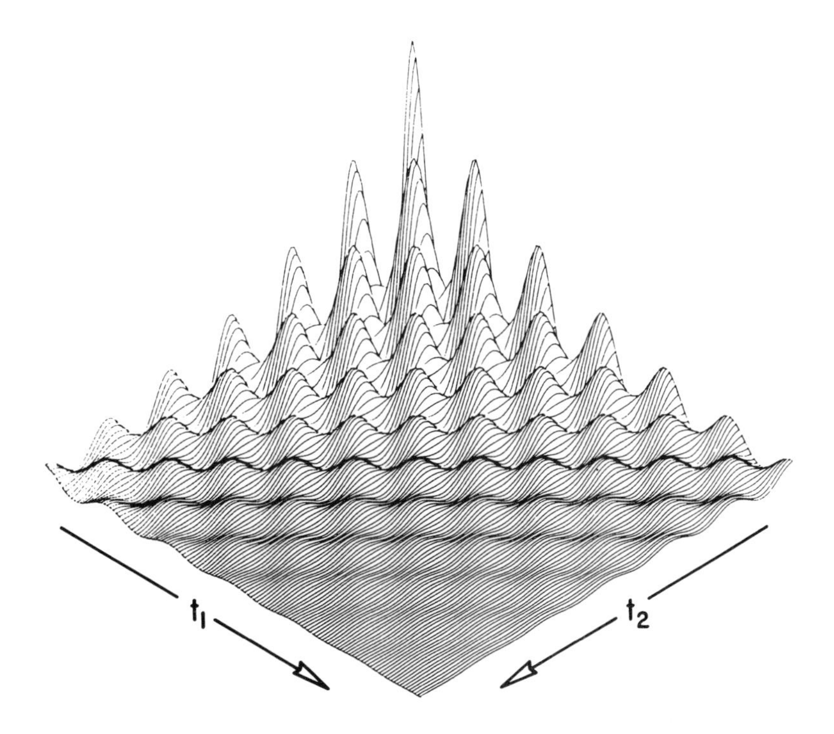

Fig. 1. Schematic diagram of the initial data matrix $S(t_1,t_2)$ consisting of a single cosine wave decaying exponentially in both time dimensions.

extract one point from each spectrum. Consequently, there is a matrix transposition stage $S(t_1,F_2) \rightarrow S(F_2,t_1)$, which then gives direct access to the time evolution of corresponding points in the F_2 spectra (Fig. 3). It is convenient to call these *interferograms* in order to differentiate them from signals which evolve in real time, such as a free induction decay. Now each interferogram may be Fourier transformed individually and the corresponding spectra replaced on the disc.

The new data matrix $S(F_2,F_1)$ or its transpose $S(F_1,F_2)$ is the required two-dimensional spectrum (Fig. 4). It is of course in this instance a trivial case of a single response with frequency coordinates (δ,δ). It happens to have the characteristic shape of a two-dimensional Lorentzian simply because the decay curves in t_1 and t_2 were assumed to be exponential.

So far, this is not very exciting. The importance of Jeener's idea stems from the fact that the conditions governing the motion of the spins during the evolution period may be different from those prevailing during the detection period. Indeed, since the receiver does not need to be switched on during t_1, there may be no detectable NMR signals at all during this period; for example it is possible to excite multiple-quantum coherence* and follow its evolution indirectly by its effect on the conventional signal observed during t_2. One very early exploitation of the new time dimension was to devise a spin-echo experiment which operated in t_1, refocusing the chemical shift but retaining

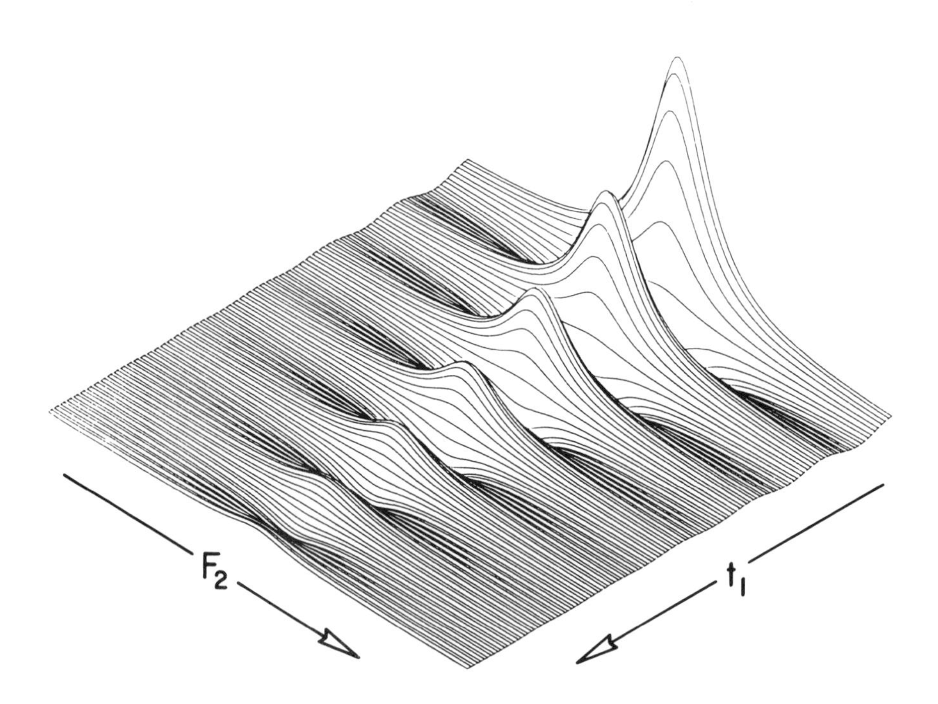

Fig. 2. Intermediate data matrix $S(t_1,F_2)$ obtained by computing the Fourier transform of each t_2 trace in Fig. 1.

spin–spin coupling. This 'J-spectroscopy'* experiment is the prototype of several two-dimensional methods which separate physical parameters in this fashion (2–5).

If the two-pulse experiment described above is extended to the case of two coupled groups A and X, as in Jeener's original experiment, the second pulse transfers some A magnetization to site X and some X magnetization to site A. In more complicated spin systems, many such coherence transfers may take place, but only between coupled spins. The result is called a chemical shift correlation* spectrum since it identifies chemically distinct sites that are related by spin–spin coupling. Note that the second pulse does not induce an *indiscriminate* exchange of coherence, it is more likely a relay race where each runner passes on a baton to a *predetermined* partner. Shift correlation (COSY) has proved to be one of the most useful forms of two-dimensional spectroscopy (2).

To date, Jeener has not published his original work, apart from an abstract and some lecture notes distributed at the summer school (1). Much of the credit for demonstrating the generality of the concept and extending it to many more varied chemical applications, is due to Ernst (2). Close parallels can be traced with the development of double irradiation methods in the nineteen-sixties; indeed many two-dimensional experiments can be thought of as the translation of double resonance techniques into the time domain.

It is still possible to make a broad division of two-dimensional experiments into the four categories:

(1) separation of parameters;
(2) correlation experiments;

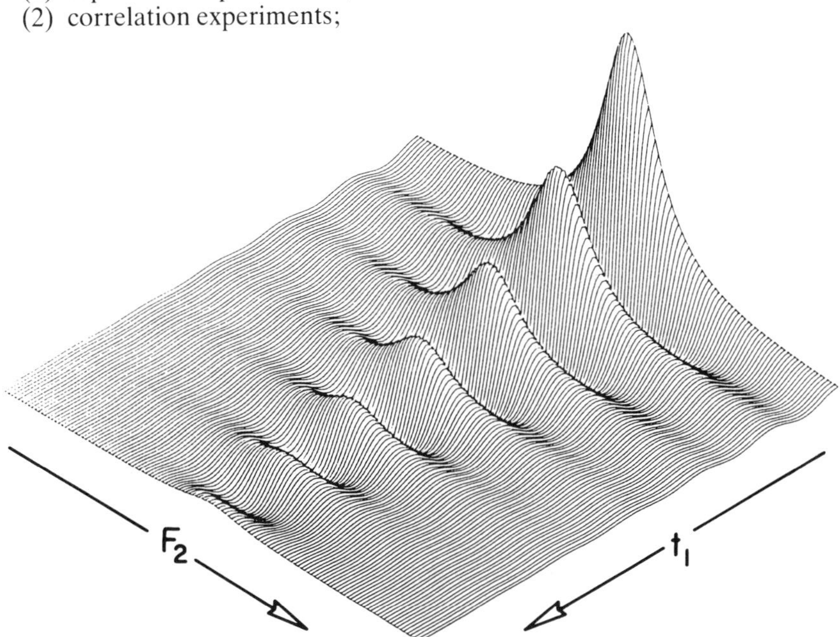

Fig. 3. Transpose of the matrix of Fig. 2. In some spectrometers this is necessary to provide rapid access to t_1 traces from the storage disc, ready for the second Fourier transformation stage.

(3) cross-relaxation and chemical exchange*; and
(4) forbidden transitions,

but it can be dangerously unproductive to draw up formal rules about two-dimensional spectroscopy. It makes surprising leaps into new fields, for example *Fourier zeugmatography*, one of the most elegant ways of coding spatial information into the NMR signal (6). Cross-relaxation or the nuclear Overhauser effect* requires only a minor modification of the COSY experiment (7), yet it opens up a wide field of investigation of biological macromolecules where this information about internuclear distances is invaluable.

Two-dimensional concepts have also had an important impact on 'conventional' high-resolution spectroscopy, stimulating the invention of several new pulse techniques. It is clear that the 'INEPT' polarization transfer* technique owes a great deal to the principles worked out for shift correlation. Similarly, the widely-used 'editing' technique for multiplicity determination* in carbon-13 spectroscopy through spin-echo modulation can be thought of in terms of a projection of the corresponding two-dimensional J-spectrum (8). Even the concept of J-scaling can be traced back to ideas first worked out in two-dimensional spectroscopy. It seems that Jeener's concept, rather tentatively proposed in 1971, has proved one of the most important stimuli for innovation in high resolution NMR spectroscopy.

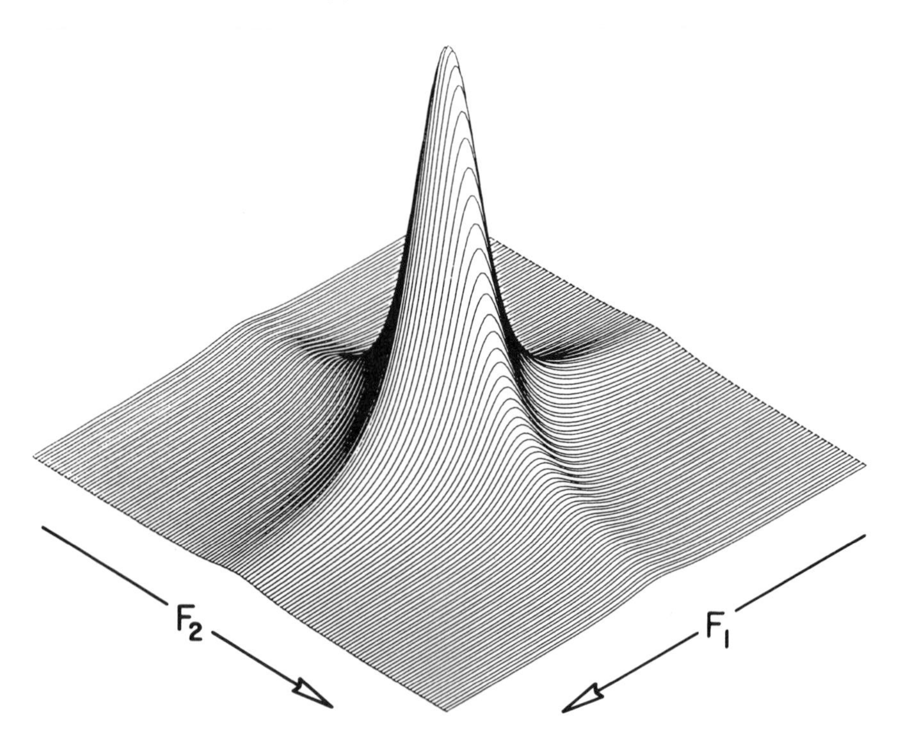

Fig. 4. Two-dimensional Lorentzian line $S(F_1,F_2)$ obtained by computing the Fourier transform of each t_1 trace of Fig. 3. The Lorentzian shape originates in the assumption of exponential decays in $S(t_1,t_2)$.

REFERENCES

1. J. Jeener, *Ampère International Summer School*, Basko Polje, Yugoslavia, 1971.
2. W. P. Aue, E. Bartholdi and R. R. Ernst, *J. Chem. Phys.* **64**, 2229 (1976).
3. R. Freeman and G. A. Morris, *Bull. Magn. Reson.* **1**, 5 (1979).
4. R. Freeman, *Proc. Roy. Soc. (Lond.)* A **373**, 149 (1980).
5. A. Bax, *Bull. Magn. Reson.* **7**, 167 (1985).
6. A. Kumar, D. Welti and R. R. Ernst, *J. Magn. Reson.* **18**, 69 (1975).
7. J. Jeener, B. H. Meier, P. Bachmann and R. R. Ernst, *J. Chem. Phys.* **71**, 4546 (1979).
8. M. H. Levitt and R. Freeman, *J. Magn. Reson.* **39**, 533 (1980).

Cross-references

Chemical exchange
J-spectroscopy
Lineshapes in two-dimensional spectra
Multiple-quantum coherence
Multiplicity determination
Nuclear Overhauser effect
Polarization transfer
Shift correlation
Spin–spin relaxation

Vector Model

Spin choreography is becoming more and more intricate. Many modern NMR experiments involve complex manipulations of populations or coherences, decoupling, refocusing, filtration, purging or phase cycling. These experiments may be treated by density matrix theory or by the product operator formalism*, but the algebra soon becomes complicated and any intuitive insight is quickly lost if there are several coupled spins or if the manipulations are too sophisticated. The vector model fills this gap by permitting a clear visualization of the process, facilitating the task of devising new sequences. It is a natural extension of the classic treatment of Bloch, embodying the transient solutions of the Bloch equations (1, 2).

Although nuclear spins obey quantum laws, the ensemble average taken over a large number of spins behaves just like a classical system, obeying the familiar laws of classical mechanics. We consider first an isolated set of spin-$\frac{1}{2}$ nuclei in a strong static field B_0 and a rotating radiofrequency field B_1. In the usual rotating frame*, synchronous with B_1, these spins are represented by a single vector M, the resultant of all the individual nuclear magnetizations within the active volume of the sample. At Boltzmann equilibrium and in the absence of any recent radiofrequency excitation, the nuclear precession phases are entirely random and there is no transverse component M_{XY}. The longitudinal (Z) component M_0 reflects the slight excess of nuclear spins aligned along B_0 compared with those aligned antiparallel to B_0. Most experiments start from this initial condition.

The vector model is very simple: if there is any residual magnetic field in this rotating frame, M simply precesses about it until the field is extinguished. Most pulse excitations are so short in comparison with relaxation times that there is no significant change in the length of the vector during the pulse. In the absence of applied radiofrequency fields, the only remaining field is ΔB, the offset from the transmitter frequency. Then M precesses about the Z axis at an angular frequency $\gamma\Delta B$ rad s^{-1}. (This happens to be the audiofrequency observed in the resulting free induction decay* since the heterodyne action within the receiver and the transformation to the rotating frame both have the effect of subtracting the transmitter frequency from the Larmor precession frequency.)

Most pulse experiments employ 'hard' or non-selective pulses such that $B_1 \gg \Delta B$ for all the chemically shifted species. During the pulse the rotation is essentially about the B_1 axis for all species, since B_{eff} is almost identical to B_1. The extent of the rotation is

$$\alpha = \gamma B_1 t_p, \qquad [1]$$

and for many experiments t_p is chosen to make $\alpha = \pi/2$ or π radians. In these simple situations the pulse sequence reduces to a set of rotations about the X or Z axes of the rotating frame, and the trajectory of a representative magnetization vector can be traced out as arcs on a unit sphere.

Spin–lattice relaxation* experiments may usually be visualized on this simple picture without further elaboration. The experiment is initiated by generating non-equilibrium magnetization M_Z which then returns to its equilibrium value M_0 aligned along +Z. In fact $(M_Z - M_0)$ decays exponentially and the time constant is T_1. The various different techniques for following spin–lattice relaxation differ mainly in how the initial non-equilibrium state is generated, for example by a 180° *population inversion* pulse.

A free induction signal is generated by transverse (XY) magnetization vectors. In the absence of applied radiofrequency fields these decay exponentially with time constant T_2, the spin–spin relaxation* time. In practice, there is usually a complication introduced by the presence of spatial inhomogeneities in the applied static magnetic field B_0. These may be treated by introducing the concept of a *spin isochromat*, a term coined by Abragam (2). The active sample volume is imagined to be broken down into a mosaic of volume elements each small enough that the B_0 inhomogeneity can be neglected over the volume of that element, but each having a slightly different natural precession frequency. The ensemble average over one such volume element is an isochromat represented by a vector m; M is now the resultant of all the individual vectors m. Immediately after an excitation pulse, all the isochromats would normally be aligned (say along the Y axis), but since they have slightly different precession frequencies they lose phase coherence and their resultant decays asymptotically to zero. This process, which is only an effect of the imperfection of the magnet, is usually assigned a 'time constant' T_2^*. It should of course be carefully distinguished from the irreversible process of spin–spin relaxation which has a time constant T_2. The concept of isochromatic vectors permits the phenomenon of spin echoes* to be given a simple pictorial description.

For the purposes of high-resolution spectroscopy where there are normally many lines in the spectrum, it is customary to extend the Bloch picture to accommodate several groups of nuclei represented by independent vectors M_A, M_B, . . ., etc., having suitable relative intensities and resonance offsets ΔB_A, ΔB_B, . . ., etc. The latter represent the different chemical shifts of the various groups. A 90° excitation pulse would align all these vectors along the Y axis, and free precession would allow them to fan out in the X–Y plane. A 180° refocusing pulse would bring them all into phase coherence again through the formation of a spin echo.

First-order coupling between an observed spin A and some other spin X subdivides M_A into n components corresponding to the n components of the A spin multiplet; they precess at frequencies which differ by J_{AX} and their centre of gravity precesses at δ_A. There is now an important new effect. These 'multiplet vectors' reflect the spin states of the coupling partner X, and if X experiences a 180° inversion pulse (which interchanges the α and β spin states) the labels on the A multiplet vectors must be reversed left-to-right so that faster vectors become slower vectors and vice versa. This is the basis for the pictorial description of spin-echo modulation and the 'INEPT' polarization transfer* experiment.

Selective or *soft* pulses (see Selective excitation*) excite one chosen group of chemically shifted nuclei (A), but approximate the condition $B_1 \ll |\Delta B_A - \Delta B_B|$, $|\Delta B_A - \Delta B_C|$, etc. for all the other groups, with $\gamma B_1/2\pi \gg |J_{AB}|$, $|J_{AC}|$, etc. This is sometimes called *semiselective* excitation. In the more extreme case, the selectivity is so high that only a single transition is excited, others (including other members of the same spin multiplet) remain essentially untouched. This requires $\gamma B_1/2\pi \ll |J_{AB}|$, $|J_{AC}|$, etc. Soft pulses are usually implemented by attenuating the B_1 intensity with a corresponding increase in the pulse length t_p to achieve the desired flip angle α. Another possibility (the DANTE sequence) employs a repeated sequence of hard pulses of small flip angle interspersed with short periods of free precession (3). The vector model provides a clear visualization of this process and shows how 'sideband' excitation conditions occur when the vector accomplishes a whole number of complete rotations between pulses.

This simple vector picture is an invaluable aid to visualizing the effect of pulse imperfections and the compensating properties of certain composite pulses*. A 'sandwich' of several contiguous radiofrequency pulses* (usually applied about different axes, X, Y, −X or −Y) causes successive rotations of the vector M on the unit sphere. An analysis of the resulting trajectories, and in particular their symmetry properties, provides direct insight into the process of compensation (4). One important application has been the evolution of new methods of broadband decoupling*, for example with the MLEV or WALTZ sequences of composite pulses (5).

Not all experiments can be described in terms of the vector model. One of the more important cases where it breaks down is in the description of the evolution of multiple-quantum coherence*. In these experiments two A vectors are prepared in an antiparallel arrangement along the ±X axis and a 90° pulse is applied *along* this axis. Clearly no rotation of the vectors occurs. But because of the effect of the 90° pulse on a coupled X nucleus a mixing of the A vectors occurs and they become locked in antiparallel pairs with the *same* precession frequency. They can therefore never induce a net voltage in the receiver coil and this is why a double-quantum coherence is not directly detectable in an NMR spectrometer. The double-quantum coherence is best visualized as a quadrupole rather than a vector.

REFERENCES

1. F. Bloch, *Phys. Rev.* **102**, 104 (1956).

2. A. Abragam, *The Principles of Nuclear Magnetism.* Oxford University Press, 1961.
3. G. A. Morris and R. Freeman, *J. Magn. Reson.* **29**, 433 (1978).
4. M. H. Levitt, *Progress in NMR Spectroscopy* **18**, 61 (1986).
5. M. H. Levitt, R. Freeman and T. Frenkiel, *Advances in Magnetic Resonance* **11**, 47 (1983).

Cross-references

Broadband decoupling
Composite pulses
Free induction decay
Multiple-quantum coherence
Polarization transfer
Product operator formalism
Radiofrequency pulses
Rotating frame
Selective excitation
Spin echoes
Spin–lattice relaxation
Spin–spin relaxation

Zero Filling

Fourier transformation of the free induction decay* is normally carried out digitally on a small computer, so that it is necessary to sample the free induction signal at discrete, evenly-spaced intervals between t = 0 and the maximum value t = T, the acquisition time. The fast Fourier transform* program requires that the size of the data table to be transformed be an integral power of 2, say $N = 2^n$. Now, in practical cases the free induction signal often decays to a low level before all N samples have been acquired, and to avoid significant contributions from noise in the tail of this signal, a weighting function is commonly used which reduces both signal and noise to a negligible level at the end of the free induction decay. It is therefore natural to complete the data table to the next power of 2 by adding zeroes.

In some other experiments, data acquisition may have to be curtailed before the free precession signal has properly decayed. This is particularly unfortunate in cases where there are two or more lines in the spectrum that are not quite resolved, because we realize that the resolving power has been unnecessarily impaired, just as in optical spectroscopy when the slit width is too large. It turns out that doubling the data table by adding an equal number of ordinates of zero intensity does in fact improve the resolution in this situation, allowing the peaks in question to be recognized as separate entities (Fig. 1). But how can adding zeroes possibly improve the quality of the information in the final spectrum? Surely, by comparison with a free induction decay followed for 2T seconds, a transient signal zero filled from T to 2T must contain *false* information? The key to this apparent paradox is that under normal conditions significant information is actually discarded and zero filling allows it to be retrieved (1).

Suppose that a free induction decay S(t) is sampled with N real coefficients from time t = 0 to t = T. In the programs normally used, Fourier transformation gives N/2 absorption-mode data points and N/2 dispersion-mode data points, which are independent data sets. Normally, only the absorption-mode part would be retained and the dispersion-mode information would be lost. Zero filling provides a method of retrieving this information and thus improving the definition and resolution of the frequency-domain spectrum. Since the function to be transformed is periodic in time (period 2T), appending N zeroes to the tail is entirely equivalent to putting N zeroes immediately in front of the free induction decay, giving a signal which runs from −T to +T, designated Z(t). By specifically defining the signal to be zero at all times prior to the radiofrequency pulse, this is said to impose *causality*.

The signal Z(t) may be decomposed (1, 2) into an even part $Z_{even}(t)$ and an odd part $Z_{odd}(t)$ as illustrated in Fig. 2.

$$Z(t) = 0.5[Z_{even}(t) + Z_{odd}(t)]. \quad [1]$$

Since the Fourier transforms of an even function must itself be even, $Z_{even}(t)$ transforms into a pure absorption-mode spectrum, whereas $Z_{odd}(t)$ gives a dispersion-mode spectrum. Each spectrum now contains N data points and each has an identical information content, having shared the information in S(t) equally. It is *as if* there had been a transfer from dispersion to absorption (and *vice versa*) by way of a Hilbert transform. The end-result is an absorption-mode spectrum with N coefficients, twice as many as the Fourier transform of the original free induction decay S(t). The 'new' values lie interleaved between the

Two spectroscopists exchanging views about zero-filling

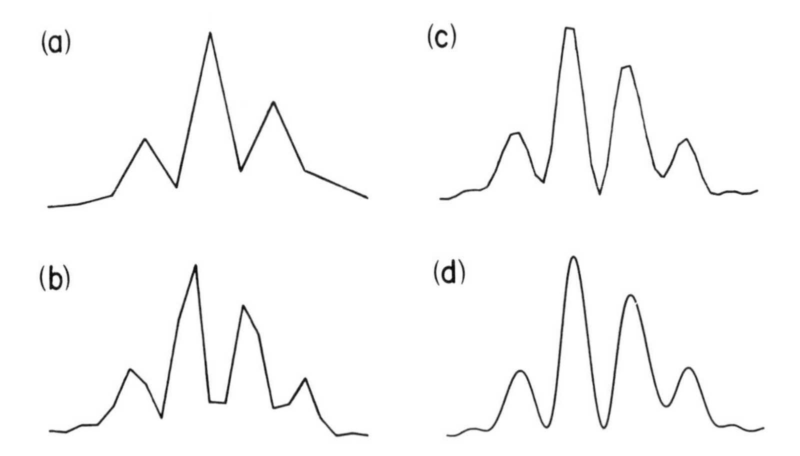

Fig. 1. Spectrum of the methylene protons of ethanol obtained by transforming a truncated free induction decay. (a) No zero-filling, eleven data points defining the multiplet. (b) Zero-filled to give 21 data points. (c) Zero-filled to give 41 data points. (d) Zero-filled to give 2561 data points. Resolution is appreciably improved by the first stage of zero-filling but further extension of the free induction signal simply improves the *digital* resolution.

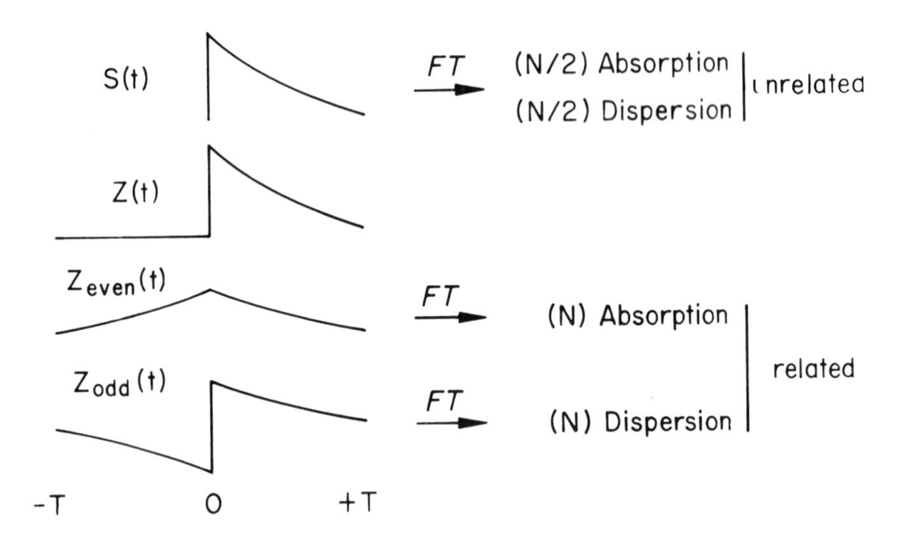

Fig. 2. With the commonly used Fourier transform algorithm, a free induction decay S(t) defined by N data points generates an 'absorption' spectrum of N/2 points while the 'dispersion' spectrum is discarded. If S(t) is zero-filled, Z(t) may be considered as the superposition of an even signal and an odd signal. Fourier transformation of $Z_{even}(t)$ gives an absorption spectrum defined by N data points, an appreciable improvement in resolution.

'old' values. The absorption-mode and dispersion-mode spectra now contain the same information and the latter may now be abandoned without regret.

The paradox is now resolved. The processing of a raw experimental free induction decay without zero filling does not use all the available information; zero filling retrieves the 'lost' data. The interesting point is that it does so automatically, achieving in one step what we might have expected to do by way of a Hilbert transform of the unused dispersion-mode spectrum. Spectroscopic resolution is improved and in marginal cases an unresolved multiplet can become resolved. It also affords a small sensitivity improvement ($2^{1/2}$) through a doubling of the number of measurements of the signal, each carrying uncorrelated noise. However, in the usual case where a sensitivity enhancement* function has been imposed on the free induction decay, these noise components have already been correlated and any sensitivity improvement anticipated.

Once the period has been extended from T to 2T by zero-filling, no new information can be retrieved by adding more zeroes. This would only improve the *digital* resolution of the spectrum interpolating between 'true' data points and giving better definition to the lineshape. In practice, it can often be important to improve the digitization* in this way, for example when fitting the lineshape or when using a computer routine to find the frequency and intensity of a given peak in the spectrum. It may, however, prove more convenient to perform the equivalent process of interpolation in the frequency domain, since this operation would be carried out only on the actual peaks, not on the entire data table.

We may summarize the conditions under which zero-filling can be advantageous:

(1) in order to complete the data table up to 2^n words for the purposes of the Cooley–Tukey algorithm*;

(2) in order to enhance the resolution* of fine structure from a truncated free induction decay by doubling the length of the time-domain data table;

(3) in order to improve the fineness of digitization in the frequency domain by zero-filling beyond the criterion of (2).

REFERENCES

1. E. Bartholdi and R. R. Ernst, *J. Magn. Reson.* **11**, 9 (1973).

2. E. Oran Brigham, *The Fast Fourier Transform.* Prentice-Hall: New Jersey, 1974.

Cross-references

Cooley–Tukey algorithm
Digitization

Free induction decay
Fourier transformation
Sensitivity enhancement
Resolution enhancement

Index

Bold entries refer to main sections of the book